Introduction to Modern Infrastructure Construction

Introduction to Modern Infrastructure Construction serves as a pivotal resource for construction management education, focusing primarily on heavy civil construction and the latest technological innovations in the field. This essential textbook is designed for both academic and professional use, thoroughly covering critical topics including earthwork, highway planning, design, asphalt production, paving, recycling technology, and transportation asset management. Additionally, it explores various aspects of infrastructure such as bridges, railways, airports, and pipelines, offering comprehensive insights beneficial to project management in these areas. Each chapter is supplemented with discussion questions or assignments to enhance educational value, and the text includes lab practice appendices to reinforce practical application.

Authored by leading experts in the field George Wang, Jennifer Brandenburg, and Don Chen, *Introduction to Modern Infrastructure Construction* draws on their extensive experience in academic teaching, research, and practical application. Their expertise provides readers with a unique blend of theoretical knowledge and real-world perspective, making this book an indispensable guide for anyone aspiring to excel in the field of infrastructure construction.

George Wang, PhD, PEng, F.ASCE, brings extensive professional experience across Australia, Canada, and the United States, with a long-standing career in both industry and academia. He currently holds the position of Gregory Poole Distinguished Professor of Construction Management at East Carolina University, where his teaching and research focus on infrastructure construction, sustainable technologies, and materials. Dr. Wang actively contributes to several technical committees, including those of ASCE, TRB, and ASTM, and is a lifetime member of ACCE.

Jennifer Brandenburg, PE, is a professional engineer with over 35 years of experience in government transportation agencies and the private sector. She holds a degree in civil engineering from North Carolina State University and has overseen numerous transportation infrastructure projects. Jennifer has also held senior management positions in the transportation industry and is currently an adjunct faculty member in construction management at East Carolina University.

Don Chen, PhD, LEED, has more than 30 years of experience in the engineering field. He is a professor at the University of North Carolina at Charlotte, teaching courses on Highway Design and Construction, Applied Analytics and Machine Learning, and Building Information Modeling (BIM). His research focuses on pavement performance evaluation and pavement management, emphasizing improving roadway network performance through local calibration, performance modeling, and deep learning.

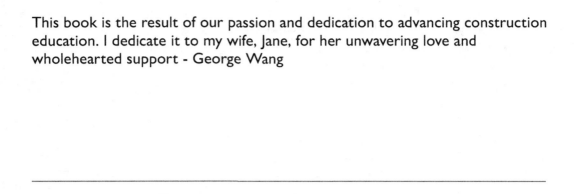

This book is the result of our passion and dedication to advancing construction education. I dedicate it to my wife, Jane, for her unwavering love and wholehearted support - George Wang

Introduction to Modern Infrastructure Construction

George Wang, Jennifer Brandenburg
and Don Chen

Routledge
Taylor & Francis Group

NEW YORK AND LONDON

Designed cover image: image credits to trekandshoot;
ShutterShock ID: 1454904092

First published 2025
by Routledge
605 Third Avenue, New York, NY 10158

and by Routledge
4 Park Square, Milton Park, Abingdon, Oxon, OX14 4RN

Routledge is an imprint of the Taylor & Francis Group, an informa business

ISBN: 978-1-032-05488-9 (hbk)
ISBN: 978-1-032-05159-8 (pbk)
ISBN: 978-1-003-19776-8 (ebk)

DOI: 10.1201/9781003197768

Typeset in Times New Roman
by Deanta Global Publishing Services, Chennai, India

Contents

Preface

In our journey of teaching construction management and engineering technology students about civil infrastructure construction, we noticed textbooks scarcely address the unique features of civil infrastructure, often termed as heavy civil or horizontal construction. Our collective backgrounds in civil engineering and construction management, spanning the spheres of industry, academia, and government, have distinctly underscored this absence. This prompted us to create a textbook aimed at students outside the traditional civil engineering discipline, as well as budding project engineers and construction managers, filling a critical knowledge void in this domain.

Over the past decade, the construction industry has witnessed a surge in innovative technological applications, a progression not adequately captured in existing literature. To bridge this gap, we conducted extensive literature reviews and collaborated with construction firms, enabling us to showcase the most current technologies employed globally in civil infrastructure construction.

This book provides an integrated perspective on the practical applications, specific topics and features, and emerging technologies in transportation infrastructure construction. It aims to prepare future construction professionals by offering a comprehensive understanding of their diverse challenges. Topics covered include soils, geotechnical aspects, earthwork, highway planning and design, asphalt paving, mix design, asset management, and various construction elements related to bridges, rail systems, airports, and pipeline infrastructure.

Our primary objective is to enhance understanding of the distinct aspects of transportation infrastructure construction and to offer practical insights into modern infrastructure construction technologies. We highlight the differences and similarities between infrastructure construction and other types, such as commercial and residential construction. Additionally, this text aims to introduce and facilitate the application of innovative technologies across various construction projects, guiding them from research and trial phases to economical commercialization.

Given the dynamic nature of civil infrastructure construction technology and management, we encourage readers to stay abreast of new technological developments in the field.

We extend our gratitude to industry and agency colleagues for their assistance. Special thanks to Mr. Michael Avery of USDOT, Mr. Taylor Wright, Ms. Ashley Melesse, Mr. Will Janning of Balfour Beatty US Civils, Mr. Scott Cooper of Caterpillar Inc., Mr. Tony Collins of NCDOT, Dr. Jim Zhang of Kiewit Corporation, Dr. Yilei Huang and Mr. Akash Waghani of East Carolina University, and Mr. Daniel Martins, BIM consultant, of Toronto for their providing project information, pictures, or review related documents.

Our appreciation also goes to the library staff from Joyner Library at East Carolina University, especially Mr. Joseph Thomas, Ms. Angela Whitehurst, and Dr. Marti Van Scott, Director of Technology Transfer, for reviewing related materials; Associate Professor Katherine Dickson, Copyright & Licensing Librarian from J. Murry Atkins Library at the University of North Carolina at Charlotte; and Mr. Ed Needle, Ms. Kirsty Hardwick, Ms. Lisa Mosier, Ms. Deepika Ashok Kumar of Routledge, Ms. Rhona Carroll, Taylor & Francis Group, and Ms. Pradiksha Dharsini of Deanta Global Publishing Services for their support throughout the manuscript preparation and production process.

George Wang and **Jennifer Brandenburg**
East Carolina University, Greenville, North Carolina
Don Chen
University of North Carolina, Charlotte, North Carolina

Acronyms and Abbreviations

Agencies and Organizations

AASHTO	American Association of State Highway and Transportation Officials <www.transportation.org>
ACI	American Concrete Institute <www.concrete.org>
ACPA	American Concrete Pavement Association <www.pavement.com>
AI	Asphalt Institute <www.asphaltinstitute.org>
APA	Asphalt Pavement Alliance <www.asphaltalliance.com>
ARRA	Asphalt Reclamation and Recycling Association <www.arra.org>
ASCE	American Society of Civil Engineers <www.asce.org>
ASTM	ASTM International <www.astm.org>
CSA	Canadian Standards Association <www.csa.ca>
DOT	Department of Transportation
ECCO	Environmental Council of Concrete Organizations
EPA	Environmental Protection Agency <www.epa.gov>
FAA	Federal Aviation Administration <www.faa.gov>
FHWA	Federal Highway Administration <www.fhwa.dot.gov>
FRA	Federal Railroad Administration <www.railroad.dot.gov>
FTA	Federal Transit Administration <www.transit.dot.gov>
IAAC	Institute of Advanced Architecture of Catalonia <www.iaac.net>
MIT	Massachusetts Institute of Technology
MOTs	Ministries of Transportation
NAPA	National Asphalt Pavement Association <www.asphaltpavement.org>
NCDOT	North Carolina Department of Transportation <www.ncdot.gov/Pages/default.aspx>
NCHRP	National Cooperative Highway Research Program <www.trb.org>
NRC	National Research Council <www.nrc.ca>
OECD	Organization for Economic Cooperation and Development <www.oecd.org>
PCA	Portland Cement Association <www.cement.org>
PIARC	Permanent International Association of Road Congresses (PIARC/AIPCR) <www.piarc.org>
SHA	State Highway Agencies
SHRP	Strategic Highway Research Program <www.infoguide.ca>
TRB	Transportation Research Board <www.trb.org>

USACE	US Army Corp of Engineers <www.usace.army.mil>
USDA	US Department of Agriculture <www.usda.gov>
USDOT	US Department of Transportation <www.transportation.gov>
USGS	US Geological Survey <www.usgs.gov>

Technical Terms

AADT	Average Annual Daily Traffic
ABC	Aggregate Base Course
ADT	Average Daily Traffic
AI	Artificial Intelligence
AR	Augmented Reality
ARRA	American Recovery and Reinvestment Act
ASR	Alkali-Silica Reactivity
BABs	Build America Bonds
BIM	Building Information Modeling
BMS	Bridge Management Software Systems
BRI	Bridge Rating Index
C&D	Construction and Demolition
CAD	Computer-Aided Design
CAE	Computer-Aided Engineering
Caltrans	California Department of Transportation
CAutoD	Computer-Automated Design
CBR	California Bearing Ratio
CCPR	Cold Central Plant Recycling
CIR	Cold In-Place Recycling
COVID-19	Coronavirus disease 2019
CP	Cold Planning or Cold Milling
CRCP	Continuously Reinforced Concrete Pavement
DBR	Dowel Bar Retrofit
DCP	Dynamic Cone Penetrometer
DHV	Design Hourly Volume
ECM	Reclaimed Concrete Material
EIA	Environmental Impact Assessments
ESA	Federal Endangered Species Act
ESAL	Equivalent Single Axle Load
EV	Electric Vehicle
FDR	Full-Depth Recycling or Full-Depth Reclamation
FWD	Falling Weight Deflectometer
GARVEE	Grant Anticipation Revenue Vehicles
GASB	Governmental Accounting Standards Board
GBE	Granular Base Equivalency
GCR	General Condition Ratings
GI	Group Index
GIS	Geographic Information System
GLONASS	Global Navigation Satellite System
GNSS	Global Navigation Satellite System

GPR	Ground Penetrating Radar
GPS	Global Positioning System
HIR	Hot In-Place Recycling
HMA	Hot-Mix Asphalt
HPMS	Highway Performance Monitoring System
HSGT	High-Speed Ground Transportation
HSIPR	High-Speed Intercity Passenger Rail Program
HSR	High-Speed Rail
HVAC	Heating, Ventilation, and Air Conditioning
IoT	Internet of Things
IC	Intelligent Compaction
ICMV	Intelligent Compaction Measurement Values
IRI	International Roughness Index
IRRM	Index of Retained Resilient Modulus
ISTEA	Intermodal Surface Transportation Efficiency Act
JMF	Job Mix Formula
JPCP	Jointed Plain Concrete Pavement
JRCP	Jointed Reinforced Concrete Pavement
LCA	Life-Cycle Assessment
LCCA	Life-Cycle Cost Analysis
LED	Layered Elastic Design
FAARFIELD	FAA Rigid and Flexible Iterative Elastic Layered Design
LiDAR	Light Detection and Ranging
LL	Liquid Limit
LLAP	Long-Life Asphalt Pavements
LOS	Level of Service
LTPP	The FHWA Long-Term Pavement Performance
MEP	Mechanical, Electrical, and Plumbing
MEPDG	Mechanistic-Empirical Pavement Design Guide
MMS	Maintenance Management Systems
MQA	Maintenance Quality Assurance
MTV	Material Transfer Vehicle
MUTCD	Manual on Uniform Traffic Control Devices
NBI	National Bridge Inventory
NBIS	National Bridge Inspection Standards
NDT	Non-Destructive Testing
NEC	Northeast Corridor
NECIP	Northeast Corridor Improvement Project
NEPA	National Environmental Policy Act
NHS	National Highway System
NMI	National Maglev Initiative
OECD	Organization for Economic Cooperation and Development
O&M	Operation & Maintenance
PCC	Portland Cement Concrete
PCI	Pavement Condition Index
PCR	Pavement Condition Rating
PDRs	Partial-Depth Repairs

PG	Performance Grade
PG	Performance Graded
PGAB	Performance Graded Asphalt Binder
PGAC	Performance Graded Asphalt Cement
PI	Plasticity Index
PL	Plastic Limit
PMA	Polymer Modified Asphalt
PMS	Pavement Management Systems
PPPs or P3s	Public Private Partnerships
PRIIA	Passenger Rail Investment and Improvement Act
PSI	Present Serviceability Index
QA	Quality Assurance
QC	Quality Control
RAP	Reclaimed Asphalt Pavement; Recycled Asphalt Pavement
RAS	Recycled Asphalt Shingles
RCA	Recycled Concrete Aggregate
RFID	Radio Frequency Identification Devices
RPD	Railroad Policy and Development
RUC	Road Usage Charge
RZBs	Recovery Zone Bonds
SEC	Southeast Corridor Commission
SGC	Superpave Gyratory Compactor
SHRP	Strategic Highway Research Program
SIBs	State Infrastructure Banks
STSFA	Surface Transportation System Funding Alternatives
SUPERPAVE	Superpave Superior Performing Asphalt Pavement System
TAM	Transportation Asset Management
TDC	Top-Down Cracking
TIFIA	Transportation Infrastructure Finance and Innovation Act
UAS	Unmanned Aerial Systems
USC	Unified Soil Classification
USCS	Unified Soil Classification System
UTW	Ultra-Thin White topping
VFA	Voids Filled with Asphalt
VMA	Voids in the Mineral Aggregate
VMT	Vehicle Miles Traveled
VR	Virtual Reality
WIFIA	Water Infrastructure Finance and Innovation Act
3D	Three-Dimensional

SI* (Modern Metric) Conversion Factors

Approximate Conversions to SI Units

Symbol	When you know	Multiply by	To find	Symbol
Length				
in	inches	25.4	millimeters	mm
ft	feet	0.305	meters	m
yd	yards	0.914	meters	m
mi	miles	1.61	kilometers	km
Area				
in^2	square inches	645.2	square millimeters	mm^2
ft^2	square feet	0.093	square meters	m^2
yd^2	square yards	0.836	square meters	m^2
ac	acres	0.405	hectares	ha
mi^2	square miles	2.59	square kilometers	km^2
Volume				
fl oz	fluid ounces	29.57	milliliters	mL
gal	gallons	3.785	liters	L
ft^3	cubic feet	0.028	cubic meters	m^3
yd^3	cubic yards	0.765	cubic meters	m^3

Note: Volumes greater than 1,000 L are shown in m^3.

Symbol	When you know	Multiply by	To find	Symbol
Mass				
oz	ounces	28.35	grams	g
lb	pounds	0.454	kilograms	kg
T	short tons (2,000 lb)	0.907	Megagrams (or "metric tons", "tonnes")	Mg (or "t")
Temperatures (Exact Degrees)				
°F	Fahrenheit	5(F − 32)/9 or (F − 32)/1.8	Celsius	°C
Illumination				
fc	foot-candles	10.76	lux	lx
fl	foot-Lamberts	3.426	candela/m^2	cd/m^2
Force and Pressure or Stress				
lbf	poundforce	4.45	newtons	N
lbf/in^2	poundforce per square inch	6.89	kilopascals	kPa

Approximate Conversions from SI Units

Symbol	When you know	Multiply by	To find	Symbol
Length				
mm	millimeters	0.039	inches	in
m	meters	3.28	feet	ft
m	meters	1.09	yards	yd
km	kilometers	0.621	miles	mi
Area				
mm^2	square millimeters	0.0016	square inches	in^2
m^2	square meters	10.764	square feet	ft^2
m^2	square meters	1.195	square yards	yd^2
ha	hectares	2.47	acres	ac
km^2	square kilometers	0.386	square miles	mi^2
Volume				
mL	millimeters	0.034	fluid ounces	fl oz
L	liters	0.264	gallons	gal
m^3	cubic meters	35.71	cubic feet	ft^3
m^3	cubic meters	1.307	cubic yards	yd^3
Mass				
g	grams	0.035	ounces	oz
kg	kilograms	2.202	pounds	lb
Mg (or "t")	Megagrams (or "metric tons", "tonnes")	1.103	short tons (2,000 lb)	t
Temperature (Exact Degrees)				
°C	Celsius	1.8C + 32	Fahrenheit	°F
Illumination				
lx	Lux	0.0929	foot-candles	fc
cd/m^2	candela/m^2	0.2919	foot-Lamberts	fl
Force and Pressure or Stress				
N	newtons	0.225	poundforce	lbf
kPa	kilopascales	0.145	poundforce per square inch	lbf/in^2

*SI is the symbol for the International System of Units. Appropriate rounding should be made to comply with Section 4 of ASTM E380.

Chapter 1

Introduction

1.1 Introduction

The definition of *infrastructure* used hereinafter is that defined by the American Society of Civil Engineers (ASCE)[1] and includes 16 categories, i.e., aviation, bridges, dams, drinking water, energy, hazardous waste, inland waterways, levees, parks and recreation, ports, rail, roads, schools, solid waste, transit, and wastewater. Infrastructure is the system of public works of a country, state, region, or municipality that relates to the work of civil engineers and construction and environmental professionals. These systems tend to be high-cost investments, which are vital to a country's economic development and prosperity.

The civil infrastructure can also be categorized into the following nine sectors:

- Transportation systems (roads, highways, railroads, bridges, canals, locks, ports, airports, mass transit, and waterways)
- Structures (including buildings, bridges, dams, and levees)
- Water supply and treatment systems
- Wastewater treatment and conveyance systems
- Solid waste management systems (collection, reuse, recycling, and disposal of waste)
- Hazardous waste management systems
- Stormwater management systems
- Parks, schools, and other government facilities
- Energy systems (power production, transmission, and distribution)

Infrastructure impacts every facet of daily life. Whether walking on sidewalks, cycling in bike lanes, using mass transit like buses or light rail systems, or driving on roads, our movements are facilitated by infrastructure. The functioning of modern society depends upon infrastructure. Additionally, the presence or absence of infrastructure can influence a municipality's residential, commercial, and industrial development, either promoting or deterring growth (Penn & Parker, 2012).

A large part of our infrastructure works behind the scenes, either hidden underground or seamlessly blended into our everyday environments, making it easy for society to take it for granted. Despite its usually hidden nature, infrastructure systems and components are vulnerable to failures. When failures occur, the effects can be widespread and occasionally disastrous. The causes of these failures are diverse and include

DOI: 10.1201/9781003197768-1

- Faulty construction practices
- Design deficiencies
- Neglect of essential maintenance
- Unanticipated changes in performance-influencing factors over time
- Problems arising from interconnected system components
- The impact of natural disasters
- Unforeseen accidents
- Acts of terrorism, vandalism, and conflicts

1.2 The Current Status of Infrastructure in the United States

Every four years, America's civil engineers provide a comprehensive assessment of the nation's 16 major infrastructure categories in ASCE's infrastructure report card. The report card examines current infrastructure conditions and needs, assigning grades and making recommendations using the following criteria:

- Capacity: Does the infrastructure's capacity meet current and future demands?
- Condition: What is the infrastructure's existing and near-future physical condition?
- Funding: What is the current level of funding from all levels of government for the infrastructure category compared to the estimated funding need?
- Future Needs: What is the cost to improve the infrastructure? Will future funding prospects address the need?
- Operation and Maintenance: What is the owners' ability to operate and maintain the infrastructure properly? Is the infrastructure in compliance with government regulations?
- Public Safety: To what extent is the public's safety jeopardized by the condition of the infrastructure, and what could be the consequences of failure?
- Resilience: What is the infrastructure system's capability to prevent or protect against significant multi-hazard threats and incidents? How can it quickly recover and reconstitute critical services with minimum consequences for public safety and health, the economy, and national security?
- Innovation: What new and innovative techniques, materials, technologies, and delivery methods are being implemented to improve the infrastructure?

According to the ASCE's 2021 Report Card on America's Infrastructure, the United States received a D+ grade on its infrastructure. Roads, aviation, transit, and bridges received grades of D, D, D−, and C+, respectively. These grades have not changed since ASCE's 2013 report, and only one, bridges, has received a higher grade since the report started in 1988. The 2017 report also states that to fix the country's entire infrastructure, it would cost more than $4.5 trillion. Every year, natural disasters such as hurricanes, fires, landslides, and floods damage our infrastructure, causing additional resources to be focused on repairs to the existing systems and taking away funding that could be used to upgrade and improve our infrastructure.

The United States was once a world leader in infrastructure, which played a considerable role in the country's development. Railroads allowed for Western expansion and led to a fast-growing infrastructure network. In 1956, President Dwight D. Eisenhower signed the National Interstate and Defense Highways Act, starting the most expensive interstate highway program that ended up stretching nearly 50,000 miles and costing more than $329 billion. However, today, these highways are crumbling; according to the Department of Transportation, as many as 14,000 deaths every year can be tied to poor road conditions. In the World Economic Forum rankings, America ranks 16th in overall infrastructure, and a 2014 White House report said that 65% of American roadways are in poor condition, plus one in four bridges require repair.

1.2.1 ASCE's Infrastructure Report Card

ASCE's Infrastructure Report Card is issued by ASCE every four years as a way of assessing the state of US infrastructure and offering recommendations for how to improve it. The report card has helped make members of the general public and their elected officials more mindful of the critical role that infrastructure plays in America's technologically advanced economy. Much work remains to be done to educate the average person regarding the importance of adequately funding and maintaining infrastructure. At national and state levels, the report card has been used by ASCE members and others to advocate for infrastructure improvements.

ASCE issued its first report card in 1998, timed to commemorate the 10th anniversary of the release of Fragile Foundations: A Report on America's Public Works. Released in February 1988, Fragile Foundations was prepared by the congressionally chartered National Council on Public Works Improvement. In a novel move, the council assigned individual grades to eight categories of public works: highways, mass transit, aviation, water resources, water supply, wastewater, solid waste, and hazardous waste. The grades ranged from a high of B for the water resources category to a low of D for hazardous waste.

For its part, the National Council on Public Works Improvement makes several recommendations, including dramatically increased capital spending by the private sector and all levels of government, greater flexibility regarding compliance with federal and state mandates, accelerated spending of existing federal trust funds for infrastructure, and greater financing of public works by those who benefit from services. Unfortunately, the Fragile Foundations report did little to change how America addressed its public works. Nearly a decade after the release of the Fragile Foundations report, ASCE was seeking to promote greater interest in infrastructure among elected officials and the general public (US GAO, 2022).

Aware of the impending anniversary of the 1988 report, ASCE took up the challenge of reviewing the state of the nation's infrastructure in 1998 and assigning its own grades to various public works sectors. The members of the policy committees and ASCE's government relations staff spent the next several months preparing the first report card. Information from 1988 to 1998 was reviewed.

The 1998 report card shown in Table 1.1 featured ten infrastructure categories, two more than its predecessor: roads, bridges, mass transit, aviation, schools, drinking water, wastewater, dams, solid waste, and hazardous waste. Highlighting a general decline in the nation's public works during the previous decade, the report card assigned an overall average grade of D for the nation's infrastructure. Grades for individual categories ranged from a C for mass transit to an F for schools.

Regarding presenting its information, the report card was formatted to be digested by readers as simply as possible. Because most Americans were familiar with an A–F grading system, ASCE chose to grade each infrastructure category in this manner. The 1998 report card included a grade, a sound bite, a paragraph, and a page. ASCE released the results of the initial report card to about 15 reporters.

The four-year release cycle for new reports has worked well. Any shorter period than that, the data is not much different and wouldn't yield any different results. Meanwhile, waiting longer than four years risks the report card losing its relevance.

Table 1.1 1988 – 2017 ASCE Report Card Summary

Category	1988	1998	2001	2005	2009	2013	2017
Aviation	B-	C-	D	D+	D	D	D
Bridges	-	C-	C	C	C	C+	C+
Dams	-	D	D	D+	D	D	D
Drinking Water	B-	D	D	D-	D-	D	D
Energy	-	-	D+	D	D+	D+	D+
Hazardous Waste	D	D-	D+	D	D	D	D+
Inland Waterways	B-	-	D+	D-	D-	D-	D
Levees	-	-	-	-	D-	D-	D
Ports	-	-	-	-	-	C	C+
Public Parks & Recreation	-	-	-	C-	C-	C-	D+
Rail	-	-	-	C-	C-	C+	B
Roads	C+	D-	D+	D	D-	D	D
Schools	D	F	D-	D	D	D	D+
Solid Waste	C-	C-	C+	C+	C+	B-	C+
Stormwater	-	-	-	-	-	-	D
Transit	C-	C-	C-	D+	D	D	D-
Wastewater	C	D+	D	D-	D-	D	D+
GPA	C	D	D+	D	D	D+	D+
Cost to Improve	-	-	$1.3T	$1.6T	2.2T	3.6T	4.59T

Table 1.2 2021 ASCE Report Card

Category	Grade	Category	Grade
Aviation	D+	Public Parks	D+
Bridges	C	Rail	B
Dams	D	Roads	D
Drinking Water	C	Schools	D+
Energy	C−	Solid Waste	C+
Hazardous Waste	D+	Stormwater	D
Inland Waterways	D+	Transit	C+
Levees	D	Wastewater	D
Ports	B−		

ASCE's release of versions of the report card for individual states has brought into focus the issues in each state, educating lawmakers and other decision-makers on the condition of our national infrastructure. The stature of ASCE and the civil engineering profession, in general, has also helped ensure a positive reception for the report card and its findings. The Infrastructure Report Card has positively impacted the nation's infrastructure; in 2021, ASCE gave an overall grade of C−, as shown in Table 1.2. While there is still much to be done, there is a desire to prioritize America's infrastructure (ASCE, 2021).

The ASCE estimates the funding needs over the next five years to be approximately $6 trillion and projects a funding gap of more than $2.5 trillion. Natural disasters, such as Hurricane Ian in 2022, the complete devastation of the island of Maui in 2023, Hurricane Helene and Milton in

2024 add significantly to that estimate. Notably, the amount of money required to raise the overall infrastructure grade to a state of good repair (a grade of "B") far exceeds the estimated budget. This shortfall in funding will, at best, result in infrastructure functioning at its current level. At worst, the level of performance will decrease, causing more and more user delays and unsafe conditions.

At the federal level, the report card has helped spur improved policies and more significant funding for various facets of infrastructure. As an example, the creation in 2014 of the Water Infrastructure Finance and Innovation Act (WIFIA) program. Administered by the US Environmental Protection Agency, WIFIA is a federal credit program that provides low-interest loans for certain drinking water and wastewater projects. By highlighting the need for improvement in these sectors, the report card "helped elected officials understand what the economic impact" of inadequately funded water and wastewater infrastructure "meant to America."

Alabama offers a recent example of just such a case. In 2019, the Rebuild Alabama Act became law, phasing in a 10-cent-per-gallon increase in the state's gas tax over three years and indexing it for inflation. Funding from the act is to be used for the state's roads and bridges as well as the dredging of the Port of Mobile. The ASCE 2015 Report Card for Alabama's Infrastructure was a "big component of the basis of information for knowing that we need the additional revenue."

Illinois offers another case in point. The Illinois Section of ASCE issued a report card for its state infrastructure in 2022, indicating the state had made significant improvements in its infrastructure with marked improvements in transit and roads. "Every person in Illinois relies on infrastructure every day – from brushing our teeth, taking the train to work or going to the grocery store – infrastructure is the backbone of our daily lives and communities," said Andrew Walton, President, ASCE – Illinois Section. He continued:

> Bipartisan collaboration resulting in Rebuild Illinois and the federal infrastructure bill signals a new era for our state's built environment, and this report card demonstrates how proper and consistent funding can ultimately produce more jobs, safer communities, and more money in taxpayers' wallets.[2]

Illinois rated an overall C−, which is above the national average of a D, showing that consistent and improved funding can lead to improvements in our country's infrastructure (ASCE, 2022).

Georgia's infrastructure has also improved with the increased attention brought to infrastructure by the ASCE Report. In 2019, for the first time ever, Georgia's rating rose to a cumulative grade of C+.[3] Since 2014, significant progress has been made in several areas, including port expansion projects, $1 billion in transportation funding increases, increased dam inspections, and transit legislation. The addition of nearly 20 stormwater utilities within Georgia represents a 44% increase in these utilities since 2014 (ASCE, 2019).

1.3 Funding Infrastructure

Funding for transportation infrastructure is constantly in the headlines. The United States is a country built on the backbone of a strong transportation network and remarkable for getting goods to market. Not only is commerce built on this nationwide transportation system, but the public also expects to move freely in any direction and at any time without disruptive delays or system outages. This combination of industry and public expectation drives the funding conversation among lawmakers at federal, state, and local levels as well as among private industry executives.

Funding for transportation comes from many different sources. Federal, state, county, and even local governments fund some portion of the transportation infrastructure in their jurisdiction. With the increased need, even private entities have stepped up to help fund projects. These Public-Private Partnerships, or P3s use private funds to build, operate, and maintain specific portions of roadways for a period of 30 or more years with the understanding that they will be reimbursed by the applicable governmental entity. The P3 is usually allowed to generate revenue by charging drivers for using the roadway, thereby creating income.

1.3.1 Federal Funding

One of the oldest and most common ways of funding transportation infrastructure is the use federal tax dollars. Federal investment comes from several sources. Congress appropriates funding to the United States Department of Transportation (USDOT) and authorizes transportation programs based on national priorities.

The USDOT has operated as a Federal Cabinet department since 1967, merging the Federal government's transportation oversight functions (e.g., Federal Aviation Agency, Bureau of Public Roads). The diversity of USDOT's jurisdiction—aviation, rail, roads, trucks, buses, maritime, pipeline, and hazardous materials, and more—promotes subject matter expertise within the operating administrations such as the Federal Highway Administration (FHWA) for roads, the Federal Aviation Administration (FAA) for airports, and the Federal Transit Administration (FTA) for public transit.

USDOT has traditionally played a critical role in maintaining and improving the transportation network across the country through Federal funding. In part, USDOT receives Congressional appropriations in support of stated national priorities and programs, for which the Secretary of Transportation delegates authority to the modal administrators (e.g., FHWA, FAA, FTA) to carry out the mission of USDOT. These appropriations are used to implement various funding and financing programs within the respective operating administrations, in addition to covering personnel, administrative, and other costs.

While Federal government spending on infrastructure, in inflation-adjusted expenditures, has steadily increased, the amount of spending on infrastructure, expressed as a fraction of total federal government spending, has remained relatively steady for the last 20 years at approximately 3% and is significantly less than it was in the 1960s and 1970s.

The cost of infrastructure components varies greatly depending on their size and complexity. One very important concept is the need for funds to not only build infrastructure improvements or replacements (i.e., capital funds) but also funds to operate and maintain projects, i.e., Operation & Maintenance (O&M) funds. Since the mid-1970s, O&M expenditures have been greater than capital expenditures. This trend will most likely continue because O&M expenses tend to increase for any individual infrastructure component as it ages and approaches the end of its design life.

The primary source of all highway funding comes from motor fuels tax. These taxes are collected at the pump when we put fuel in our cars. In 2021, the federal motor fuel tax collected on a gallon of gasoline was $0.184 per gallon, a number that has not changed since 1990. The Department of Transportation reports that in 2016 federal fuel taxes raised $36.4 billion in Fiscal Year 2016, with $26.1 billion raised from gasoline taxes and $10.3 billion raised from taxes on diesel and special motor fuels.

This revenue is distributed between the states to be expended mainly on interstates and primary routes. There are some smaller federal grant programs that local entities are eligible to use on local routes.

1.3.2 State Funding

State governments traditionally fund their state highway networks, including additional funding on interstate and US routes in their states, through state motor fuel taxes and other transportation-related fees. The funding sources listed below are primarily dedicated to highways but vary by state and may also fund bridges, rail, and ports (Dierkers & Mattingly, 2009).

1.3.2.1 Fuel Taxes

All states have some kind of motor fuel tax. Nationally, state motor fuel taxes averaged 21.72 cents per gallon for gasoline (ranging from 7 to 32 cents per gallon), 22.62 cents per gallon for diesel, and 21.54 cents for gasohol.

1.3.2.2 Vehicle Registration Fees

All states collect some vehicle registration fee, which amounted to a total collection of almost $20 billion in 2008.

1.3.2.3 Traditional Bond Proceeds

Nearly all states have transportation bonding authority. Transportation bonds are fixed-rate bonds issued by local, regional, state, and federal government agencies to fund projects in the transportation sector. These can include initiatives such as the construction and improvement of highways, bridges, ports, airports, rail lines, and public transit systems. With traditional bonds, states repay bondholders from user revenues, including taxes, vehicle-related fees, and toll receipts.

1.3.2.4 Tolls

There are approximately 150 toll roads, bridges, and tunnels in the United States that operate in 27 states. These toll facilities are administered by state operating authorities who collect the revenue and operate and maintain the facility.

1.3.2.5 General Funds

Thirty-two states have general fund revenues that collectively account for approximately 6% of total state highway funding. State general funds are established through income taxes, sales taxes, property taxes, and other state and local fees. A number of states use dedicated state transportation trust funds to manage and disperse some or all of their transportation funds.

1.3.2.6 Other Sources (Fees, Taxes, and Other Funds)

Twenty states use one or more other sources of funding, including inspection fees, driver's license fees, advertising, a rental car tax, state lottery/gaming funds, oil company taxes, vehicle excise taxes, vehicle weight fees, investment income, and other licenses, permits, and fees revenue.

States are also using nontraditional and innovative approaches to funding and financing, including new sources of revenue, new financing mechanisms, new funds management techniques, and new institutional arrangements. These might include new sources for bond repayment or electronic road tolls that charge based on usage time. Although traditional sources still produce the majority of state transportation revenues, new and innovative approaches have generated billions of dollars to fund state transportation projects over the past decade. Leading categories of new and innovative transportation financing being used in the states or overseas include options like bonds, federal credit assistance, and state infrastructure banks. The descriptions of each are shown below:

1.3.2.7 GARVEE Bonds

Grant Anticipation Revenue Vehicles, or GARVEE bonds, are any debt financing instrument (bond, note, certificate, mortgage, or lease) that a state issues whose principal and interest are repaid primarily by future federal aid funds. Before their creation in 1995, states could not use federal-aid funds to support bonding.

1.3.2.8 ARRA Bonds

The American Recovery and Reinvestment Act of 2009 (ARRA) provided for two new transportation bonds, Build America Bonds (BABs) and Recovery Zone Bonds (RZBs). In the first several months of availability, public issuers sold nearly $8 billion in BABs, including a successful $1.375 billion issue by the New Jersey Turnpike Authority. ARRA established a $10 billion national bond cap for RZBs.

1.3.2.9 Federal Credit Assistance

Through the Transportation Infrastructure Finance and Innovation Act (TIFIA) loan program, the federal government provides states with direct loans, loan guarantees, and lines of credit for major transportation infrastructure projects. Traditionally, these types of projects were supported by federal grants.

1.3.2.10 State Infrastructure Banks

State Infrastructure Banks (SIBs) are revolving loan funds to finance highway and transit projects. Fifteen SIBs are in place in 35 states, although more than 95% of the funding is concentrated in eight states. These became widespread in 1998 when the federal government expanded eligibility and provided $150 million in seed funding for initial capitalization.

1.3.2.11 Congestion and Cordon Pricing

Congestion pricing is designed to shift demand to less-congested areas or time periods by charging motorists for road use or varying charges, during times of peak demand. Virginia uses this congestion pricing model on their Interstate toll lanes in the Washington, DC area. VDOT states that

> Congestion Pricing aims to reduce rush-hour traffic by providing a financial incentive to use alternative transportation modes (like carpools and transit) or to travel during off-peak periods. If traffic is heavy, the toll is higher—if traffic is light, the toll is lower.[4]

Congestion pricing can be set in advance, or it can be set "dynamically," meaning tolls increase or decrease every few minutes to ensure that the lanes are fully used without a breakdown in traffic flow.

Cordon pricing similarly charges users for entry into a congested area, such as a city center, during some portion of the day. A few states use congestion fees and none use cordon pricing, but these tools are used in a number of countries as a means of both demand mitigation and revenue generation (for example to help fund transit options). The United Kingdom, Norway, and Sweden have been operating successful congestion and cordon pricing schemes for several years; Singapore created the first congestion pricing program in the 1970s.

1.3.2.12 Public-Private Partnerships

Public-private partnerships (PPPs) establish a contractual agreement between a public agency and a private sector entity to collaborate on a transportation project. Twenty-six states have some sort of PPP enabling legislation, and 24 states have used some form of public-private partnership for surface transportation, including roads, freight facilities, and transit, for a total of 71 projects. PPP activity is much greater outside the United States, where partnerships have been used to fund more than four times as many projects as have been undertaken here.

1.3.2.13 Vehicle Miles Traveled Fees

Vehicle miles traveled (VMT) fees charge drivers directly for each mile traveled; the fees replace a traditional motor fuel tax. States are beginning to examine using VMT fees as a way to stop the decline in transportation revenue as gas-powered vehicles become more efficient and electric vehicles become more prevalent. Many states are now piloting a variety of VMT taxes, and the federal highway administration is developing guidance. VMT-based fees are in place for trucks in Germany, Switzerland, and Austria.

1.3.2.14 Other Sources (Impact Fees, Traffic Camera Fees, Container Fees, and Emissions Fees)

Other types of new or innovative vehicles or user fees are also employed by states and internationally to generate revenue. Twenty-three states and several European countries are using impact fees to help fund new infrastructure and transportation projects. Meanwhile, 23 states and many European countries use traffic cameras to generate revenue for surface transportation. Several European and Asian countries rely on vehicle emissions fees, which are currently unavailable in the United States.

1.3.3 Local Government Funding

Counties and cities fund transportation in their jurisdictions in much the same way as states. Local governments build, operate, and maintain their own network of streets, bridges, subways, and transit systems and must find ways to provide the financing for these. Local funding sources include local highway user taxes (motor fuel and motor vehicle), general fund appropriations, property taxes, other local sources dedicated to roads and streets, and miscellaneous receipts such as investment income, traffic fines, and parking fees.

Local governments may also have more flexibility in using private funding sources such as donations in cash or the transfer of real property, the construction of facilities, and the performance of support services (surveys or preliminary and construction engineering).

Additionally, state funds, including state highway user taxes (motor fuel, motor vehicle, and motor carrier), state general funds, and other state funds (state bond proceeds, state lottery proceeds, and imposts such as severance taxes, cigarette taxes, etc.) may be transferred to local governments depending on how state legislation is written.

Federal funding for local transportation includes FHWA funds and funds from all other federal agencies, such as Federal Transit Administration, US Department of Agriculture, Housing and Urban Development, National Highway Traffic Safety Administration, US Army Corps of Engineers, Bureau of Indian Affairs, Bureau of Land Management, Federal Emergency Management Administration, National Park Service, and others.

1.3.4 Funding Challenges

With the increase in electric vehicles and more fuel-efficient gasoline engines, the traditional funding mechanisms for transportation have been eroding steadily over the years. States will continue to face big hurdles in finding funding sources to maintain and repair roads and bridges, build new facilities, and maintain existing infrastructure so it remains safe and efficient.

The fuel efficiency of cars today is drastically different from just a decade ago. National fuel economy standards were 27.5 miles per gallon in 1982, 30.2 miles per gallon in 2011, 37.7 miles per gallon in 2019, and they will go up to 49.7 miles per gallon by 2026 for passenger vehicles and light-duty trucks.[5]

According to a 2018 report by the Iowa Department of Transportation studying the impact of electric vehicles (EV) on highway funding, "Even the most cautious forecasts of long-term EV growth estimate EV market share to grow beyond 10% of the total light-duty fleet." A 2018 report by the Edison Electric Institute projected electric vehicle sales to exceed 3.5 million in 2030 and more than 20% of annual vehicle sales.

Americans' transportation habits are also changing. With the COVID-19 worldwide pandemic in 2020, social distancing and working from home became a reality. Businesses were forced to shut down in-person offices and find other ways to get the job done, and telecommuting became a very real option. The US Department of Transportation reports that 2020 vehicle miles traveled nationwide dropped by 11%. The aftermath of the pandemic has brought workers back to the office in lower numbers, and VMT continues to be lower than it was in 2019.

New mobility options have also changed the fuel tax-based mindset. Uber, Lyft, and other shared transportation services continue expanding throughout the United States, changing the paradigm that everyone needs to own a car and drive themselves from place to place. These rideshare services mean fewer vehicles being purchased and paying user fees. Urban centers are flooded with electric bicycles and scooters, shifting how urban Americans get from point A to point B. As technology continues rapidly advancing, the deployment of autonomous vehicles will undoubtedly alter travel and commuting patterns, and transportation agencies need to be cognizant of the impact on transportation funding.

The federal government is also investigating ways to replace the motor fuels tax for revenue. The Highway Trust Fund – the major source of federal road and bridge funding – largely supported by gas taxes, has seen declining revenues as fuel economy standards increase and more alternative fuel vehicles travel the roads. A Federal Highway Administration program provides funding for states to pilot alternatives to gas tax revenues. Most states chose to test systems that charge for road use per mile driven.

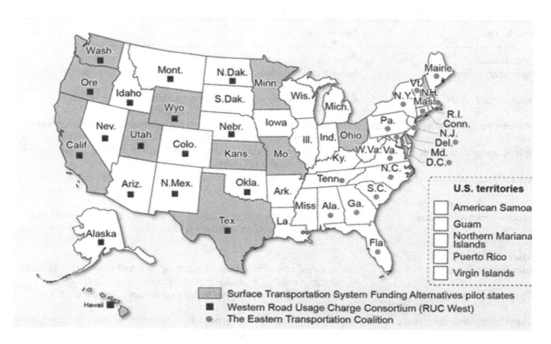

Figure 1.1 States using STSFA Funds to Pilot Alternative User Fees

Source B Courtesy of GAO analysis of Federal Highway Administration Data and Ma Resources; GAO-22-104299

The US Governmental Accountability Office found that since fiscal year 2016, 13 states, including 2 multistate coalitions, have used Surface Transportation System Funding Alternatives (STSFA) funds to pilot and research user-based funding alternatives to the fuel tax. Figure 1.1 shows the use of STSFA funds to research user-based fees. Projects in these states are at various stages of development and implementation. Most of these states explored mileage fee systems, which track miles driven, with the goal of ensuring that drivers pay for their use of the roads. For example, California used STSFA funds to test various technologies to record and report miles driven as part of its mileage fee system pilot. Driver participation in STSFA projects is voluntary to date, has ranged from about 100 to 5,000 participants per pilot.

Recognizing these issues, all states have begun to investigate alternatives to their own motor fuel taxes to increase revenue. States such as North Carolina have tasked their Blue Ribbon Committees with identifying possible future funding methods. In January 2021, the NC FIRST Commission published its final report outlining recommendations and options. The report included options for immediate impact, long-term modernization, local government funding, and other future considerations. This report was delivered to the NC Legislature, which is now tasked with writing and passing legislation to improve the state's infrastructure (McFarlane & Nye, 2021). In 2020, the Washington State Transportation Commission recommended the enactment of a small-scale road usage charge (RUC) program as a first step in a gradual transition away from taxing motor fuel to fund the upkeep of state roads and bridges (WSDOT, 2020).

Table 1.3   AASHTO Road Classifications

Classification	Description
Rural Roads	
Major access roads	Serve a dual function: access to abutting land and connecting service to other roads.
Minor access roads	Almost exclusively used for access to abutting land – roads are typically short – they may be dead ends – speeds are low.
Industrial/commercial access roads	Provide access to factories or commercial outlets – high proportion of truck traffic – roads are typically short and dead-ended.
Agricultural access roads	Provide access to fields – traffic often consists of large, heavy, and/or slow agricultural equipment.
Recreational and scenic roads	Serve specialist land uses such as parks and campsites – roads have low truck traffic but typically cater to large, slow recreational vehicles.
Resource recovery roads	Serve mining or logging operations – traffic has a high proportion of large, heavy resource industry trucks.
Urban Road	
Major access streets	Serve a dual function: access to abutting land and connecting service to other roads – typically shorter than rural equivalents.
Urban residential streets	Access to single and multiple-family residences – large trucks are rare.
Industrial/commercial streets	Serve developments that generate a substantial number of large and/ or heavy trucks – a typical function is to provide a connection between a factory and a highway.

Source: AASHTO, Guidelines for Geometric Design of Very Low-Volume Local Roads (AADT < 400), American Association of State Highway and Transportation Officials (AASHTO), Washington, DC, 2001

1.3.5 The US Road System

The US Road system is provided in the *Guidelines for Geometric Design of Very Low-Volume Local Roads*. Rural and urban roads with an average annual daily traffic (AADT) of less than 400 vehicles/day are classified according to function, as shown in Table 1.3.

1.4 Sustainability and Resilience of Future Infrastructure Construction

The pattern of shifting environmental conditions is a critical challenge human beings are facing on Earth in the 21st century. Construction is one of the largest industries in the world. The construction industry constitutes around one-tenth of the gross domestic product worldwide. In the United States, the construction industry is a major player in the nation's economy, contributing over $1 trillion, including $770.4 billion of private construction and $316.6 billion in the public sector as of December 2008. Construction activities consume tremendous amounts of materials, including cement, concrete, aggregate, steel, nonferrous metals, and other inorganic nonmetallic materials, which are made from winning and processing natural ores and minerals. During this process, a huge amount of energy is used, and greenhouse gas emissions take place. Reusing and recycling industrial and municipal coproducts and by-products can significantly save natural resources and contribute to sustainability in environmental, social, and economic aspects. The benefits of sustainable design and construction offer the potential to change the way in which we, as humans face challenges in the future. These challenges are not insignificant.

The concept of sustainability has gained popular momentum over the last 20 years. The goals of sustainability are to enable all people to meet their basic needs and improve their quality of life, while ensuring that the natural systems, resources, and diversity upon which they depend are maintained and enhanced, for both their benefit and that of future generations. The infrastructure construction industry is beginning to adopt the concept of sustainability in all construction activities, including various recycle, reuse, and reduce technologies in highway pavement construction and rehabilitation, and has a significant opportunity to mitigate environmental problems associated with construction activities while contributing to a high quality of life for its clients (Wang, 2016).

The demand for resources is greater now than ever before in world history. The world's population has doubled since 1950, the urbanization trend is growing rapidly, and the amount of resources consumed per person has also increased dramatically. Since resource demand is a function of these two variables (population and per capita consumption), the increase is compounded. For example, world energy consumption has doubled since the mid-1970s. Similarly staggering numbers can be found that quantify recent increases in water consumption, mining activity, waste production, vehicle miles traveled, cropland converted to development, and so on. In light of these increases, the need to ensure that sufficient resources are available for future generations is greater than ever. In an increasingly global political and economic environment, we must all be concerned not only with sustainability within the United States, but throughout the world. Motivation to do so may come from a sense of ethical duty or humanitarian desire, or more pragmatically, from security concerns; a lack of resources and political instability in some countries, such as Somalia and Yemen, are contributing factors to the rise of terrorism.

1.4.1 Three Rs Consideration Driven by Natural Disasters

Natural disasters affect the United States more frequently than ever before. From devastating hurricanes like Harvey, Ian, Katrina, Sandy, Helena, and Milton to major flooding events and tornadoes, to wildfires and even volcanic eruptions, states are charged with getting the transportation network back online as quickly as possible. After the storm passes, the time comes to evaluate the damage to roads, bridges, airport runways, stormwater systems, urban planning, and other transportation infrastructure. This damage assessment is based on replacing the infrastructure that was there, not making it better. This process needs to be re-examined, not just focusing on short-term fixes but rebuilding with the future in mind, designing and constructing a more resilient infrastructure.

Many of the data that are available for urban planners, construction companies, and governments are dated. Some are more than 50 years old. Getting this data updated entails a lengthy conversation, but it does make one consider the first steps that should be taken when rebuilding the infrastructure after natural events.

Three areas of focus provide good first steps when it comes to addressing infrastructure needs: resiliency, risk, and recycling. The three Rs can restore our transportation infrastructure.

1.4.1.1 RESILIENCY

When looking to restore the infrastructure or build new infrastructure, our roads and bridges need to withstand current (and projected) use. Conventional designs and planning specifications need one eye looking at the present and one looking into the future. Today's residential communities, business districts, highways, and aviation facilities were built using data from decades ago. They don't reflect today's severe weather patterns. (Wang, 2016)

Due to shifting weather patterns, historical climate data is no longer a reliable predictor of future impacts in designing structures to withstand local conditions. Resiliency requires planners to anticipate and prepare for future weather conditions.

1.4.1.2 RISK

A better understanding of the multiple risks posed by environmental changes and the benefits of avoiding damages to coastal infrastructure requires quantification and evaluation of potential impacts. For instance, governments need to know the risks involved if floods and heavy winds could cause structures and businesses damage that would negatively impact and endanger citizens. The chemical plant in Texas and the explosions that resulted from Harvey's impact are a good example. Current highway systems, particularly in large, urban areas, weren't built to handle current traffic loads. Risks involved in new construction to handle the new loads and the effects of this new construction will also need to be addressed.

1.4.1.3 RECYCLE

A *Washington Post* story claimed that Hurricane Harvey "unloaded 33 trillion gallons of water in the United States." This same piece states that the Chesapeake Bay has 18 trillion gallons of water. As mentioned before, current paving technologies mainly consider horizontal drainage, therefore preventing water from being absorbed into the ground. New technologies for porous pavement using recycled materials, including various slags, scrap rubber, and other available recycled materials, can provide new, sustainable, and cost-effective innovations for future transportation construction, which allow for the easier vertical drainage of water, as opposed to runoff. The introduction of these materials into any infrastructure rebuild will help to mitigate the standing water damage that results from weather events.

Now is the time for governments to reinvent themselves regarding future infrastructure needs. Vulnerability and emerging risk analyses should be conducted. Results will highlight new and future threats. If a new approach is not taken, our infrastructure will continue to be unsustainable, less resilient, and more vulnerable (Wang, 2017).

1.5 Questions

1. How does the American Society of Civil Engineers (ASCE) define "Infrastructure"?
2. What are the 16 categories of infrastructure as defined by the ASCE?
3. Why might society often take infrastructure for granted?
4. According to the ASCE's 2021 Report Card on America's Infrastructure, what grade did the United States receive? and how does it compare to the 1998 report?
5. How do states traditionally fund their state highway networks?
6. How many states in the United States have used public-private partnerships (PPPs) for surface transportation projects?
7. How do counties and cities typically fund their transportation needs within their jurisdictions?
8. How have increased electric vehicles and fuel-efficient gasoline engines impacted traditional transportation funding mechanisms?
9. How are rural and urban roads with an AADT of fewer than 400 vehicles/day classified?

10. What is the significance of average annual daily traffic (AADT) in the classification of roads?
11. How can reusing and recycling industrial and municipal coproducts contribute to sustainability?
12. How has the concept of sustainability evolved over the past 20 years in infrastructure construction?
13. How have natural disasters impacted the transportation network in the United States in recent years?

Notes

1 The American Society of Civil Engineers (ASCE), founded in 1852, is the nation's oldest and largest professional society. It has more than 150,000 members in 177 countries from private practice, government, industry, and academia who are dedicated to advancing the science and profession of civil engineering.
2 ASCE Illinois News Release, American Society of Civil Engineers, 2022.
3 ASCE Georgia Section of the American Society of Civil Engineers Infrastructure Report Card, American Society of Civil Engineers, 2019.
4 Virginia Tolling: Then and Now, Virginia Department of Transportation, 2023.
5 Final rule published by the Environmental Protection Agency and the National Highway Traffic Safety Administration in 2012.

References

ASCE (2019). *2019 Report Card for Georgia's Infrastructure*. At https://www.infrastructurereportcard .org/wp-content/uploads/2016/10/American Society of Civil Engineers (ASCE)_Brochure%E2%80 %94GA2019.pdf, American Society of Civil Engineers (ASCE), Reston, VA.

ASCE (2021). *2021 Report Card for America's Infrastructure*. At https://infrastructurereportcard.org/, American Society of Civil Engineers (ASCE), Reston, VA.

ASCE (2022). News Release. *Illinois Earns a C- on its 2022 Infrastructure Report Card*. At https://infrastruct urereportcard.org/illinois-earns-c-on-its-2022-infrastructure-report-card-while-making-strides-on-roads -and-transit/, American Society of Civil Engineers (ASCE), Reston, VA.

Dierkers, G. & Mattingly, J. (2009). *How States and Territories Fund Transportation, An Overview of Traditional and NonTraditional Strategies*. NGA Center for Best Practices Environment, Energy & Natural Resources Division.

Goldberg, L. (1996). *Local Government Highway Finance Trends: 1984–1993*. Public Roads Magazine Summer 1996. At https://highways.dot.gov/public-roads/summer-1996/local-government-highway -finance-trends-1984-1993, FHWA, Washington, DC.

Government Accountability Office (2022). *HIGHWAY TRUST FUND Federal Highway Administration Should Develop and Apply Criteria to Assess How Pilot Projects Could Inform Expanded Use of Mileage Fee Systems*. US GAO. At https://www.gao.gov/products/gao-22-104299

Landers, J. (2020). Advocating for Infrastructure. *Civil Engineering*, November 2020: 60–65.

McFarlane, N. & Nye, W. (2021). *NCF1RST Commission Final Commission Report*. January 8, 2021. At https://www.ncdot.gov/about-us/how-we-operate/finance-budget/nc-first/Pages/final-report.aspx,   NC Department of Transportation, Ralegn, NC.

Penn, M.R. & Parker, P.J. (2012). *Introduction to Infrastructure – An Introduction to Civil and Environmental Engineering*. John Wiley & Sons, Inc, Hoboken, NJ.

Wang, G. (2016). *The Utilization of Slag in Civil Infrastructure Construction*. Elsevier, London, UK and Cambridge, MA.

Wang, G. (2017). *Reinventing for future infrastructure needs*. American City and County, October 18, 2017. At http://americancityandcounty.com/blog/reinventing-future-infrastructure-needs

WSDOT (2020). *Road Usage Charge Assessment Final Report*. January 2020. https://wstc.wa.gov/studies -surveys/road-usage-charge-assessment/, Washington State Transportation Commission (WSDOT), Olymoia, WA.

Chapter 2

Review of Soils

Properties, Investigation, and Compaction

2.1 Introduction

Civil infrastructure construction encompasses a wide range of earthwork activities. In general, earthwork includes the excavation and placement of various materials, such as soil, granular substances, rock, and miscellaneous materials. The processes under earthwork span from geotechnical investigations to subgrade preparations, grading, and excavation. It further extends to embankment construction, compaction, trenching, pipe-laying, and the crucial tasks of erosion and sediment control (Figure 2.1).

2.2 Soil Classification and Properties

Soil can be described as an accumulation of particles resulting from disintegrated rocks. These soils primarily originate from three fundamental rock types: igneous, sedimentary, and metamorphic. Soil formation at or near the Earth's surface occurs through various processes. Mechanical weathering, such as freezing and thawing, and chemical weathering, influenced by factors like the presence of water or natural acids, play crucial roles in this transformation.

2.2.1 Soil Formation and Classification

Igneous rocks form as magma, a high-temperature solution of rock-forming elements under intense pressure, cools and solidifies. Granite, gabbro, and basalt are some common examples. Over time, when rocks are exposed to atmospheric conditions, they undergo wear and tear due to weathering and erosion. As a result of the actions of water and wind, these weathered materials get redeposited, eventually forming sedimentary rocks. One of the most recognizable features of sedimentary rocks is their layered or stratified appearance. Examples include shale, sandstone, and limestone. These can vary in hardness, from being very soft to being as rigid as certain igneous rocks. Metamorphic rocks emerge when igneous or sedimentary rocks undergo changes in texture or mineral composition due to heightened heat or pressure. This transformation can be gradual, with all transitional stages present. Sometimes, the metamorphism becomes so profound that any trace of the original rock disappears. Typically, metamorphic rocks are harder than their precursors. For instance, gneiss evolves from granite, while marble forms from limestone.

From an engineering perspective, soils are chiefly categorized into gravel, sand, silt, and clay based on grain size and consistency. There are multiple classification systems in place, including the Unified Soil Classification System (USCS), the American Association of State Highway and

DOI: 10.1201/9781003197768-2

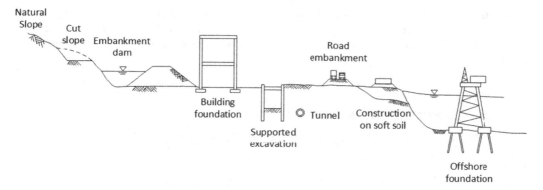

Figure 2.1 Types of Earthwork in Infrastructure Construction

Source: Courtesy of Atkinson. Adapted from Atkinson, J., The Mechanics of Soils and Foundations, 2nd ed. Taylor & Francis, 2007

Table 2.1 Soil Grain Sizes

Soil Type	USCS Symbol	Grain Size Range (mm)			
		USCS	*AASHTO*	*USDA*	*MIT*
Gravel	G	76.2–4.75	76.22	>2	>2
Sand	S	4.75–0.075	2–0.075	2–0.05	2–0.06
Silt	M	Fines <0.075	0.075–0.002	0.05–0.002	0.06–0.002
Clay	C		<0.002	<0.002	<0.002

Transportation Officials (AASHTO) system, the US Department of Agriculture (USDA) system, and the Massachusetts Institute of Technology (MIT) system.

Table 2.1 delineates these soil types according to their grain sizes. To ascertain particle size exceeding 0.075 mm and its distribution, mechanical sieve analysis is employed, as outlined by standards like ASTM D 422 or AASHTO T88.

2.2.1.1 Sieve Test

A dried soil sample undergoes sifting through a series of standard sieves, each having a distinct mesh size. The weight of the soil retained on each sieve is determined and expressed as a percentage of the sample's total weight. This data is then plotted on a cumulative curve, illustrating the grain size distribution of the sample.

2.2.1.2 Sedimentation Test

For distinguishing between silt and clay, one can use either the wash method (as per ASTM D1140 – Standard Test Methods for Amount of Material in Soils Finer Than the No. 200 or 75μm) or the hydrometer analysis (as outlined in ASTM D7928-16 Standard Test Method for Particle-Size

Distribution of Fine-Grained Soils Using Sedimentation Analysis). In the sedimentation test, the soil sample undergoes a mixing process with water and specific chemicals. After thorough mixing, a hydrometer measures the solution's density. It's advisable to conduct a sedimentation test when the combined quantity of clay and silt surpasses a set threshold, such as 15%, (Figures 2.2 and 2.3).

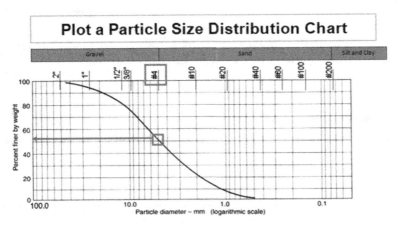

- Use the sieve numbers instead of the actual particle size
- Particle size distribution chart shows:
  - ❖ Range of particle sizes found in this soil
  - ❖ Weight percentages of all particle sizes

Figure 2.2 Particle Distribution Chart

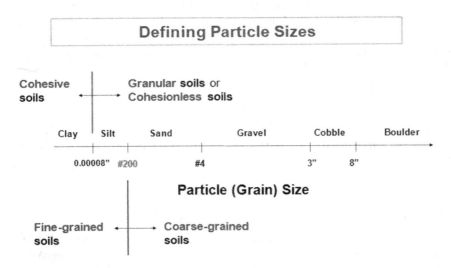

Figure 2.3 Particle Size Definition

2.2.1.3 Consistency

The consistency of fine-grained soils is a vital attribute, often varying from soft to firm to hard based on their water content. The mechanical properties of the soil shift in tandem with these consistency changes.

Standardized laboratory compaction tests commonly classify fine-grained soils. Two key parameters assessed are the liquid limit (LL) and the plastic limit (PL). The difference between the LL and the PL yields the plasticity index (PI), which gauges the soil's plasticity level. Soil exhibiting a low PI is termed as low plastic soil, characterized by its acute sensitivity to water content variations (Figure 2.4).

2.2.2 Soil, Water, and Drainage

Soil is composed of a three-phase system, and the presence of water complicates its behavior and the intricacies of construction (Figure 2.5).

Water accumulation in subgrade soils beneath a foundation or structure can lead to numerous problems. For instance, when a roadbed is saturated either fully or partially, the imposition of dynamic loads can heighten pore pressures. This, in turn, diminishes internal friction, reducing shear resistance, or the buoyant effect of water can compromise the friction between soil particles. Water can cause expansive soils to swell noticeably, resulting in differential swelling and undermining pavement structures and foundations. During adverse weather, the freeze-thaw dynamics related to water presence can lead to volume changes, culminating in pavement damage and the formation of potholes. Thus, before commencing any construction, it's crucial to establish effective drainage systems to prevent water accumulation and potential damage.

To ascertain the seepage volume or the amount of water flowing through the soil over a specific period, and to determine uplift pressure, a flow net analysis might be necessary.

Range of water content over which the soil remains plastic

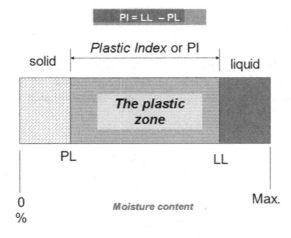

Figure 2.4 Atterberg Limits and Consistency

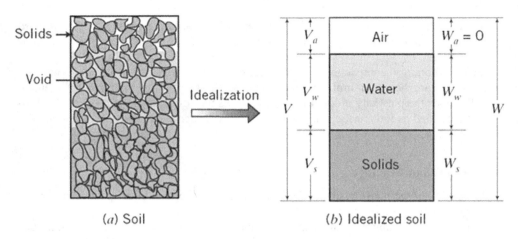

Figure 2.5 The Three-Phase System of Soil

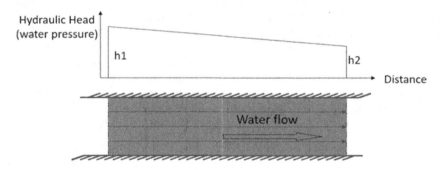

Figure 2.6 A Simpler Flow Chart

In Figure 2.6, the hydraulic gradient reduces from h1 to h2, meaning water flows from the left towards the right. The flowline represents the water's pathway, while the equipotential line indicates consistent water pressure across its length.

Quicksand typically forms in sandy or non-cohesive silt areas, as depicted in Figure 2.7. The force from upward water seepage suspends soil particles, making them unstable. In certain conditions, like during an earthquake, liquefaction causes soils to mimic quicksand behavior temporarily.

2.2.2.1 Dewatering during Construction

Dewatering is often essential during construction to create a suitable working area for tasks such as excavation, foundation construction, and backfill. Various solutions for dewatering include:

- Sump Pump: This method is used primarily for shallow depths. It collects water in a ditch and pumps it out. Being the simplest and most economical option, it's ideal for minor dewatering tasks.

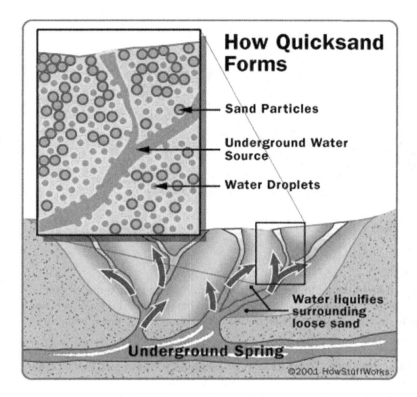

Figure 2.7 Formation of Quicksand

- Well-Point System: Consisting of a series of wells spaced between 3 to 10 feet apart, the well-point system is effective for intermediate depths, catering to dewatering depths of up to approximately 20 feet. This system is set up beneath the water table. Water is drawn out through a "riser pipe" and subsequently collected via a "header pipe."
- Multistage Well-Point: Suitable for deeper dewatering requirements that exceed 20 feet, the multistage well-point system can overcome the constraints of centrifugal pumps, methodically lowering the water table (Figure 2.8).
- Vacuum Dewatering: This method employs a vacuum on a sealed well-point system. Riser pipes in this system are sealed off, necessitating closer well-point spacing. Vacuum dewatering is particularly effective for low-permeable soils, as the conventional well-point system, which relies on gravity to channel water, doesn't perform optimally with less permeable terrains.
- Electro-Osmosis: Here, direct current electricity is applied to a well-point system, essentially "driving" the water out. This method becomes relevant when neither gravity (as seen in the conventional well-point) nor vacuum methods prove effective. It's particularly suitable for clayey terrains and is employed when other methods are ineffective.

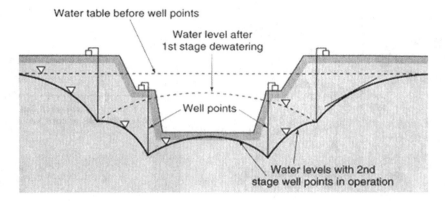

Water table before well points
Water level after
1st stage dewatering
Well points
Water levels with 2nd
stage well points in operation

Figure 2.8 Multiple Well-Point Setting

2.2.2.2 Permanent Drainage

Permanent drainage is crucial for ensuring foundation stability, mitigating erosion due to groundwater flow, and preventing seepage or leakage in structures such as basements. Alongside structural frameworks, it's essential to incorporate effective permanent drainage systems. These encompass:

- Foundation Drains (French Drain): This drainage system acts as a barrier against significant water buildup adjacent to structures. A perforated pipe, commonly known as a French drain, is situated alongside the foundations. Surrounding this pipe, coarse filter materials are placed to assist in water filtration. The water then gets directed through the pipe, ultimately draining into a storm drain, pool, or sump pump.
- Blanket Drain: Designed to alleviate uplift pressure beneath structures, blanket drains incorporate a layer of fine gravel or crushed rock placed under the structure, facilitating effective drainage.
- Interceptor Drain: This system acts as a barrier, intercepting and redirecting the flow of water, preventing accumulation at unwanted sites.

Effective drainage systems that decrease the water table play a pivotal role in preventing water accumulation beneath pavement surfaces, warding off issues stemming from vapor or capillary actions. Such drainage measures ensure minimal water retention in the subgrade, which otherwise tends to linger, persisting even in dry conditions. Additionally, drainage aids in preventing porous material pockets from transforming into hidden water reservoirs.

2.3 Soil Investigation

Soil investigation primarily aims to identify the vertical and horizontal extents of the various soil and rock layers that lie beneath a proposed construction site. Another fundamental objective is to ascertain the position of the groundwater table. Additionally, it seeks to assess the engineering properties of the subsurface materials. These objectives are typically met through testing samples collected during the investigative process.

2.3.1 Before Field Sampling

Prior to conducting site boring and sampling, it's essential to collect and examine all accessible information. This includes the site's background data, layout plan, local planning documents, key plan, past reports, geological charts, hydrogeological and hydrology maps, and overburden and bedrock contour maps.

2.3.2 Site Exploration and Drilling

The principal procedures and considerations in the on-site investigation include the following.

2.3.2.1 Test Hole Arrangement

- Project managers utilize specific methods for determining the count and depth of test holes based on factors like budget, the proposed structure's dimensions and purpose, the anticipated water table depth, and suspicions of contaminants.

2.3.2.2 Equipment for Drilling

- Drill rig (or well rig), backhoe, and hand auger probes

2.3.2.3 Site Services and Preparations

- Scheduling utility location appointments (e.g., phone, electric, water, sewage, cable, gas)
- Clearing the site
- Setting up drilling protocols for bedrock coring
- Conducting methane gas evaluations
- Employing the right technical staff, such as a drilling supervisor and, if needed, a surveying crew

2.3.2.4 Sampling Tools and Materials

- Ensure the toolkit, which houses all necessary equipment for special fieldwork or testing, is complete and matches the provided checklist

2.3.2.5 Site Overview

- General Topography: Understand if the area is mountainous or level, if coastlines are even or jagged, and whether the project site is located at a hill's base, center, or peak.
- Detailed Site Analysis: Provide a descriptive account of the site's appearance, highlighting features like sudden steep inclines in a primarily sloped area or the inverse. If feasible, offer a rough contour map, document any pre-existing structures; capture photographs; gather feedback from the local population.

2.3.2.6 Documentation and Testing

- Utilize field boring forms
- Follow initial soil sampling methodologies
- Conduct in-situ tests and determine the N value
- Determine the borehole sequence
- Accurately measure depths
- Execute laboratory examinations
- Develop a soil profile
- Engage in air quality monitoring
- Administer standard penetration assessments
- Implement field vane evaluations
- Utilize pocket penetrometers
- Check for combustible gases within boreholes
- Test the headspace of soil and water
- Install monitoring wells and observe artesian conditions
- Utilize pressure meters
- Conduct packer evaluations
- Undertake field permeability assessments using standpipes
- Execute aquifer pumping evaluations
- Conduct cone penetration examinations
- Undertake specialized tests
- Measure the thickness of the topsoil and pavement
- Monitor water levels

2.3.2.7 Finalizing the Site

- Clean and restore the site post-exploration

2.3.3 Laboratory Preparation Procedures

2.3.3.1 Sample Handling

2.3.3.1.1 ENVIRONMENTAL SAMPLES

- Ensure proper storage in sealed containers to prevent contamination
- Appropriate labeling with relevant data such as location and date
- Temperature-controlled storage is necessary when necessary

2.3.3.1.2 GEOTECHNICAL SAMPLES

- Prioritize care in handling to retain its natural state
- Use protective gloves to prevent cross-contamination
- Storage recommendations: cool and dry locations

2.3.3.1.3 OTHER SAMPLES

- Adhere to standard procedures specific to the sample type
- Adequate labeling and storage based on type-specific requirements

2.3.3.2 Lab Testing Schedule

- Water Contents: Determination of water quantity in the sample
- Density: Assessment of the sample's mass per unit volume
- Atterberg Limits: Consistency limits evaluation
- Sieve Analysis: Particle size distribution metrics
- Hydrometer Test: Analysis of fine-grained particle distribution

2.3.3.3 Special Testing

2.3.3.3.1 ENVIRONMENTAL TEST

- Contaminant analysis, such as heavy metals, organic compounds, etc.
- Adherence to environmental regulations and standards
- Structured reporting of results, especially highlighting potential environmental concerns

2.3.4 Soil Investigation Report

2.3.4.1 Site Details

- Information: Comprehensive data regarding the location and characteristics of the site
- Purpose: The main objective of the investigation
- Borehole Logs: Detailed records of the drilled boreholes
- Rock Logs: Detailed records of rock layers and their characteristics
- Recommendations: Advised actions or precautions based on the findings
- Borehole Distribution Plan: A detailed map showcasing the distribution of boreholes

2.3.4.2 Investigation Process

2.3.4.2.1 PRELIMINARY STUDY (STAGE I)

- History of the site
- Review of existing geotechnical data
- Analysis of aerial and satellite photographs
- Examination of geological literature, including maps and books
- Study of topographic maps

2.3.4.2.2 FIELD INVESTIGATION (STAGE II)

- No fixed criteria for soil exploration programs; they vary based on project requirements
- Heavy structures like garages might necessitate more extensive exploration compared to lighter buildings

Dam investigations differ from road projects. Considering the scope, it's essential to evaluate:

- Proposed Construction Details: type, weight, size, functionality, and construction style
- Specific Requirements: design data, settlement limits, etc.
- Costs associated with the investigation
- Pre-Existing Data: previous boring records, neighboring building performances, etc.

- Soil nature and characteristics
- Management and disposal of excavated spoil

2.3.4.2.3 BORING SPECIFICS

- Extend borings through unsuitable layers until reaching a "stable" stratum
- For projects involving piles, the boring depth should go beyond the pile tip elevation, preferably 20% deeper
- Determining boring spacing and numbers can be challenging without initial fieldwork. Larger projects might commence with a preliminary investigation, followed by a more detailed and structured program (Tables 2.2–2.7)

2.3.4.3 California Bearing Ratio (CBR) Test

CBR value is used for pavement design. Based on CBR tests from samples taken during an initial soil survey, unsuitable subgrade soils are deemed deficient. These soils also encompass areas rich in organic matter or those too saturated for effective compaction (Table 2.8).

Table 2.2 Special Terms Used in Soil Samples Description Based on Amounts by Weight within the Respective Grain-Size Fractions

noun	gravel, sand, silt ,clay	>35% and main fraction
"and"	and gravel, and silt, etc.	>35%
adjective	gravelly, sandy, silty, clayey, etc.	20–35%
"some"	some sand, some silt, etc.	10–20%
"trace"	trace sand, trace silt, etc.	1–10%

Table 2.3 Use Liquid Limit to Classify Clays and Silts as to Degree of Plasticity

Low	Degree of plasticity	LL < 30
Medium	Degree of plasticity	30 < LL < 50
High	Degree of plasticity	LL > 50

Table 2.4 Approximate Consistency of Cohesive Soils

Consistency	Field Identification
Very Soft	Easily penetrated a couple of inches by the fist
Soft	Easily penetrated a couple of inches by the thumb
Firm	Can be penetrated a couple of inches by the thumb with moderate effort
Stiff	Readily indented by the thumb but penetrated only with great effort
Very Stiff	Readily indented by the thumbnail
Hard	Indented with difficulty by the thumbnail

Table 2.5 Compactness Condition of Sands from Standard Penetration Tests

Compactness Condition	SPT N-Index (Blows/3 Feet)	Angle of Internal Friction
Very Loose	0–4	<30
Loose	4–10	30–35
Compact	10–30	35–40
Dense	30–50	40–45
Very Dense	Over 50	>45

Table 2.6 Consistency of Cohesive Soils from Standard Penetration Tests

Compactness Condition	SPT N-Index (Blows for 2nd and 3rd 6"Penetration)	P.P. Value
Very Soft	0–2	0–0.25
Soft	2–4	0.25–0.5
Firm	4–8	0.5–1
Stiff	8–15	1–2
Very Stiff	15–30	2–4
Hard	>30	>4.0

Table 2.7 Depth and Spacing of Boreholes

Type of Project	Depth (m)	Spacing (m)
Roads	1.5–2	50
Sewers	1.5 m below invert	30–50
Parking Lots	1.5–2	50
Light Buildings	3–6	30
Medium Buildings	7–12	25–30
Heavy Buildings	12–20	15–25
Bridges	10–20	At least 1 at each abutment and pier
Dams	At least to 1.5 times the height of the dam	

It should be noted that if adverse subsoil conditions are encountered, the number and depth of borings should be increased

Table 2.8 Soil Support Categories

CBR	Support Category
15+	Excellent
10–14	Good
6–9	Fair
5 or less	Poor

(Source of data: Barksdale & Takefumi, 1991)

2.4 Soil Compaction

Compaction refers to the method of enhancing a material's density and load-bearing capabilities by applying static or dynamic external pressures. Within construction, the bearing strength, stability, impermeability, and load-bearing capacity of subgrade soil are directly related to its proper compaction. This segment focuses on soil compaction techniques.

2.4.1 Techniques for Compacting Soil and Fill Material

The primary principles underlying soil and aggregate compaction equipment are static pressure, vibratory forces, and impact. Several factors, such as soil type, moisture content, rigidity of the underlying layer, and duration of the compaction process, play a role in choosing the appropriate compaction method and its subsequent outcome.

2.4.1.1 Static Compaction

In static compaction, machines utilize their inherent weight to exert pressure on the surface, compressing the filled substance. Due to the rapid decrease in static pressure as depth increases, static compactors are effective up to a limited depth, necessitating the compaction of material in shallower layers. In this method, the surface solely experiences applied pressure.

To modify the static pressure applied to the surface, one can either change the equipment's weight or its contact area. The compaction efficacy also depends on the speed of the compactor and the number of repetitions.

Traditional static compactors, which have been used, include

- Smooth-Wheeled Rollers: These are large, heavy cylinders that can be either static or vibratory. They are particularly effective for compacting granular soils like sand and gravel.
- Sheepsfoot Rollers: Recognizable by their "lugs" or "feet." They are more effective for compacting cohesive soils, such as clay. The protrusions or "feet" exert pressure points into the soil, which aid in the compaction process.
- Pneumatic Tired Rollers (PTR): These are rollers with rubber tires. The machine's weight is transferred to the soil through the tires, giving a kneading action effective for both cohesive and non-cohesive soils.
- Tamping Rollers or Pad Foot Rollers: Similar to sheep foot rollers but with padded feet. They are designed to provide a kneading action and compact silty and clayey soils.
- Grid Rollers: These have a cylindrical surface comprising a grid of steel bars, beneficial for compacting weathered rock and well-graded coarse soils.

The choice of roller largely depends on the type of soil being compacted, its moisture content, and the required depth of compaction. Properly compacting the soil ensures stability and provides a firm foundation for construction.

Caterpillar's one of the new single-drum vibratory soil compactors line with the introduction of the new 84-in drum width Cat GC Series, built for simple operation, high reliability, and low cost-per-hour performance. With both smooth and padfoot drum models available for the compaction of granular and cohesive soils, the new models deliver reliable performance for applications such as road building, site preparation, large earth fill applications, etc. (Figure 2.9).

Figure 2.9 11-ton CS11 GC Introduced by Caterpillar Inc.

2.4.1.2 Vibratory Compaction

Vibratory compactors utilize a rapid series of impacts on the surface, producing pressure waves that permeate the fill, causing soil particles to move. This movement diminishes the internal friction, enabling particles to settle into tighter configurations. The heightened particle-to-particle contact subsequently enhances load-bearing capabilities.

Predominantly, vibratory compaction is most effective on coarse-grained soils. Medium to heavy vibratory compactors adeptly compress such soils, even those with minimal inherent cohesion, in expansive layers. While its influence is more subdued on fine-grained fills, vibratory compaction remains one of the premier compaction methods.

When compared to static compaction, vibratory methods achieve superior densities and have a more profound depth effect on all fill types. The ability to reach optimal density with fewer passes makes vibratory equipment both efficient and cost-effective, often outperforming heavy static machines in various scenarios.

2.4.1.3 Impact Compaction

This method involves leveraging robust impact forces. Such forces create a pressure wave within the soil, intensifying the pressure deeper into the layer.

Achieving a substantial impact on the surface can be accomplished by dropping a weight, typically hoisted by a crane.

Tamping machines, with their extensive stroke height, produce significant impact forces. This results in an impressive depth effect, making them more adept at compacting cohesive soils compared to vibratory plates. Static tamping rollers, specifically designed for cohesive soils, operate at such velocities that their feet deliver impactful strikes to the soil.

Occasionally, compactors equipped with triangular, rectangular, or pentagonal drums are employed. Their design leads to uncompacted spaces between impacts, thus necessitating multiple passes for consistent compaction.

To maximize their potential, impact rollers should operate at speeds notably greater than their static or vibratory counterparts. They present the best economic value when utilized over extensive terrains.

In the realm of impact compaction, the tamper's substantial stroke height ensures a dominant downward thrust, leading to notable compaction depths.

2.4.2 Compaction Equipment

When selecting compaction equipment, it's vital to consider the type of fill material, layer thickness, compaction standards, and job size. Above all, the machine's capability to meet compaction standards cost-effectively is paramount. Several machines are available for soil compaction, and some of the most prevalent ones are detailed below.

2.4.2.1 Self-Propelled, Single-Drum Vibratory Rollers

- Features one vibrating drum and pneumatic drive wheels
- Effective on rockfill and soil
- Padfoot models excel on clay
- Weight range: 4–25 tons

2.4.2.2 Vibratory Tandem Rollers

- Typically equipped with vibration and drive on both drums
- Suitable for soil (mainly subbases and bases) and asphalt
- Weight range: 1–18 tons

2.4.2.3 Vibratory Plate Compactors

- Moves forward due to the machine's vibratory action. Models with reversible plates are also available
- Weight range: 40–800 kg

2.4.2.4 Walk-Behind Rollers/Trench Compactors

- Feature two drums with either a rigid or an articulated frame
- Weight range: 400–2000 kg

2.4.2.5 Static Three-Wheel Rollers

- Equipped with two driving steel drums and a steering drum
- Weight range: 8–15 tons

2.4.2.6 Tampers

- Produces a high compaction force suitable for various soil types
- Weight range: 40–100 kg

2.4.2.7 Pneumatic-Tired Rollers

- Typically comes with 7–11 pneumatic tires, overlapping in the front and rear
- Weight range: 10–35 tons

2.4.2.8 Static Tamping Rollers

- Features four padfoot drums and articulated steering
- Weight range: 15–35 tons

For each equipment type, the primary compaction parameters include static linear or wheel load, amplitude, frequency, and speed. The compaction effectiveness is usually enhanced by a higher load and specific amplitude. Generally, speeds above 6 km/h for most machines and 10 km/h for Static Tamping Rollers may reduce the compaction effectiveness.

2.4.2.9 Compaction Properties of Different Soil and Rock Fill Materials

The choice of compaction equipment hinges on various factors, like the job type, site size, soil type, underlying layer's stiffness, standards, capacity needs, and weather conditions. Vibration compaction is generally more effective for granular soils like sans and gravel. For cohesive soils, a combination of kneading and vibration can be effective, if the soil is near its optimum moisture content (OMC).

2.4.2.10 Rock Fill (Boulders and Cobbles)

- Encompasses various rock sizes, potentially as large as 1–1.5 m, with minimal fines
- Properly compacted rock fill prevents undesirable settlements
- Vibratory equipment is optimal for rock fill compaction, whereas static and impact compactors might not be ideal. Using a heavy falling weight may risk crushing the rocks
- Heavy-duty vibratory machines are essential for relocating larger boulders and ensuring density and stability
- Given the intense loads during rock fill compaction, it's critical to utilize equipment specifically tailored for the job

2.4.3 Compaction for Minor Tasks

Various light compaction tools exist for minor tasks and constricted spaces. This range includes vibratory plate compactors, tampers, double drum walk-behind rollers, and trench compactors. These instruments are ideal adjuncts to larger rollers, especially in spaces where larger machines are impractical. These smaller machines play a pivotal role in ensuring the safety, quality, and longevity of structures. The criteria for materials and compaction degree often align with the demands of major projects.

2.4.3.1 Minor Roadwork and Infrastructure

Vibratory plate compactors and diminutive vibratory rollers are optimal for compacting embankments, subbase, and base layers. The scope of these tasks can range from small potholes to areas covering approximately a thousand square meters. These machines are also quintessential for constricted spaces and refurbishment tasks.

For compacting coarse-grained soils in slender layers, light vibratory plate compactors are ideal. Denser layers of such soils and semi-cohesive soils demand larger plate types.

For narrower areas, vibratory plate compactors with curved bottom plates are recommended because of their user-friendliness.

2.4.3.2 Foundations and Slab Underlays

To forestall settling, slabs and ground floors require proficiently compacted fills. In constrained spaces, the optimal tools are forward and reverse vibratory plate compactors. Lighter variants serve well for superficial compaction, leveling, and compacting fills adjacent to foundations and basement walls.

2.4.3.3 Trench Operations

Uncompacted trench backfills exert undue strain on pipes or culverts. Consistent, quality compaction mitigates this pressure, which is vital around metallic culverts.

For trench operations, especially for compacting the pipe bed, light compaction gear is pivotal. Trench or vibratory plate compactors over 100 kg are generally chosen. Proper embedding requires alternate filling and compaction on the pipe or culvert's flanks.

Preventing future settling of backfill over pipelines, especially in road crossings and beneath structures, is crucial. Vibratory plate compactors and tampers are optimal choices, with tampers preferred in narrow spaces and cohesive material contexts.

Specialized trench compactors work for granulated and cohesive soils and can be remotely controlled.

2.4.3.4 Bridge Foundations

Surface ridges, resulting from insufficient compaction, often form where road embankments merge with bridge platforms. To attain desired densities, forward and reverse vibratory plates weighing a minimum of 400 kg are typically employed.

2.4.3.5 Dam Foundation Construction

Within dams, top subgrade fill material compaction is essential for impermeability. The subgrade may be slightly uneven, necessitating thin layer placements for leveling. Given that the fill materials are generally semi-cohesive or cohesive, vibratory plate compactors of at least 400 kg or tampers are chosen.

Light compaction tools are also indispensable for soil compression adjacent to metal sheet walls and concrete structures in dam construction and foundational tasks.

2.4.4 Field Compaction Monitoring Methods

To ensure soil compaction meets the specified requirements, several techniques are employed in the field. Among the spot measurement techniques are density assessments, load-bearing tests, leveling evaluations, and more. Additionally, there is the roller-integrated compaction monitor, which, when paired with a documentation system, offers continuous oversight of the compaction process and its outcomes.

2.4.4.1 Sand Replacement Technique

This method involves creating a shallow excavation in the soil, weighing its contents, and then determining the volume by filling the hole with calibrated dry sand or using a water balloon.

2.4.4.2 Tube Core Extraction

Particularly for fine-grained soils like clay, a tube is inserted into the terrain to extract a core specimen, which is then used for density evaluations.

2.4.4.3 Nuclear Density Gauge

This tool quickly indicates the compacted soil's density and its moisture content. It functions on the premise that material density proportionally attenuates radiation from a radioactive isotope. This method yields optimal results when applied to homogeneous soils.

2.4.4.4 Static Load Deformation Test

Carried out on the compacted material's surface, this method determines the deformation underneath a known plate and load. The results can then be used to derive the elasticity modulus of the compacted soil. The measurements might be influenced by the bearing capacity of the layers beneath, with the degree of influence depending on the compacted layer's thickness.

2.4.4.5 Soil Penetration Evaluations

There are multiple penetration tests, with the California Bearing Ratio (CBR) test being a standard. The CBR test, though arbitrary, measures the resistance of soil against the penetration of a standard cylindrical piston. The soil's resistance is then benchmarked against standard values, and the ratio between the soil's resistance and the benchmark determines the CBR. This test is typically used for fine-grained soils.

2.4.4.6 Following-Weight Assessments

These tests employ units that quickly gauge the bearing capability of on-site construction layer surfaces. Typically managed by one operator, the device records the surface deflection caused by a dropped weight and derives a dynamic elasticity modulus. Both lightweight and heavyweight units are available.

2.4.4.7 Surface Settlement Leveling

Primarily used for rock fills, this method involves a leveling tool to monitor the elevation of various reference markers before and after compaction.

2.4.4.8 Surface Indentation Test (Proof Rolling)

In this test, a significantly heavy roller with air-filled tires traverses the compacted terrain. The indentations formed must remain within predefined limits.

2.4.5 Advanced Intelligent Compaction

Intelligent compaction (IC) embodies an innovative approach to road material compaction, utilizing state-of-the-art vibratory rollers outfitted with an integrated measurement system, a sophisticated onboard computer reporting mechanism, global positioning system (GPS)-driven mapping, and optional adaptive feedback mechanisms. The prime advantage of IC rollers lies in their capacity for instantaneous compaction surveillance and immediate process adjustments, harnessing an amalgamation of measuring, documenting, and adaptive systems. These rollers offer an uninterrupted visual record through color-coded plots, granting users access to detailed depictions of roller location, pass counts, and material stiffness metrics.

Figure 2.10 illustrates the IC work principles. The intelligent compaction measurement values (ICMV) are parameters used to measure the compaction level of road material layers during roller compaction in the field. During the compaction process, the compaction roller is controlled by an operator to achieve the targeted degree of compaction for the material. The roller operator obtains real-time visualization of the compaction from the compaction measurement system.

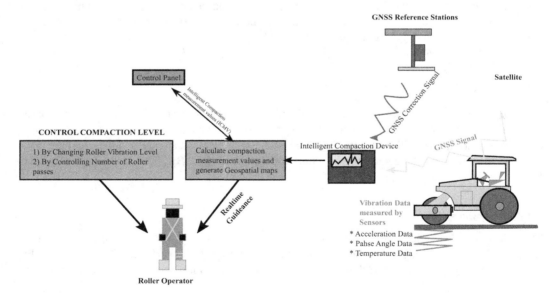

Figure 2.10 Illustration of Intelligent Compaction Technology

Source: Courtesy Ranasinghe, R., Sountharajah, A. & Kodikara, J. (2023)

It is common knowledge that geomaterials have ideal densities for ensuring proper support, resilience, and structural integrity. With traditional compaction equipment, there's a risk of obtaining inconsistent or insufficient material densities. Such inconsistencies could lead to the premature degradation of pavements. IC plays a pivotal role in addressing this concern, enhancing the compaction process's efficiency.

In essence, IC rollers are vibratory rollers fortified with a system that channels data into a real-time documentation and adaptive control mechanism. Critical parameters like the roller's exact position, speed, and pass count over specific spots are plotted using advanced GPS or similar technologies, which are conventionally employed for grading and equipment oversight.

To gauge compaction efficacy, specialized compaction meters or accelerometers are strategically positioned in or around the drum. These devices scrutinize the compaction effort's magnitude, rhythm, and the ensuing material response. The algorithms driving these measurements, albeit proprietary in nature, often conceptualize soil reactions through compaction indices or stiffness/modulus figures. To draw a parallel between the stiffness/modulus values and conventional on-site metrics such as material modulus or density, project test samples are frequently employed.

In the realm of asphalt IC rollers, additional temperature sensors are pivotal in tracking the external temperature of the asphalt paving material. This monitoring is imperative since vibratory compaction within specific temperature ranges can spawn detrimental outcomes.

2.4.5.1 Documentation and Adaptive Feedback

A distinctive characteristic of IC rollers is their embedded system, capable of real-time data acquisition, processing, and interpretation. This feature manifests as a visual, color-coded map for the operator, detailing the roller's position, pass counts, cumulative compaction activity, and material response in real-time. Through optional adaptive feedback mechanisms, the roller can adjust its force and frequency, ensuring maximum compaction efficiency. The intuitive IC roller display signals the operator once optimal compaction is reached, obviating superfluous passes. Should any compaction challenges emerge, the system highlights the areas where additional passes would prove fruitless and records these spots for subsequent examination.

2.5 Aggregates in Infrastructure Construction

Aggregates serve as fundamental raw materials for infrastructure construction. Every city, whether a bustling metropolis or a quaint town, and the roads connecting them rely on aggregates for construction and maintenance. Most foundation bases utilize unbound granular materials. Specifically, 90% of asphalt pavements and 80% of concrete are composed of aggregates.

Infrastructure, including roads, airports, utilities, and other facilities, is essential for the growth and prosperity of any populated region. A significant portion of the US infrastructure was erected during the 1950s and 1960s and has since shown signs of wear and tear. Especially in areas experiencing rapid population growth, there's a pressing need to upgrade and expand the infrastructure.

Sand, gravel, and crushed stone are some of the most common aggregates. In 2014, around 6,600 active operations across the 50 states produced sand and gravel, amassing an estimated output of 911 million tons valued at over $7 billion. That same year, approximately 1.26 billion metric tons of crushed stone were produced from 4,100 active quarries, valued at $12.8 billion.

Recently, the trend of using recycled aggregates sourced from demolished infrastructure and industrial byproducts has gained traction. Old roads and buildings have become significant sources of these recycled materials. In some scenarios, recycled aggregates have matched the price and quality of new aggregates.

2.5.1 Critical Role of Aggregate in Construction System

Aggregates are foundational to the nation's infrastructure, serving as a cornerstone in the construction industry. They are the primary components of vital materials such as hot-mix asphalt and Portland cement concrete and form the bedrock of various construction projects.

Properties: The properties of aggregates play a decisive role in determining the longevity and efficacy of the infrastructure systems in which they are incorporated. Often, the longevity of a road or a building hinges on the appropriate selection and application of aggregates. Inadequate or improper use of these materials can lead to premature system failures, highlighting the importance of understanding their characteristics thoroughly.

Quality Control: Ensuring the high quality of aggregates is crucial for their application in construction. Stringent quality control measures are necessary to guarantee that these aggregates exhibit the desired traits, such as size, shape, strength, and durability. These characteristics are essential to ensure that the aggregates perform their intended function and contribute to the completed structure's longevity and resilience.

Binder Compatibility: The interaction between aggregates and binders, such as asphalt or Portland cement, is fundamental to the performance of construction materials. The compatibility of aggregates with these binders affects everything from the structural integrity to the durability of the construction system. A deep understanding of how aggregates interact with different binders enables the creation of materials that are robust, durable, and suited to a variety of environmental conditions.

In summary, aggregates are not just a part of construction materials; they are a critical factor in the success and sustainability of infrastructure projects. The quality, properties, and compatibility of aggregates with binders form the triad of considerations that determine the effectiveness of these essential construction components.

2.5.1.1 Aggregates Use and Contribution

The intended use of the aggregates dictates the necessary properties. Different uses require aggregates with distinct characteristics.

Aggregates play a crucial role in a variety of construction and infrastructure projects. They contribute significantly to the overall performance and durability of materials such as concrete and asphalt in several ways:

Distributing Surface Pressures and Preventing Deflection: Aggregates help in evenly distributing the weight and pressures exerted on the surface. This is particularly important in road construction, where aggregates contribute to the stability of the pavement, preventing deflection and deformation under the weight of traffic.

Resisting Weather and Chemical Damage: The inherent properties of aggregates make them resistant to damage caused by weather elements like rain, heat, frost, and cycles of freezing and thawing. They also show resistance to chemical attacks, which is essential in maintaining the integrity of construction materials.

Counteracting Applied Loads: Aggregates absorb and disperse the loads applied to the concrete or asphalt, enhancing the material's ability to withstand various stresses, such as compression, tension, and shear forces.

Negating Internal Forces: By providing structural integrity, aggregates help balance and negating internal forces within the composite material. This reduces the likelihood of cracking and other forms of structural failure.

Maintaining Compatibility with Binders: Aggregates are selected for their compatibility with binders like cement or asphalt. This compatibility is essential for ensuring that the aggregate and binder work together effectively to form a cohesive and durable composite material.

Upholding Performance Standards under Traffic: In roadway constructions, aggregates contribute to the pavement's ability to maintain performance standards despite continuous traffic stress. This includes enduring the load and wear from vehicles over prolonged periods.

Enhancing Surface Properties: Aggregates influence various surface properties of roads and pavements. For example, they can enhance skid resistance, which is crucial for the safety of vehicles. The texture of aggregates affects the noise level produced by traffic, potentially reducing noise pollution. Additionally, certain aggregates can minimize glare reflection, improving driver visibility and safety.

Overall, the selection and use of the right type of aggregate are pivotal in determining the quality, durability, and performance of construction materials. Engineers and builders must carefully consider the properties of aggregates to ensure they meet the specific requirements of each project.

2.5.2 Aggregate Types and Processing

Aggregates are essential components in construction and building materials, comprising a mixture of different elements such as sand, gravel, crushed stone, and slag. These materials are fundamental in creating concrete, asphalt, and other composite materials used in various construction projects. Aggregates serve as a reinforcement to add strength to the overall composite material. The sourcing of aggregates is diverse, allowing for a wide range of applications and characteristics.

Natural Deposits: Many aggregates come from natural deposits. These are typically sourced from riverbeds, beaches, or quarries where sand and gravel are found in their natural state. The advantage of naturally occurring aggregates is their rounded shape and smooth texture, which can be beneficial for certain applications.

Rock Crushing: This involves breaking down larger rocks into smaller, usable sizes. Aggregates obtained from rock crushing are often used in construction, especially for producing concrete and asphalt. Crushed stone, derived from limestone, granite, or other hard rock sources, is a prime example.

Waste Materials: In an effort to promote sustainability and reduce waste, some aggregates are now sourced from industrial by-products or waste materials. For instance, slag, a by-product of steel manufacturing, is repurposed as an aggregate. It's known for its strength and durability, making it a popular choice in heavy construction.

Processed Materials: These are specifically engineered for particular uses. Processed aggregates are created by altering the physical properties of natural aggregates to meet specific requirements. This can include washing to remove impurities, crushing, and screening to achieve uniform sizes.

Recycled Pavements: With the growing focus on recycling and sustainable construction, old pavements are increasingly being recycled into aggregates. This process involves crushing and reprocessing pavement materials, such as concrete or asphalt, to produce recycled aggregate for new construction projects. This not only helps reduce waste but also conserves natural resources.

Each aggregate source has unique properties and suitability for different construction applications. The choice of aggregate depends on factors such as the required strength of the material, its durability, the environmental impact of its sourcing, and the specific needs of the construction project.

2.6 Questions

1. Define earthwork in the context of civil infrastructure construction.
2. Describe the Unified Soil Classification System (USCS) and its purpose.
3. Explain the consequences of water accumulation in subgrade soils.
4. Describe the significance of effective drainage systems before construction.
5. Explain the purpose of permanent drainage.
6. Describe the significance of the plasticity index (PI) in determining soil properties.
7. What is the primary purpose of soil investigation?
8. Why are borehole and rock logs vital in soil investigation reports?
9. Based on Table 2.2, what term would describe a soil sample with 25% silt content?
10. What is the support category for soil with a CBR of seven based on Table 2.8?
11. Why are soils too saturated for adequate compaction deemed unsuitable?
12. How are subgrade soil's bearing strength and load-bearing capacity related to compaction?
13. How does soil type influence the choice of compaction method?
14. On which type of soil is vibratory compaction most effective?
15. What tools are available for light compaction tasks in constricted spaces?
16. List the techniques employed in the field to ensure soil compaction meets specified requirements.
17. Describe the purpose of the California Bearing Ratio (CBR) test.
18. What distinguishes intelligent compaction (IC) from traditional compaction methods?
19. Why are aggregates considered fundamental for infrastructure construction?
20. What are the sources from which aggregates can be obtained?

References

Atkins, H.N. (2003). *Highway Materials, Soils, and Concretes*. 4th ed. Prentice Hall, Upper Saddle River, NJ.

Barksdale, R.D. & Takefumi, T. (1991). *Design, Construction and Testing of Sand Compaction Piles*. American Sociey for Testing and Materials (ASTM). Philadelphia, PA.

Capachi, N. (2005). *Excavation and Grading Handbook*. Craftsman Book Company. P.O. Box 6500, Carlsbad, CA.

Carthy, D.F. (2007). *Essentials of Soil Mechanics and Foundations*. 7th ed. Pearson Prentice Hall.

Das, B.M. (2013). *Fundamentals of Geotechnical Engineering*. 4th ed. Cengage Learning. Boston. MA.

Derucher, K.N., Korfiatis, G.P., & Ezeldin, A.S. (1998). *Materials for Civil & Highway Engineers*. 4th ed. Prentice Hall, Upper Saddle River, NJ.

Dynapac Compaction Equipment AB, Sweden (2007). MM Communications AB, Sweden.

NSSGA (2013). *The Aggregates Handbook*, 2nd ed. The National Stone, Sand and Gravel Association, Alexandria, VA.

Ranasinghe, R., Sountharajah, A. & Kodikara, J. (2023). An Intelligent Compaction Analyzer: A Versatile Platform for Real-Time Recording, Monitoring and Analyzing Road Materials Compaction. *Sensors*, 23(17). At https://doi.org/10.3390/s23177507

Chapter 3

Highway Pavement Type, Structure, and Selection

3.1 Introduction

Highway pavement infrastructure is a cornerstone of a nation's economic machinery, enabling the swift movement of goods, services, and individuals across extended distances. In 2020, a telling statistic highlighted the pivotal role of transportation in daily life. Transportation, encompassing elements from private vehicles to public transit, made up a significant portion of an average household's annual expenses, as shown in Figure 3.1. This data points to families' substantial financial dedication to transportation, including fuel, maintenance, insurance, or transit tickets.

Further exploring the economic and logistical dimensions, highways' role in freight transport cannot be overstated. A 2017 report illustrated that a vast majority of the total freight shipment value depended on highways, as detailed in Figure 3.2. This indicates that a significant portion of goods – ranging from tech gadgets and apparel to essentials like food – heavily relied on road infrastructures. This not only showcases the efficacy and dominance of highways in freight transit but also accentuates their crucial function in sustaining a lively and interconnected economy. The Department of Transportation (DOT) characterizes the National Highway System (NHS) as encompassing not only the Interstate Highway System but also other roads pivotal to the nation's economic vitality, defense, and overall mobility. This significant undertaking of crafting the NHS was a collaborative effort between the DOT, state bodies, local authorities, and metropolitan planning organizations. Components of the NHS include the interstate, principal arterials, strategic highway network, major strategic highway network connectors, and intermodal connectors. Collectively, these highways form a robust network, bridging the gap between major intermodal facilities and the diverse subsystems of the National Highway System. A comprehensive depiction of the NHS can be found in Figure 3.3.

Selecting the appropriate pavement type for highways is paramount. This crucial step ensures that scarce financial resources are judiciously utilized to bring the greatest societal advantage.

3.2 Pavement Type

Road pavements, when considering their foundational bases, can generally be classified into two primary types: flexible and rigid. Flexible pavements typically consist of a layer of hot-mix asphalt (HMA) placed on top of the base and subbase layers supported by the subgrade layer, which is a layer of compacted soil. On the other hand, rigid pavements pertain to surfaces constructed using Portland cement concrete (PCC). Rigid pavements consist of PCC layers placed on the subgrade with or without a middle base layer. Pavements made of concrete are renowned for their

DOI: 10.1201/9781003197768-3

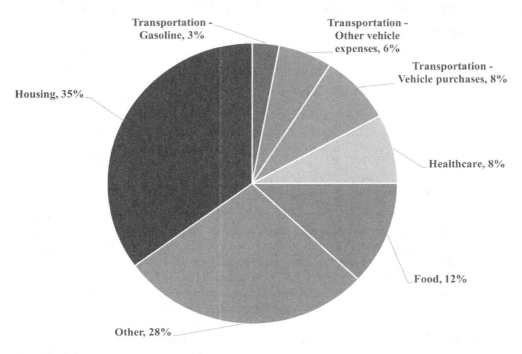

Figure 3.1 Transportation-Related Household Expenditures

Source: Bureau of Labor Statistics, Consumer Expenditures Survey

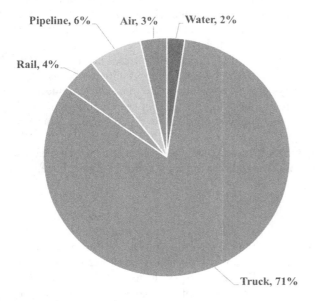

Figure 3.2 Value of Freight Shipments by Transportation Mode

Source: US Department of Transportation, Bureau of Transportation Statistics and Federal Highway Administration, Freight Analysis Framework, Version 4.5, 2019

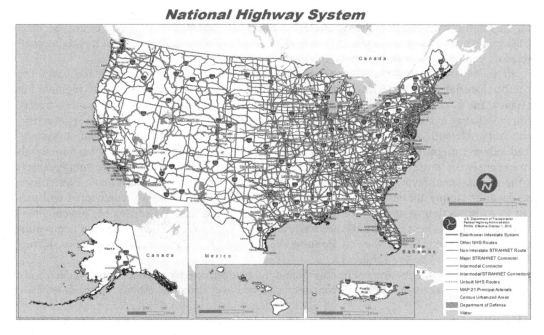

Figure 3.3 National Highway System

Source: Figure Courtesy of Federal Highway Administration

considerable flexural resilience. This inherent strength allows them to act akin to beams, bridging minor undulations or imperfections in the underlying base or subgrade.

A flexible pavement is designed to maintain close contact with, and distribute loads effectively to, the underlying subgrade. It relies on the interlocking of aggregates, friction between particles, and cohesiveness to ensure stability. Typically, flexible pavements are constructed from a combination of granular layers capped by one or more layers of hot-mix asphalt. This includes a top layer of high-quality, thin asphalt for wear and tear. The fundamental components of this structure encompass a wearing surface, an asphalt binder layer, an asphalt base course, a granular foundation, a granular subbase, and the foundational subgrade. Depending on its intended use, the wearing surface's thickness can vary. For instance, it can be less than 1 in (25 mm) for a bituminous surface treatment tailored for light-traffic, cost-effective roads or up to 6 in (150 mm) or more for asphalt concrete meant for roads with high traffic volume. It's imperative that this wearing surface can endure the abrasive impact of moving vehicles and possess the stability to prevent deformation, like shoving and rutting, due to traffic loads.

Furthermore, it plays a crucial role in blocking excessive amounts of surface water from seeping into the base and subgrade from above. For certain high-traffic highways, a top layer ranging from ½ in to ¼ in (13 mm to 18 mm) of an open-graded friction course with high drainage capacity is applied over the wearing course. This layer is specifically designed to enhance skid resistance, reduce the risks of hydroplaning at high velocities, and boost visibility during wet nighttime conditions.

The base consists of one or more layers characterized by their superior stability and density. Its primary function is to distribute the stress induced by wheel loads on the wearing surface, ensuring that the transmitted stresses to the subgrade don't lead to substantial deformation or displacement of the foundational layer. Additionally, the base's composition should be such that it remains unaffected by capillary water and potential frost action.

Often, locally sourced materials are predominantly utilized in constructing the base. The specific materials chosen for this purpose can differ considerably across various regions. For instance, the base might consist of gravel, crushed stone, or even granular substances treated with stabilizing agents like asphalt, cement, or lime-fly ash. In areas with severe frost action, weak subgrade soil, or where a construction working platform is required, a granular or stabilized subbase might be employed. Economically, in places where high-quality base materials are expensive, a more cost-effective subbase material might be preferred. The subgrade acts as the foundational layer, bearing the weight of all loads applied to the pavement. Sometimes, this layer is simply the untouched natural ground. More commonly, it's compacted soil present in a cut section or atop an embankment. Central to the function of flexible pavements is the combined thickness of the subbase, base, and wearing surface. Their collective depth must adequately minimize the stresses on the subgrade, ensuring there's no undue warping or shifting of the subgrade soil layer (Table 3.1).

Composite pavements are commonly found in reconstructed or rehabilitated older roads. Typically, they consist of a Portland concrete layer laid over damaged asphalt or, in some cases, the reverse.

Data from 2020 reveals that the total centerline distance of pavements extended to 2,907,358 miles in rural regions, as shown in Table 3.2, and 1,222,128 miles in urban settings, referenced in Table 3.3.

Table 3.1 Comparison of Flexible and Rigid Pavement

	Flexible Pavement	Rigid Pavement
1.	It consists of a series of layers with the highest quality materials at or near the surface of the pavement.	It consists of one layer of Portland cement concrete slab with relatively high flexural strength.
2.	It reflects the deformations of the subgrade and subsequent layers on the surface.	It is able to bridge over localized failures and areas of inadequate support.
3.	Its stability depends upon the aggregate interlock, particle friction, and cohesion.	Its structural strength is provided by the pavement slab itself by its beam action.
4.	The subgrade strength greatly influences pavement design.	The flexural strength of concrete is a major factor for design.
5.	It functions by way of load distribution through the component layers	It distributes the load over a wide subgrade area because of its rigidity and high modulus of elasticity.
6.	Temperature variations due to changes in atmospheric conditions do not produce stresses in flexible pavements.	Temperature changes induce heavy stresses in rigid pavements.
7.	Flexible pavements have self-healing properties due to heavier wheel loads that are somewhat recoverable.	Any excessive deformations due to heavier wheel loads are not recoverable, i.e., settlements are permanent.

Table 3.2 Length of Rural Pavements

Ownership/Functional System	Unpaved	Paved				Total
		Flexible	Concrete	Composite	Total	
Rural:						
State Highway Agency:						
Interstate	20	16,072	6,002	5,838	27,912	27,932
Other Freeways & Expressways	0	3,208	1,724	1,394	6,326	6,326
Other Principal Arterial	340	70,862	4,661	13,813	89,336	89,676
Minor Arterial	97	100,219	1,756	17,019	118,994	119,091
Major Collector	1,576	197,490	873	8,776	207,139	208,715
Subtotal	2,032	387,852	15,016	46,839	449,707	451,739
Other Jurisdictions:						
Interstate	–	610	147	286	1,043	1,043
Other Freeways & Expressways	–	70	93	–	163	163
Other Principal Arterial	14	620	246	11	877	891
Minor Arterial	348	8,004	10	337	8,351	8,700
Major Collector	34,522	124,860	6,194	6,873	137,927	172,448
Subtotal	34,884	134,164	6,690	7,506	148,361	183,245
Federal Agency:						
Interstate	–	–	–	–	0	0
Other Freeways & Expressways	–	–	–	–	0	0
Other Principal Arterial	–	36	0	–	36	36
Minor Arterial	–	1,059	0	–	1,059	1,059
Major Collector	426	1,611	8	36	1,655	2,081
Subtotal	426	2,706	8	36	2,750	3,176
Subtotal	37,343	524,722	21,714	54,382	600,818	638,160
Minor Collector	79,911				179,818	259,729
Local	1,166,525				842,944	2,009,469
Total Rural	1,283,778				1,623,580	2,907,358

3.3 Pavement Structures

The primary function of the pavement structure is to distribute vehicle wheel loads over a large area, hence reducing the resulting stresses and strains to a level that does not exceed the supportive capacity of each layer during the design life of the pavement. Another function is to provide a level and safe surface for the traveling public.

Flexible and rigid pavements distribute wheel loads differently (Figure 3.4). Flexible pavements distribute the load over a cone-shaped area under the wheel, reducing the magnitude of imposed distress as depth increases. Rigid pavements distribute the load uniformly over the area of the pavement slab.

Figure 3.5 shows the structure of a typical flexible pavement. It includes the asphalt-wearing surface, the base, the subbase, and the subgrade.

Table 3.3 Length of Urban Pavements

Ownership/Functional System	Unpaved	Paved				Total
		Flexible	Concrete	Composite	Total	
Urban:						
State Highway Agency: (3)						
Interstate	–	7,699	5,054	4,857	17,610	17,610
Other Freeways Expressways	–	5,175	3,050	2,364	10,589	10,589
Other Principal Arterial	1	31,951	4,599	12,488	49,038	49,040
Minor Arterial	17	28,102	1,187	8,023	37,313	37,329
Major Collector	5	21,911	1,966	2,526	26,403	26,408
Minor Collector	19	2,378	34	259	2,670	2,690
Subtotal	42	97,216	15,891	30,516	143,623	143,665
Other Jurisdictions:						
Interstate	–	639	160	300	1,099	1,099
Other Freeways Expressways	–	507	495	31	1,033	1,033
Other Principal Arterial	3	13,429	1,987	1,489	16,905	16,908
Minor Arterial	234	50,740	6,260	5,921	62,921	63,155
Major Collector	1,023	77,210	5,661	5,862	88,733	89,756
Minor Collector	542	14,763	497	727	15,986	16,529
Subtotal	1,802	157,288	15,059	14,331	186,677	188,479
Federal Agency:						
Interstate	–	–	–	–	–	–
Other Freeways Expressways	–	39	0	2	41	41
Other Principal Arterial	–	8	–	5	13	13
Minor Arterial	–	144	0	1	146	146
Major Collector	3	186	6	2	193	196
Minor Collector	20	54	0	–	54	74
Subtotal	23	431	6	10	447	469
Subtotal	1,867	254,934	30,956	44,857	330,747	332,613
Local	40,258				849,257	889,515
Total Urban	42,125				1,180,004	1,222,128

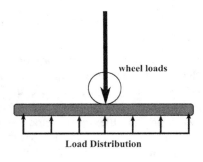

(a)   Rigid Pavement

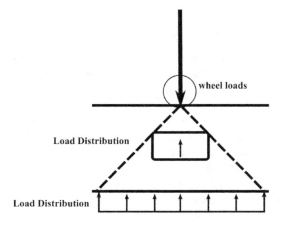

(b)   Flexible Pavement

Figure 3.4 Load Distributions

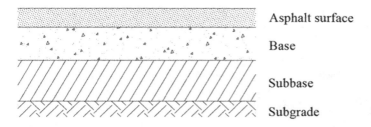

Figure 3.5 Structure of Flexible Pavements

3.3.1 Rigid Pavement

While concrete usually incurs a higher upfront cost than asphalt, it tends to have a longer lifespan and is more cost-effective in terms of maintenance. Nonetheless, in certain situations, design or construction mishaps or the use of unsuitable materials can significantly diminish pavement durability. Hence, it's crucial for pavement engineers to grasp the nuances of material selection, mix ratio, design details, drainage, construction methodologies, and the overall performance of the pavement. Additionally, comprehending the theoretical foundation behind prevalent design approaches and recognizing the boundaries of these methods' applicability is vital.

3.3.1.1 Brief History

The first concrete pavement was constructed in Bellefontaine, OH, in 1891 by George Bartholomew. Notable early pavements include Chicago's Front Street in 1905, which endured for 60 years, and Detroit's Woodward Avenue in 1909, marking the first mile of concrete road. While South Fitzhugh Street in Rochester, NY, laid in 1893, might be the first Portland cement concrete pavement, Bellefontaine remains recognized as the first long-lasting successful concrete pavement.

The rise of automobiles spurred the demand for paved roads. In 1913, a 23-mile concrete path called the "Dollarway" was built near Pine Bluff, AR, costing $1 per foot. By 1914's end, the United States had 2,348 miles of concrete pavement. Figure 3.6 shows the concrete pavement structure (Delatte, 2014).

3.3.1.2 Types of Concrete Pavements

There are two distinct features of concrete pavement: first, it bears traffic loads via concrete flexure, using reinforcement primarily for controlling cracks rather than bearing load. Second, it is subject to contraction from the concrete's drying shrinkage and expansion or contraction from thermal variations. These movements necessitate mitigation, and various pavement types employ joints, reinforcing steel, or a combination of both.

Conventional pavements (JPCP, JRCP, and CRCP) make use of several types of transverse and longitudinal joints. Transverse contraction joints are used in JPCP and JRCP, usually with dowels. At the end of each day of paving, or after a significant delay, transverse construction joints are placed, generally at the location of a planned contraction joint for JPCP or JRCP. Transverse expansion or isolation joints are placed where expansion of the pavement would damage adjacent bridges or other structures. Longitudinal contraction joints are created where two or more lane

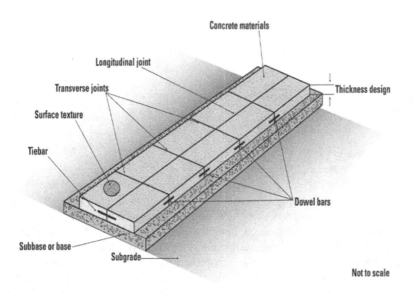

Figure 3.6 Concrete Pavement Structure

widths or shoulders are paved at the same time. In contrast, longitudinal construction joints are used between lanes or shoulders paved at different times.

The performance of concrete pavements depends to a large extent upon the satisfactory performance of the joints. Most jointed concrete pavement failures can be attributed to failures at the joint, as opposed to inadequate structural capacity. Distresses that may result from joint failure include faulting, pumping, spalling, corner breaks, blowups, and mid-panel cracking. Characteristics that contribute to satisfactory joint performance, such as adequate load transfer and proper concrete consolidation, have been identified through research and field experience. The incorporation of these characteristics into the design, construction, and maintenance of concrete pavements should result in joints capable of performing satisfactorily over the life of the pavement. Regardless of the joint sealant material used, periodic resealing will be required to ensure satisfactory joint performance throughout the life of the pavement. Satisfactory joint performance also depends on appropriate pavement design standards, quality construction materials, and good construction and maintenance procedures.

Jointed Plain Concrete Pavement: Jointed plain concrete pavement, or JPCP, consists of unreinforced concrete slabs 3.6–6.0 m (12–20 ft) in length with transverse contraction joints between the slabs. The joints are spaced closely enough so that cracks should not form in the slabs until late in the life of the pavement, as it approaches failure. Therefore, for JPCP, the pavement expansions and contractions are addressed through joints. JPCP is the most commonly used type of concrete pavement because it is usually the cheapest to construct. It is economical because there is no need to pay for any reinforcing steel in the slabs or for labor to place the steel. In most regions, contractors will also have more familiarity with JPCP than with other types of concrete pavement. In those regions where corrosion of steel is a problem, the absence of steel reinforcement means less concern for steel corrosion issues, although the steel dowels can still corrode.

Main performance issues of JPCP include

- Initial pavement smoothness, which is a function of construction practices
- Adequate pavement thickness is necessary to prevent mid-slab cracking
- Limiting the joint spacing also to prevent mid-slab cracking
- Adequate joint design, detailing, and construction

Jointed Reinforced Concrete Pavement: Jointed reinforced concrete pavement, or JRCP, is distinguished from JPCP by longer slabs and lighter reinforcement in the slabs. This light reinforcement is often termed temperature steel. JRCP slab lengths typically range from 7.5 to 9 m (25–30 ft), although slab lengths up to 30 m (100 ft) have been used. With these slab lengths, the joints must be doweled. The slab steel content is typically in the range of 0.10–0.25% of the cross-sectional area, in the longitudinal direction, with less steel in the transverse direction. Either individual reinforcing bars or wire fabrics and meshes may be used. Because the steel is placed at the neutral axis or midpoint of the slab, it has no effect on the flexural performance of the concrete and serves only to keep cracks together.

Although JRCP was widely used in the past, it is less common today. The only advantage that JRCP has over JPCP is fewer joints, and this is outweighed by the cost of the steel and the poor performance of the joints and the cracks. Because the joints are spaced further apart than JPCP, they open and close more, and load transfer suffers as joints open wider. JRCP joints always use dowels. Furthermore, even though the slabs are longer, the cracks still form at approximately the same interval as JPCP; therefore, JPCP slabs generally have one or two interior cracks each. The light steel reinforcement across these cracks is often not enough to maintain load transfer, and the cracks may fault.

Continuously Reinforced Concrete Pavement: Continuously reinforced concrete pavement, or CRCP, is characterized by heavy steel reinforcement and an absence of joints except for construction and isolation joints. Much more steel is used for CRCP than for JRCP, typically on the order of 0.4–0.8% by volume in the longitudinal direction. Steel in the transverse direction is provided in a lower percentage as temperature steel.

Well-designed and constructed CRCPs accomplish the following:

- Eliminate joint-maintenance costs for the life of the pavement, helping meet the public's desire for reduced work zones and related traveler delays.
- Provide consistent transfer of shear stresses from heavy wheel loads, resulting in a consistently quiet ride and less distress development at the cracks.

Such pavements can be expected to provide over 40 years of exceptional performance with minimal maintenance.

Cracks form in CRCP approximately 0.6–2 m (2–6 ft) apart. The reinforcement holds the cracks tightly together and provides for aggregate interlock and shear transfer. CRC pavements require anchors at the beginning and end of the pavement. The anchors keep the ends from contracting due to shrinkage and to help the desired crack pattern develop.

The use of CRCP dates back to the 1921 Columbia Pike experimental road in Virginia, with over 22,500 km (14,000 miles) of two-lane highway built in the United States by 1982. CRCP has also been used for major airports around the world. Because of the steel reinforcement, CRCP costs more than JRCP and is thus used less frequently in most regions. However, it provides a smoother ride and a longer life than any other type of pavement and is therefore preferred type in

Texas and Illinois. In many regions, the performance of CRCP has been excellent. The Long-Term Pavement Performance (LTPP) database to investigate 14 CRCP sections in the states of Alabama, Florida, Georgia, Mississippi, North Carolina, and South Carolina. At the time of the study, the sections were 21–30 years old and had carried heavy traffic but were generally in very good-to-excellent condition. With three exceptions, the pavements had serviceability indices of 4 or better despite the fact that they had already exceeded their 20-year design lives.

Key performance considerations for CRCP include

- Initial pavement smoothness
- Adequate pavement thickness is necessary to prevent excessive transverse cracking
- Adequate reinforcing steel to hold cracks together and prevent punchouts is essential. Punchouts are a distress mechanism distinct to CRCP

3.3.1.3 Overlays

Concrete pavements may also be used as overlays for either existing asphalt or concrete pavements. For each of the two existing pavement types, there are two overlay classifications based on whether the overlay is bonded to the existing pavement or whether bond is either ignored or prevented, and thus not considered in the design.

The performance of pavements is significantly determined by the underlying support layers, notably in terms of stability, bearing strength, time-dependent consolidation, and vulnerability to moisture. Often, one or multiple layers are introduced between the soil subgrade and the pavement itself.

Concrete pavements disperse loads more broadly than asphalt ones, resulting in reduced pressures on the subbase and subgrade. Consequently, the bearing capability of underlying layers is less of a concern, and there's typically no requirement for rigid base materials, except for roads bearing the most substantial loads. While it's uncommon for concrete highways to possess both a base and a subbase layer, airfield pavements accommodating large aircraft often do. Ensuring meticulous design and construction of subgrades and subbases is pivotal for the structural integrity and smoothness of all pavements. The specifications for concrete pavements can vary widely, influenced by factors such as soil type and environmental conditions.

3.3.2 Flexible Pavement

The hot-mix asphalt layer consists of a layer of asphalt mixture. It is designed to withstand high tire pressure, resist traffic abrasive forces, provide a skid-resistant driving surface, resist surface distresses, minimize traffic noise, and allow for the rapid drainage of surface water to prevent surface water from penetrating into the underlying layers. Depending on the traffic volume, the thickness of the surface layer can range from 3 inches to more than 6 inches.

The base course serves as the primary structural layer in flexible pavements, effectively dispersing wheel loads to ensure that the strength limits of the subbase and subgrade aren't surpassed. Typically, this layer comprises granular elements, such as crushed stone, slag, gravel, or sand. It's crafted to be both dense and stable to ward off fatigue cracking and structural warping. As a result, the criteria for the base course, encompassing plasticity, gradation, and strength, are more stringent compared to the subbase layer. If certain materials don't align with these standards, they can still be utilized if properly stabilized with components like Portland cement, asphalt, or lime.

Figure 3.7 Illustration of Dense, Open, and Gap Mixes

The subbase course typically comprises materials of slightly inferior quality compared to those used for the base. Serving as a structural component of the flexible pavement, it additionally helps distribute wheel loads to the subgrade. The materials chosen for this layer should also align with specifications concerning gradation, plasticity, and strength. Another crucial role of the subbase is to shield the subgrade from potential harm inflicted by construction and site vehicles during the building process. If the subgrade is robust and stable, a subbase might not be necessary.

The subgrade is usually the in-situ natural soil or a layer of well-compacted borrow material that serves as the foundation of the pavement structure.

Flexible pavement incorporates various mix types. As depicted in Figure 3.7, these include dense grade, open graded, and gap grade HMA mixes.

3.4 Pavement Selection Process

In pavement design, it's crucial to focus not just on technical aspects like thickness and structure. Design considerations should extend beyond height to compare alternative options in terms of economic benefits. Initial capital costs are necessary, but it's equally essential to factor in long-term maintenance and life-cycle costs. Therefore, a comprehensive cost analysis should be conducted to select the most cost-effective option that meets technical requirements.

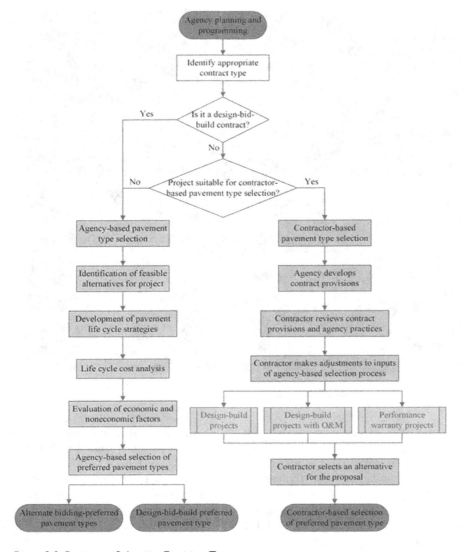

Figure 3.8 Pavement Selection Decision Tree

Selecting pavement type is an integral part of a pavement management system. The traditional method is to perform a life-cycle cost analysis (LCCA) to compare the costs of alternatives during a performance period. Recently, pavement type can be selected through the alternate design or alternate bidding procedure that involves the bidding contractors' opinions. The selection process is challenging because short-term and long-term performance, initial and life-cycle costs, and highway user impacts must be considered and balanced. As shown in Figure 3.8, pavement selection can be either agency-based (the agency makes the decision) or contractor-based (the decision is made by the contractor) depending on feasible pavement alternatives, as well as economic and non-economic factors (Hallin, Sadasivam, Mallela, Hein, Darter, & Von Quintus, 2011).

Typical economic factors include

- Initial costs
- Rehabilitation costs
- Maintenance costs
- User costs
- Life-cycle costs

Typical non-economic factors include

- Roadway/lane geometrics
- Continuity of adjacent pavements and lanes
- Traffic during construction
- Availability of local materials and experience
- Conservation of materials/energy
- Local preference
- Stimulation of competition
- Noise
- Safety
- Subgrade soils
- Experimental features (e.g., the performance of new materials or design concepts)
- Future needs
- Maintenance capability
- Sustainability

Agency-based pavement selection is preferred for design-bid-build projects. For design-build projects, pavement types can be selected using either agency-based or contractor-based processes. Feasible pavement alternatives can be identified from an approved list of pavement types. LCCA is performed using either a deterministic or probabilistic approach. The cost-effectiveness of alternatives is evaluated, and the feasible ones are then compared using non-economic factors. If an alternative is cost-effective and has no non-economic disadvantages, this alternative is considered the most preferred pavement type. If there is more than one qualified alternative, they are evaluated using the screening matrix (Table 3.4).

3.5 Life-Cycle Cost Analysis

Pavement management systems (PMS) are pivotal in refining the methodologies associated with highway construction, upkeep, and restoration. By integrating an exhaustive compilation of past data on construction materials, ongoing on-site road assessments, and cutting-edge software tools, PMS delivers critical insights, enabling informed policymaking for government entities. The predictions and recommendations generated by PMS for routine maintenance, major maintenance, or rehabilitation will be used in LCCA. PMS allows agencies to consider a spectrum of investment strategies, striving to enhance road longevity and quality within set financial parameters. With this system in place, authorities can precisely ascertain the funds needed to hit particular performance standards, thus equipping decision-makers with a grounded understanding of feasible outcomes. This topic is explored comprehensively in Chapter 12, Transportation Asset Management.

Table 3.4 Pavement Selection Screening Matrix Example – Group Structure and Weights

Group	Description Example	Weight Example
Group A	Economic factors	50%
Factor A1	Initial costs	
Factor A2	Rehabilitation costs	
…	…	
Factor An	User costs	
Group B	Construction factors	25%
Factor B1	Continuity of adjacent lanes	
Factor B2	Traffic during construction	
…	…	
Factor Bn	Lane geometrics	
Group C	Local factors	10%
Factor C1	Availability of local materials	
Factor C2	Local preferences	
…	…	
Factor Cn	Stimulation of competition	
Group D	Other factors	15%
Factor D1	Noise	
Factor D2	Subgrade soils	
…	…	
Factor Dn	Experimental features	

In the realm of new pavement construction, or old pavement reconstruction or rehabilitation, and selecting pavement alternatives, mere adherence to technical specifications and initial investment scrutiny is insufficient. It is imperative that all endeavors, be it new construction, revamping, restoration, or routine maintenance, incorporate an economic assessment of diverse design options. This ensures the identification of the most economically viable approach and optimal execution timeline over an extended service duration. In this context, life-cycle cost analysis stands out as a trusted tool, aiding in assessing long-term investment efficacy. LCCA, rooted in economic analysis, evaluates the long-term economic efficiency of various investment avenues. Beyond initial costs, it factors in future discounted expenses from agencies, users, and other relevant stakeholders over an investment's lifespan. The objective is clear: to identify investments that achieve desired performance at the most cost-effective rate (JEGEL, 1997).

LCCA is defined by the FHWA as an analysis technique that builds on the well-founded principles of economic analysis to evaluate the overall long-term economic efficiency between competing alternative investment options. It does not address equity issues. It incorporates initial and discounted future agency, user, and other relevant costs over the life of alternative investments. It attempts to identify the best value (the lowest long-term cost that satisfies the performance objective being sought) for investment expenditures (FHWA report FHWA-SA-98-079).

LCCA serves as an invaluable decision-making tool when determining pavement selection, structural and mix choices (for flexible pavements), construction methodologies, as well as maintenance and rehabilitation tactics. Generally, LCCA is executed through the following fundamental stages:

- Setting Initial Strategies and Analysis Parameters: Before diving deep into LCCA, it's essential to set preliminary decisions, projections, and assumptions to set the groundwork for the analysis.

- Projecting Costs: It's crucial to calculate the costs related to both the agency in charge and the end-users for each option available.
- Evaluating Options: This step usually requires representing each choice through a universally accepted measure, like net present value or the benefit-cost ratio, making it easier to draw comparisons.
- Interpreting Outcomes and Reassessing Choices: Post-analysis, it's essential to delve into the results, pinpointing the dominant costs, elements, and assumptions. Sensitivity analyses often aid in achieving this clarity. Based on this detailed assessment, initial design strategy options should be revisited to enhance the cost-effectiveness of each solution.

3.5.1 Steps in the LCCA Process

3.5.1.1 Establish Alternative Design Strategies

The LCCA process begins pavement design when the various design alternatives have been identified sufficiently so that their costs can be reasonably estimated. Each design alternative is a combination of initial pavement design and associated future preservation, rehabilitation, and reconstruction activities. Typically, two or three design alternatives will be considered, all of which all meet technical requirements; and there is a need to compare them economically. For consistency, all strategies must be set forth for the same analysis period (FDOT, 2019).

It is recommended that a 35–40-year analysis period be used, a length of time usually sufficient to include several major activities in the rehabilitation and reconstruction cycle. The pavement's design service life does not need to coincide with the analysis period. All pavement designs must consider work zone requirements during initial construction and any subsequent activities. Traffic control, detours, temporary drainage, environmental monitoring and control, and other efforts may add to the agency's costs and significantly impact the traveling public through user costs.

3.5.1.2 Determine Performance Periods and Activity Timing

The performance life of both the initial construction and any subsequent preservation, rehabilitation, and reconstruction activities has a significant impact on the life-cycle cost analysis results that need to be determined. If an initial pavement construction, at year 0, is expected to last 15 years, with an overlay to be placed after that performance period, it follows that the first overlay is planned for year 15.

A pavement management system (PMS) can provide data on the status of the road network and performance prediction to determine an optimum long-term pavement maintenance and rehabilitation program. A proper PMS can assess the condition of each road, predict the year in which each road section will fall below a minimum pavement quality index, and suggest the most cost-effective rehabilitation strategies.

3.5.1.3 Estimate Agency and User Costs

These costs include planning, design, contract administration, construction, traffic control, environmental monitoring and control, and any other costs from the project budget. Note that these costs are present not only for initial construction but also for future preservation, rehabilitation, and reconstruction activities.

3.5.1.4 Develop Cash Flow Diagrams

Cash flow diagrams serve as straightforward visual tools that depict the expected timeline for different cash flows linked with an option. Typically, the commencement of the cash flow duration is designated as year 0. The chronological sequence of years is plotted on a horizontal axis, and every notable cash flow is represented by an arrow, indicating the specific year of its occurrence. Even though each cash flow transpires over a continuous span, for simplicity and visualization purposes, it's conventionally considered to manifest at a singular time.

3.5.1.5 Compute the Present Worth of Costs for Each Alternative

This method allows alternatives with different cash flow values and patterns to be compared in a way that reflects the time value of money. The result can be thought of as the amount that would be needed at the beginning of the project to fund the initial and future net costs if any unspent amounts were invested at the given discount rate until needed. The alternative with the lowest present worth is considered the most economically attractive.

3.5.2 Components of Pavement Life-Cycle Cost Analysis

3.5.2.1 Present Worth Analysis

For a comprehensive evaluation of structurally similar design alternatives that utilize different materials for pavements, one must look beyond just the immediate costs associated with each option. A thorough assessment should encompass the cumulative costs over the design's lifespan. Sometimes, the alternative that is most affordable upfront might not prove to be the most economical choice when other aspects, like maintenance, rehabilitation, inflation, and the opportunity cost of capital (essentially, the benefits of investing money today for potential returns in the future, or interest), are factored in. Life-cycle cost analysis stands out as the most reliable approach to gauge the economic viability of different asphalt pavement designs.

The present worth technique has garnered widespread acceptance among organizations that employ life-cycle cost analyses. Accurate forecasting of the life-cycle expenses of each alternative using this method necessitates a clear understanding of the inflation rate, interest rate, and discount rate.

It's crucial to note that the inflation rate (which denotes the proportional escalation in prices for goods, such as construction materials) and the interest rate (which signifies the potential earnings from investments) are influenced by the prevailing economic conditions. The discount rate, which captures the nominal appreciation of money's value over time, is deduced from both the interest and inflation rates, as represented by the following formula:

$$Discount\,rate(\%) = \frac{interest\,rate(\%) - inflation\,rate(\%)}{1 + inflation\,rate(\%)/100}$$

The present worth technique balances the current and future costs of each alternative, including associated maintenance and rehabilitation expenses, over the project's duration. This discounting principle facilitates a comparison of options with expenditures spread over a lengthy time frame, enabling designers to account for the intertwined effects of interest rates and inflation on the total project cost.

Choosing the right discount rate is a critical step in the life-cycle cost analysis. While there have been instances when the gap between inflation and interest rates was notably wide, these episodes were typically short-lived. Some projects have utilized a 6% discount rate, coupled with sensitivity analysis to gauge the influence of fluctuating discount rates. Generally speaking, elevated discount rates lean towards alternatives that either have reduced costs or necessitate heightened maintenance, especially when such expenses accrue toward the project's latter stages. It's vital to note that the chosen discount rate should mirror the funding source of the project – whether it's a governmental or private sector loan. Financial institutions usually offer somewhat more favorable rates to government bodies compared to private entities.

When it comes to examining how shifts in the discount rate influence the life-cycle cost analysis for diverse pavement designs, the relationship is quite straightforward. An upward adjustment in the discount rate means that pavement designs with a reduced upfront cost tend to showcase more favorable life-cycle costs, even if they demand increased and more regular maintenance. Marginal fluctuations (in the ballpark of 1–2%) in the discount rate typically don't drastically alter the comparative rankings of different pavement options.

Several critical cost components shape the results of a life-cycle cost analysis. These components were meticulously evaluated during the assessment of pavement design alternatives that are structurally on par. Beyond the considerations of inflation, interest, and discount rates, other vital components include

- Initial Costs: Encompasses all expenses related to the construction of the chosen pavement design, often referred to as the capital expenditure.
- Maintenance Costs: Expenditures associated with regular maintenance activities, such as crack sealing, which extend the pavement's functional lifespan.
- Rehabilitation Costs: Costs linked to significant maintenance endeavors like resurfacing, reconstruction, or enhancements, mandated when the pavement quality (such as ride comfort) drops to a certain serviceability threshold. This threshold typically aligns with the road or highway's classification.
- Residual Value: Represents the remaining utility or service life from any maintenance or rehabilitation activity by the analysis period's conclusion. This is generally interpreted as a negative cost or a credit.
- Salvage Value: The worth of any components that can be repurposed post the analysis period. This is usually accounted for as a negative cost or credit. A prime example is reclaimed asphalt pavement (RAP), which boasts a recyclability rate of 100%.
- User Delay Costs: This pertains to a range of user expenses: from vehicle operational costs, the monetary value of user travel time, and construction-induced traffic delays, to costs resulting from accidents and discomfort-related costs. Quantifying these user costs can be challenging, and they should be framed as additional expenses beyond what's typically expected.

By considering these components, a more comprehensive and accurate life-cycle cost analysis can be achieved, ensuring that all potential costs and benefits are captured in the decision-making process.

Table 3.5 shows the factors that influence the life-cycle cost analysis results.

Table 3.5 Factors Affecting LCCA

Item	Relative Influence on Alternative Selection
Inflation Rate	Low
Interest Rate	Low
Discount Rate	High
Initial Costs	Moderate to High
Maintenance Costs	Moderate
Rehabilitation Costs	Moderate
Residual Value	Low
Salvage Value	Low
User Delay Costs	Low to Moderate

3.5.2.2 Initial Costs

The initial costs for pavement projects can be determined by analyzing the bid unit costs (prices) for typically similar-sized projects in the same jurisdiction. As such, this cost information may not directly apply to specific projects affected by local materials, project scale, contractor pricing, and project-specific conditions. The initial construction cost of each alternative was then estimated using typical prices for each primary construction material, component, or process. The unit costs can be the average of the three lowest bid prices or the average of all the bid prices for pavement construction and rehabilitation projects. As the life-cycle cost comparison was intended to determine the relative cost of each alternative, minor differences in unit prices do not significantly affect the analysis results. This can be confirmed through a sensitivity analysis.

3.5.2.3 Maintenance Costs

In calculating the life-cycle costs of various asphalt pavement alternatives, the timing of regular maintenance activities, such as pothole repairs, spray patching, localized base restorations, hot-mix patching, and pavement marking, can be referenced from estimations made available by authoritative bodies like the Transportation Research Board (TRB), the National Asphalt Pavement Association (NAPA), and other State Highway Agencies (SHA).

Regular maintenance measures can be planned throughout the lifespan of an asphalt pavement. For instance, a structured regimen for routing and sealing can be set in motion, where every pavement crack is duly routed and sealed at regular intervals, either after the initial construction or following the installation of an overlay, and periodically thereafter. Prompt sealing of these fissures is among the most economically viable preventive maintenance practices. Undertaking this practice in accordance with endorsed protocols can prolong the pavement's service life by a duration ranging between two to five years.

3.5.2.4 Rehabilitation Costs

Major rehabilitation or maintenance is typically necessary for asphalt pavements to ensure they remain at or surpass a set minimum serviceability threshold. Additionally, these measures serve to prolong the lifespan of the asphalt pavement options throughout the duration of the considered life-cycle cost analysis. The timing for such interventions is contingent upon the type of asphalt pavement materials used and the routine maintenance strategy implemented. It's imperative that both the materials and the processes are of high quality.

To forecast the optimal schedule for pivotal maintenance and rehabilitation tasks, it's advisable to reference performance metrics for individual asphalt pavement elements available in scholarly publications. These can be further enriched with data from transportation authorities. Accurate predictions regarding the longevity of new pavements, as well as the duration of the effects of pavement maintenance and rehabilitation efforts, are pivotal to a comprehensive life-cycle cost analysis.

3.5.2.5 Residual and Salvage Values

In addition to the initial construction costs and the ongoing expenses for maintenance and rehabilitation throughout the pavement's lifespan, residual and salvage values are crucial components in a comprehensive life-cycle cost analysis. The salvage value refers to the benefit or value of a material that can be repurposed or reused after its initial service life. On the other hand, the residual value denotes the remaining service life or 'value' of the rehabilitation effort as the analysis period concludes.

For the asphalt pavement design alternatives explored in this study, a baseline assumption was that the pavement condition rating (PCR) would consistently be upheld at an acceptable threshold. To achieve this, it's essential to incorporate some rehabilitation measures, such as an asphalt overlay, ensuring the pavement's service life extends to the terminal point of the life-cycle cost analysis duration. Depending on the particular alternative, the maintenance and rehabilitation initiatives could possess residual value extending beyond the analysis timeframe. Notable rehabilitation efforts, like asphalt overlays, were assigned specific residual values. In the context of the present worth analysis, these values, after discounting, were incorporated as credits.

The salvage value for each asphalt pavement design alternative was assessed at the conclusion of a 40-year life-cycle analysis duration. As the life-cycle cost analysis period extends, the influence of the salvage value on the present worth of every pavement option diminishes. At the end of the analysis duration, a standard salvage value, equivalent to 5% of the initial capital expenditure for constructing the pavement structure, was designated.

3.5.2.6 User Delay Costs in Analysis

In the life-cycle cost analyses, user delay costs are integrated to account for the disruption faced by roadway users due to scheduled maintenance or rehabilitation undertakings. The monetary value of the delay is ascertained by quantifying the costs associated with the time the pavement is inaccessible to the public. Several organizations have formulated standard values for user delays, which should be adjusted for inflation and incorporated into the life-cycle cost evaluation.

Modern contracting approaches recognize the effect of maintenance or rehabilitation on road users. Major maintenance and rehabilitation tasks are typically planned for post-peak traffic hours and usually do not require continuous lane occupation.

These considerations are normally mirrored in the calculated user delay costs. Maintenance tasks, for instance, are presumed to create delays primarily outside of peak traffic times, affecting a minor portion (about 10%) of the average annual daily traffic (AADT). In contrast, rehabilitation tasks, while also assumed to cause delays during non-peak times, span a more extended duration compared to maintenance tasks.

Vehicle operation costs, encompassing factors like augmented fuel usage and vehicle upkeep expenses, can also be incorporated into a pavement life-cycle cost assessment. This analysis operates on the premise that all considered asphalt pavements will be maintained at analogous ride quality standards. As a result, the influence of vehicle operation costs on the distinct asphalt pavement design life-cycle expenses is negligible. Consequently, this study has chosen to exclude vehicle operation costs.

For this analysis, a streamlined approach was employed to integrate user-delay costs. Typically, all maintenance tasks were given equal weight, implying they all induced similar delays. Conversely, all rehabilitation tasks were seen as equivalent but causing longer delays than their maintenance counterparts.

3.6 Perpetual Pavement

A perpetual pavement is defined as an asphalt pavement designed and built to last longer than 50 years without requiring major structural rehabilitation or reconstruction, needing only periodic surface renewal in response to distresses confined to the top of the pavement.

Perpetual pavements, or long-life asphalt pavements, are not new. Full-depth and deep-strength asphalt pavement structures have been constructed since the 1960s, and those that were well-designed and well-built have been very successful in providing long service lives under heavy traffic. According to the 1990 European asphalt study tour (FHWA/AASHTO/NAPA), perpetual or long-life flexible pavements had been widely used in Europe. New technologies have contributed to the development of long-life pavement, which includes new mix designs, stone mastic asphalt (SMA), for instance; porous asphalt and high-tech surface treatments; new attitudes toward quality and longevity in pavements – "perpetual" pavement; SHRP activities, Superpave (construction equipment, April 1997), AASHTO 93 – pavement structure – serviceability and fatigue (if checked), AASHTO 2002 (mechanistic) fostering in the United States and Canada, which are straightforward to implement.

International technical literature has presented long-life asphalt pavements (LLAP) and distress type(s) newly identified in recent years, such as top-down cracking (TDC) in the United States. International practical experience in France, Belgium, and the United Kingdom, for instance, indicates that the deterioration of thick, well-constructed asphalt concrete pavements is not generally structural and that the deterioration generally starts at the surface as cracking, i.e., TDC, and surface course asphalt concrete rutting, such as non-structural deterioration. This implies that an adequately designed flexible pavement built with appropriate strength and quality will remain structurally serviceable for a lengthy design life, provided a pavement management strategy/system (PMS) and maintenance management systems (MMS) are implemented to ensure that systematic maintenance and pavement preservation, such as crack sealing and surface course HMA renewal (3 to 4 cm, on about a 20-year cycle, are adapted to detect/monitor and mitigate non-structural deterioration (cracks and rutting) before it impacts structural performance. Good pavement foundations (strong, drained, stable, and consistent) are essential to preclude deep distress within the pavement structure (APA, 2001, 2002, 2004).

3.6.1 Pavement Structure Design

The long-life flexible (asphalt) pavement structure design and associated technology involved the following work to meet the components completion timeline, particularly subgrade, base and subbase, pavement surface, and warranty period. The design of recommended long-life flexible pavement structures included assistance with site CBR, Falling Weight Deflectometer (FWD), and Dynamic Cone Penetrometer (DCP) testing and characterization of representative subgrade soil (fill) samples, granular materials, and asphalt concrete. The activities involved preliminary design(s), materials characterization, final design(s), complete life-cycle performance, and life-cycle cost analysis comparisons of the recommended long-life flexible pavement structures with the current semi-rigid asphalt pavement structure, including a value engineering comparison.

The basic premise of obtaining a long pavement life is that an adequately thick HMA pavement placed on a stable foundation will preclude distresses that originate at the bottom of the pavement and eventually require expensive reconstruction to correct properly. Structurally, the pavement must have the proper combination of thickness and stiffness to resist deformation in the foundation material or the underlying subgrade. Likewise, the HMA layers must be thick enough and have the right properties to resist fatigue cracking originating at the bottom of the structure. Currently, most pavement design procedures do not consider each pavement layer's contribution to resisting fatigue, rutting, and temperature cracking in the structure. Since each pavement layer has its unique part to play in performance, an improved structural design method is needed to analyze each pavement layer.

The foundation of the pavement is fundamental to the establishment and efficiency of perpetual pavement. During its installation, this foundation offers a sturdy platform that accommodates heavy dump trucks and paving machinery used to lay the HMA layers. Moreover, it provides resistance against deflection from rollers, ensuring the top layers of the pavement are compressed adequately. As the pavement goes into its operational phase, the foundation plays an indispensable role in bearing traffic loads and minimizing inconsistency in support across different seasons, especially with factors like freeze-thaw and moisture variations. Ensuring the foundation's proper design and build is paramount to counter volume changes due to cycles of wetting and drying in expansive clays or freeze-thaw shifts in frost-prone soils.

Perpetual pavement design, among other design alternatives, must be subject to a life-cycle cost analysis, just like other pavement designs. The design that offers the lowest life-cycle cost will be earmarked as a suitable choice. The concept of perpetual pavement design is illustrated in Figure 3.9, while Figure 3.10 draws a comparison between perpetual and standard pavement designs.

3.7 Reconstruction and Rehabilitation Design Selection

Similar to the new pavement design, for reconstruction and rehabilitation design, the life-cycle cost analysis needs to be used to evaluate long-term cost-effectiveness. There is an increasing demand to make better use of shrinking resources for rehabilitation and maintenance, and a variety of strategies are available to meet these demands. Selecting the most cost-effective strategy for pavement rehabilitation continues to be a significant challenge to the transportation professional.

3.7.1 Systematic Strategy Selection

Rehabilitation design for flexible and rigid pavement must account for all applicable parameters and their impacts on choosing alternatives. These parameters may be both pavement and non-pavement related. Agency experience, initial cost of rehabilitation, anticipated maintenance, and future rehabilitation requirements also influence strategy selection.

Policy decision-making that advocates applying the same standard fixes to every pavement does not produce successful pavement rehabilitation. Successful rehabilitation depends on decisions based on the specific condition and design of individual pavements.

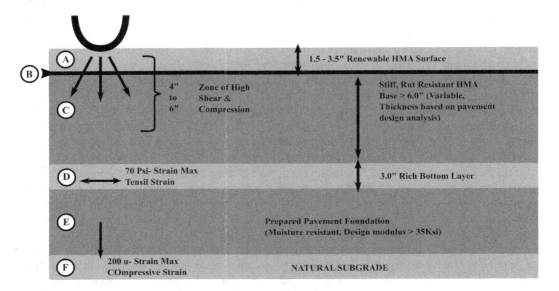

Figure 3.9 Illustration of Perpetual Pavement Design

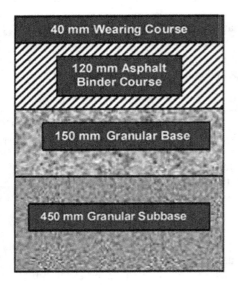
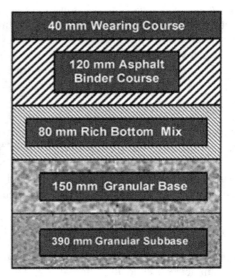

Figure 3.10 Comparison between Standard and Perpetual Pavement Designs

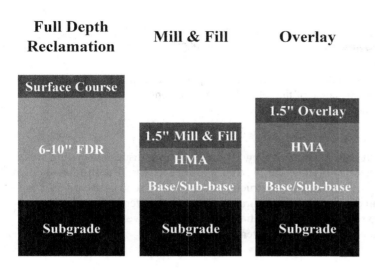

Figure 3.11 Flexible Pavement Rehabilitation Methods

Assessing the effectiveness of rehabilitation alternatives must include examining project conditions, life-cycle costs, and constraints on strategy selection.

3.7.2 Project Information

Detailed project information is important for the selection of the preferred rehabilitation strategy. Five basic types of information are necessary: design, construction, traffic, environmental, and pavement condition.

3.7.2.1 Design Data

The first information an engineer needs is the design parameters for the existing pavement. This data includes the pavement type and thickness.

Detailed information about the pavement components, including layer materials and strengths, joint design, shoulder design, drainage system, and previous repair or maintenance activities.

3.7.2.2 Construction Data

Investigation of the conditions during the construction of the original pavement can often give insight into the causes of existing pavement distress. Field books or daily log notes are excellent sources of construction information. Notes of problems or weather conditions are particularly helpful.

3.7.2.3 Traffic Data

Strategy selection requires past, current, and expected traffic growth. This helps determine the remaining effective structural capacity of the existing pavement. Knowing the structural capacity helps narrow the number of potentially effective rehabilitation options.

3.7.2.4 Environmental Data

Precipitation, temperature, and freeze-thaw conditions are important factors. Each of these factors acts on the materials and layers of a pavement system to influence material integrity, structural capacity, and rideability.

Understanding how these factors have affected the existing pavement can help a designer prevent potential problems in the performance of the rehabilitated pavement.

3.7.2.5 Distress/Condition Data

A condition survey should report the type, severity, and quantity of each distress. Isolating areas of distress can help pinpoint different solutions for different project sections. The distress should be characterized as load- or climate related. Non-destructive testing (NDT) and destructive testing (i.e., cores and borings) can help determine the structural condition and material properties below the surface. A visual or photo log survey is also essential. The main objective is to collect enough information to be able to characterize the type and extent of deterioration.

3.7.3 Evaluation

Pavement evaluation is a crucial step to draw from all project information to determine the cause of pavement distress. These factors are measurable from actual responses to destructive and non-destructive testing methods. Using assumed values provides no insurance that the rehabilitation strategy will actually meet design expectations. An engineering evaluation must address several key issues, such as functional and structural conditions, materials conditions, drainage conditions, and lane condition uniformity.

3.7.3.1 Functional and Structural Condition

In evaluation, the engineer must determine the structural condition of the pavement. Structural deterioration is any condition that reduces the load-carrying capacity of a pavement. Evaluating the level of structural capacity requires a thorough visual survey and material testing.

3.7.3.2 Materials Condition

Material problems such as asphalt aging, stripping, raveling, and rutting, for flexible pavement or alkali-silica reactivity (ASR) and D-cracking for rigid pavement can significantly reduce the integrity of the pavement. The severity of material distress may impact the number of possible candidate rehabilitation procedures.

3.7.3.3 Drainage Condition

Drainage problems usually manifest in noticeable pavement distress. Any presence of pumping joints or severe asphalt rutting indicates drainage deficiencies. The condition of ditches and drain system outlets is also important indicator.

3.7.3.4 Lane Condition Uniformity

On many four-lane routes, the outer truck lane deteriorates at a more rapid pace than the inner lane or shoulders. This condition is necessary to note during an evaluation. Significant savings may result from repairing only the pavement lane that requires treatment.

3.7.4 Rehabilitation Choices

For concrete pavement restoration, include partial-depth repair, full-depth repair, dowel bar retrofit, diamond grinding, joint and cracking resealing, cross stitching, slab stabilization, resurfacing, overlay, whitetopping, ultra-thin whitetopping (UTW), and asphalt surfacing.

For flexible pavement, the methods include cold milling, resurfacing, full-depth reclamation, and cold in-place recycling with lime, cement, and fume asphalt stabilization.

Repair work is almost always necessary before placing an overlay on existing pavement. The quantity of repair before overlay may affect the required thickness of the overlay. In some situations, better and more extensive pre-overlay repair permits a reduction in overlay thickness.

Reconstruction involves the complete removal of the pavement structure, typically but not always including the base layer(s). The structure is replaced with a new pavement or inlay. Reconstruction techniques offer the choice of using virgin or recycled materials. The use of recycled material can often lower project costs.

Rehabilitation timing is a fundamentally important ingredient in pavement rehabilitation. As pavement deteriorates, the type of rehabilitation that is most appropriate changes. It should be understood that each strategy has its most appropriate application during a particular portion of the pavement life. Proper timing of rehabilitation techniques, along with proper design and construction, is essential for good performance.

3.8 Questions

1. Define "flexible pavement" and "rigid pavement".
2. Point out two areas of distinct difference between rigid and flexible pavements.
3. What are the two major aspects of geometric design?
4. Why is flexible pavement constructed in several layers?
5. What are the main differences between the two types of pavement structure?
6. Explain rigid pavements: JPCP, JRCP, and CRCP. Which one is more commonly used? Which one is less used now?
7. Describe the concept of LCCA – life-cycle cost analysis.
8. What is the significance of selecting the appropriate pavement type for highways?
9. Why is the wearing surface's thickness crucial in constructing flexible pavements?
10. Describe the importance and role of the subgrade in a pavement structure.
11. Why is it important to focus on more than just the technical aspects, such as thickness and structure, in pavement design?
12. Beyond initial capital costs, what other types of costs should be factored into pavement selection?
13. What is the primary purpose of pavement management systems (PMS) in the context of highway construction and maintenance?
14. How does the life-cycle cost analysis (LCCA) aid in evaluating long-term investment efficacy for pavement construction and maintenance?

15. Why is it important to consider both the inflation rate and the interest rate when performing LCCA?
16. Why are good pavement foundations essential for the pavement structure?
17. What components are considered in the long-life flexible pavement structure design?
18. What is the importance of the foundation in the establishment and efficiency of perpetual pavement?
19. How do freeze-thaw and moisture variations impact the foundation of the pavement?
20. What types of information are crucial for the selection of the preferred rehabilitation strategy?

References

APA (2001). *Perpetual Pavement, Structure for the Future*. Asphalt Pavement Alliance (APA), Lexington, KY.

APA (2002). *Perpetual Pavement – A Synthesis*. Asphalt Pavement Alliance (APA), Lexington, KY.

APA (2004). *Pavement Type Selection Processes*. Asphalt Pavement Alliance (APA), Lexington, KY.

Delatte, N.J. (2014). *Concrete Pavement Design, Construction, and Performance*. 2nd ed. New York: Taylor and Francis Group.

JEGEL (1997). *Life Cycle Cost Effectiveness in Hot-Mix Asphalts Pavements in Ontario*. Ontario Hot-Mix Producers Association (OHMPA), Mississauga, Ontario.

FDOT (2019). *Pavement Type Selection Manual*. Florida Department of Transportation (FDOT), 605 Suwannee Street, Tallahassee, FL.

Hallin, J.P., Sadasivam, S., Mallela, J., Hein, D.K., Darter, M.I., & Von Quintus, H.L. (2011). *Guide for Pavement-Type Selection*. NCHRP Report 703. The National Academies Press.

Chapter 4

Transportation Planning, Geometric Design of Road, and Drainage

4.1 Comprehensive Transportation Planning: A Multifaceted Approach

Transportation planning is a complex and multi-faceted process that requires thoroughly analyzing a region's current transportation network while strategically preparing for its future needs. This field encompasses an array of factors, including budgetary constraints, overarching objectives, and prevailing policy frameworks, all of which play crucial roles in guiding the development trajectory of communities and cities. It encompasses various elements, from highways, streets, and maritime transport to public transit systems and dedicated bike lanes. The ripple effects of transportation planning are far-reaching and extend to various sectors, such as commerce and leisure, ultimately playing a significant role in shaping the quality of life experienced by residents.

4.1.1 Key Elements of Transportation Planning

4.1.1.1 Data Collection and Public Participation

Transportation planning requires establishing a solid foundation of accurate information, practical tools, and active mechanisms for public engagement. This underpinning is crucial to improving the overall efficiency and performance of the transportation system.

4.1.1.2 Influencing Choices

The valuable insights from transportation planning are pivotal in influencing many decisions. This includes the development and formulation of policies, the selection of strategic alternatives, the establishment of priorities, and the allocation of necessary funds.

4.1.1.3 Comprehensive Planning Framework

The transportation planning process extends beyond its traditional boundaries to incorporate a range of external factors, such as

- Land Use: Conforming to relevant state and local regulations
- Air Quality: Adhering to the mandates of the Clean Air Act and achieving air quality standards
- Environmental Policy: Observing the directives outlined by the National Environmental Policy Act (NEPA)

DOI: 10.1201/9781003197768-4

- Environmental Justice: Abiding by the provisions set forth in Title VI
- Accessibility: Ensuring compliance with the Americans with Disabilities Act

4.1.1.4 Demand Analysis

This facet involves a meticulous examination of prevailing travel patterns and commuter behaviors, with the aim of accurately identifying the specific transportation needs of particular regions or communities.

4.1.1.5 Infrastructure Assessment

This involves a comprehensive evaluation of the existing transportation infrastructure, including but not limited to roads, bridges, and public transit systems, to gauge their current state and effectiveness.

4.1.1.6 Traffic Optimization

Implementing measures such as refining traffic signal timing, effective lane management, and strategies for alleviating congestion.

4.1.1.7 Enhancement of Public Transportation

Concentrating on the improvement and expansion of public transit options, including buses, subways, light rail systems, and commuter trains.

4.1.1.8 Road Safety Measures

Tackling road safety concerns through design improvements, stringent traffic enforcement, and comprehensive public awareness initiatives.

4.1.1.9 Addressing Environmental Impact

Evaluating the environmental consequences of transportation initiatives and promoting eco-friendly alternatives.

4.1.1.10 Harmonizing Transportation and Land Development

Aligning transportation plans with regional or urban land-use strategies to support unified and balanced growth.

4.1.1.11 Financial Planning and Allocation

Identifying viable funding sources and meticulously developing budgets for future transportation projects.

4.1.1.12 Development and Implementation of Policies

Crafting and enforcing transportation policies and rules that guide the planning process.

4.1.1.13 Engaging with the Community

Proactively include the public and essential stakeholders to incorporate their views, needs, and concerns into the planning process.

4.1.2 Objectives of Transportation Planning

The primary goals of transportation planning include

- Establishing clear, forward-looking objectives and aspirations
- Assessing both the current and projected future conditions
- Conducting a thorough analysis of transportation needs
- Developing and implementing effective strategies for development
- Identifying, prioritizing, and securing funding for transportation projects
- Continuously evaluating the efficacy of the transportation infrastructure

At its core, comprehensive transportation planning lays the foundation for thriving, efficient, and sustainable communities, creating an environment that seamlessly integrates business, leisure, and daily life.

In the context of smart transportation, resilient infrastructure, and intelligent urban planning, transportation planning encompasses a multifaceted approach that incorporates numerous elements. Here is a detailed breakdown of the crucial considerations in each area.

4.1.2.1 Smart Transportation

- Integrated Systems: Leverage the Internet of Things (IoT) to connect and synchronize various transportation modes, traffic management systems, and user-based applications
- Data Analytics: Utilize big data and machine learning to analyze traffic patterns, predict peak periods, and optimize routes
- Autonomous Vehicles: Design infrastructure to accommodate self-driving cars and other autonomous modes of transportation
- E-Mobility: Facilitate the adoption of electric vehicles (EV) by providing charging infrastructure and incentives
- Multimodal Transportation: Encourage and plan for diverse modes of transportation, including walking, cycling, buses, and trains
- User Experience: Develop applications and services that allow users to plan routes, monitor real-time traffic conditions, and pay for public transportation

4.1.2.2 Resilient Transportation

- Climate Adaptation: Ensure transportation infrastructure is resilient against extreme weather events and related challenges

- Redundancy: Construct multiple routes and modes to maintain a continuous flow of transportation, even when one mode or route is compromised
- Rapid Recovery: Devise strategies and resources for swiftly restoring transportation services following disruptions
- Security: Implement measures to prevent and respond to various security threats, including physical (e.g., terrorist attacks) and cyber (e.g., hacking traffic management systems)
- Flexible Design: Embrace design principles that accommodate future modifications as needs and challenges evolve

4.1.2.3 Smart Cities

- Connectivity: Prioritize high-speed internet and other communication technologies to facilitate a connected urban ecosystem
- Sustainability: Promote green transportation options such as EVs, eco-friendly buses, and bike-sharing programs while ensuring urban designs support walkability
- Urban Planning: Integrate transportation planning with land use, housing, and public space design to foster holistic, user-centric urban environments
- Citizen Engagement: Leverage digital platforms to gather feedback and engage citizens in transportation and urban planning decisions
- Public-Private Partnerships: Foster collaboration with private entities, particularly tech start-ups, to harness innovative solutions for transportation and other urban challenges

A holistic approach is essential as we plan for innovative and resilient transportation within the realm of intelligent cities. This approach should focus on current needs and address future challenges, embrace technological advancements, and adapt to changing climate patterns. Transportation planners need to be forward-thinking, adaptable, and open to integrating innovative solutions into the urban fabric.

4.2 Standards for Geometric Design of Roads

Geometric design plays a crucial role in shaping the dimensions and layout of various highway elements, including vertical and horizontal curves, cross-sections, truck climbing lanes, bicycle paths, and parking facilities. This design process relies on in-depth traffic analysis that considers the behavior and characteristics of drivers, pedestrians, and vehicles, as well as the physical attributes of the road itself. These analyses then inform the determination of physical dimensions for each highway component. For instance, calculations for lengths of vertical curves or radii of circular curves are carried out to guarantee that drivers have the minimum stopping sight distance given the highway's design speed. The ultimate aim of geometric design is to establish a smooth and safe highway infrastructure that is achieved through adhering to a standardized design approach that caters to the requirements of both drivers and their vehicles (Underwood, 1991).

The American Association of State Highway and Transportation Officials (AASHTO) plays a critical role in developing guidelines and standards essential to the geometric design of highways. AASHTO comprises delegates representing each state's highway and transportation department across the United States alongside representatives from the Federal Highway Administration (FHWA). The association conducts meetings with various technical committees to evaluate design practices submitted by individual states. Following validation, these practices are incorporated into AASHTO's publication "A Policy on Geometric Design of Highways and Streets," a comprehensive guide that ensures consistency and coherence in national design standards. Although AASHTO's guidelines are available for adoption by individual state transportation departments, it's worth noting that the most recent edition (2018) doesn't deem existing streets and highways unsafe if designed based on previous editions, nor does it mandate initiating improvement projects.

This chapter aims to elucidate the principles underpinning the design of horizontal and vertical alignments and the current standards utilized in geometric design, as recommended by AASHTO. Our objective is to familiarize the readers with the various factors that influence highway design.

Highway geometric design is rooted in specific design standards and controls contingent upon several roadway system factors, including

- Functional classification
- Design hourly traffic volume and vehicle mix
- Design speed
- Design vehicle
- Cross-sections of the highway, such as lanes, shoulders, and medians
- Presence of heavy vehicles on steep grades
- Area topography
- Level of service
- Available funds
- Safety
- Social and environmental considerations

These factors often intersect and influence one another. For example, design speed is contingent upon functional classification, which typically correlates with expected traffic volume. Similarly, design speed may be influenced by topography, especially in cases where funding is limited. Typically, the principal factors dictating the standards for a specific highway design are expected traffic volume, level of service, design speed, and vehicle design. These factors, alongside essential characteristics of the driver, vehicle, and road, inform the establishment of standards for the geometric characteristics of the highway, including cross-sections and horizontal and vertical alignments, to maintain a desired level of service for a known proportional distribution of different vehicle types.

4.2.1 Geometric Design Standards

The initial step in any highway design process involves selecting the appropriate geometric design standards, as no single set of standards can be universally applied to all highways. Each highway's characteristics must be considered to ensure the selected geometric standards are suitable. For instance, standards appropriate for a scenic mountain road with low average daily traffic (ADT) would not suffice for a freeway managing heavy traffic (AASHTO, 2018).

4.2.1.1 Design Hourly Volume

The design hourly volume (DHV) is an essential metric used in highway design, representing the projected volume of traffic during the busiest hour. This volume is generally derived from a percentage of the expected ADT on the highway. The DHV is crucial for decisions related to roadway structural and capacity design, given the fluctuation of traffic volume throughout the day and year.

4.2.1.2 Design Speed

Design speed is a specific speed used to determine the various geometric features of a roadway. It is influenced by factors such as the highway's functional classification, the topography of the area, traffic volume, and land use in adjacent areas. Topography is typically categorized into level, rolling, and mountainous. While level terrain offers long sight distances with minimal construction challenges, rolling terrain features natural slopes that may restrict standard alignments. Conversely, mountainous terrain often requires extensive hillside excavations to achieve acceptable alignments.

The selected design speed should align with the speed at which motorists are expected to drive. For instance, a low design speed would be inappropriate for a rural collector road in flat topography, as drivers are likely to travel at higher speeds. The average trip length on the highway is another consideration when selecting the design speed, with highways accommodating longer average trips typically requiring higher design speeds.

Design speeds typically range from 20 to 70 mph, with increments of 5 mph. However, design elements display significant differences at 15 mph or more increments. Generally, freeways are designed for speeds between 60 and 70 mph, while other arterial roads range from 30 to 60 mph.

4.2.1.3 Design Vehicle

The design vehicle is a representative vehicle chosen based on its weight, dimensions, and operating characteristics to establish the highway's design standards. It is typically the largest vehicle likely to use the highway regularly. The design vehicle informs critical features such as radii at intersections, turning roadways, and highway grades.

When selecting a design vehicle, the following guidelines should be considered:

- A passenger car may be used as the primary traffic generator for designs involving a parking lot or a series of parking lots.
- A two-axle, single-unit truck is suitable for intersections of residential and park roads.
- A city transit bus may be used at intersections of state highways and city streets, primarily serving buses and relatively few large trucks.
- For collector streets and other facilities where single trucks are likely, a three-axle truck may be used.

4.2.2 Design Approach

The process of geometric road design encompasses a range of key elements that collectively aim to provide a safe and efficient means of transportation. These integral aspects include alignment, cross-section, intersection and interchange design, along with considerations of sight distance, superelevation and banking, roadway geometrics, grade separation, and facilities for pedestrians and cyclists. Additionally, attention must be given to access management, roadside design, environmental factors, and adherence to geometric standards. Specific projects may necessitate further specialized design elements.

4.2.2.1 Profile View

The three-dimensional highway alignment can be distilled into a two-dimensional representation, as depicted in Figure 4.1. Here, the horizontal alignment correlates with the x and z coordinates.

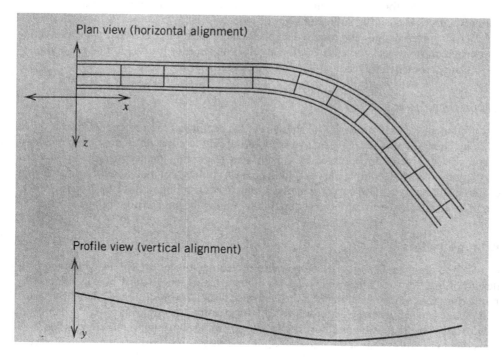

Figure 4.1 Highway Alignment in Two-Dimensional Views

Source: Principles of Highway Engineering and Traffic Analysis by Mannering, K.L. & Kilareski, W.P., John Wiley & Sons, Inc.

In contrast, the vertical alignment corresponds to the length of the highway at a constant elevation, referenced by the y coordinates.

4.2.2.2 Vertical Alignment

A profile view serves to illustrate the vertical alignment of a highway, detailing the elevation of each point measured along the road's length, which is gauged at a constant elevation. In this case, the x and z coordinates are substituted with stationing distances, each equivalent to 1 km of highway alignment. This distance is measured along the highway's centerline from a specified origin point, with the stationing notated such that a point 1258.5 m from the origin is labeled as station $1 + 258.500$.

This stationing, in conjunction with the plan view's directional guidance (horizontal alignment) and the profile view's elevations (vertical alignment), precisely identifies every point on the highway, virtually equivalent to employing x, y, and z coordinates.

4.2.2.3 Sag and Crest Vertical Curves

Both sag and crest vertical curves are integral to highway design, with the primary focus on the transition between different roadway elevations. Sag curves, distinguished by their appearance, must be evaluated for comfort, sight distance, and adequate drainage. In contrast, the main consideration for crest curves is ensuring sufficient sight distance.

Factors such as grade changes and required sight distances must be considered when determining the length of a crest vertical curve. This often involves an iterative process to find the optimum grade line, balancing practicality, economics, and the desired vertical curve.

4.2.2.4 Horizontal Alignment

Horizontal alignment focuses primarily on the curve that facilitates the directional transition of the roadway in a horizontal plane. Critical to a vehicle's ability to navigate a curve, this transition must accommodate a diverse range of vehicle capabilities. Regarding connecting straight sections of roadways, options include simple curves with a constant radius, compound curves comprising multiple simple curves, and spiral curves with a continuously changing radius (Austroads, 1993).

Figures 4.2(a) and 4.2(b) present the coordination of alignments and terrain fitting.

4.3 Drainage Issues

Ensuring adequate drainage is a crucial component in selecting a location and designing the geometric design of highways. Drainage facilities on highways and streets must be capable of efficiently redirecting water away from the pavement surface through well-designed channels. A lack of proper drainage can significantly deteriorate the highway infrastructure over time. Furthermore, accumulated water on the pavement can result in slower traffic movement and increase the risk of accidents due to hydroplaning and reduced visibility caused by splash and spray. The vital role of water management in highway construction is reflected in the budget allocation, with approximately 25% of highway construction funds dedicated to erosion control and the installation of drainage structures, including culverts, bridges, channels, and ditches.

The highway engineer primarily addresses two water sources. The first, surface water, manifests as rainfall or snowfall. The soil absorbs some of this water, while the remainder collects on the ground and must be diverted from the highway pavement, constituting surface drainage. The second source, groundwater, moves through subterranean channels and can pose a challenge in highway cuts or areas where the water table lies close to the pavement structure, resulting in subsurface drainage.

Outlined below are the fundamental design principles for managing surface and subsurface drainage. These principles incorporate essential hydrology concepts required to comprehend rainfall as a crucial water source.

4.3.1 Surface Drainage

Surface drainage encompasses comprehensive surface water management within the highway or street's pavement and right-of-way. An appropriately designed highway surface drainage system aims to efficiently intercept all surface and watershed runoff, guiding it into channels and gutters for eventual discharge into natural waterways. Preventing water infiltration through pavement cracks and shoulder areas, which can lead to significant roadway damage, hinges primarily on managing surface runoff. A well-designed surface drainage system plays a crucial role in minimizing such damage.

The surface drainage system for rural highways should feature adequate transverse and longitudinal slopes on both the pavement and shoulder, ensuring positive runoff. It should also incorporate

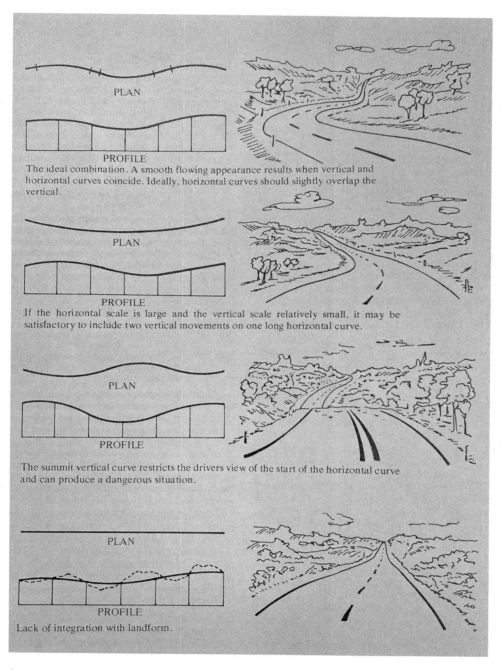

PLAN

PROFILE

The ideal combination. A smooth flowing appearance results when vertical and horizontal curves coincide. Ideally, horizontal curves should slightly overlap the vertical.

PLAN

PROFILE

If the horizontal scale is large and the vertical scale relatively small, it may be satisfactory to include two vertical movements on one long horizontal curve.

PLAN

PROFILE

The summit vertical curve restricts the drivers view of the start of the horizontal curve and can produce a dangerous situation.

PLAN

PROFILE

Lack of integration with landform.

Figure 4.2(a) Coordination of Alignments and Terrain Fitting

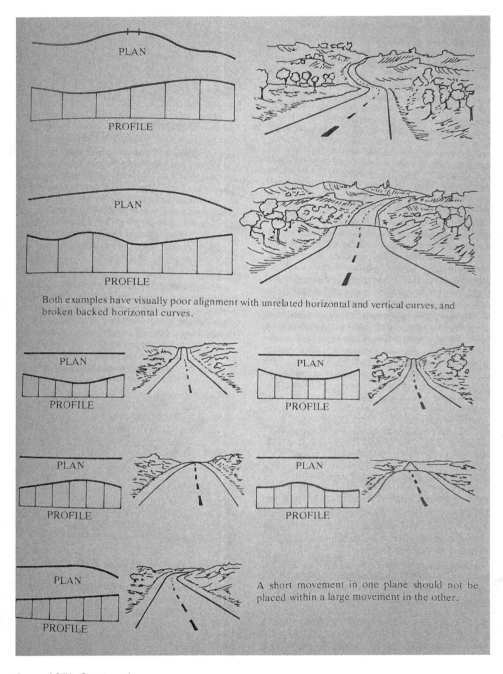

Both examples have visually poor alignment with unrelated horizontal and vertical curves, and broken backed horizontal curves.

A short movement in one plane should not be placed within a large movement in the other.

Figure 4.2(b) Continued

longitudinal channels (ditches), culverts, and bridges to facilitate surface water discharge into natural waterways. In rural divided highways, storm drains and inlets can be found in the median. The surface drainage system in urban areas similarly includes longitudinal and transverse slopes.

Longitudinal drains, typically in the form of underground pipe drains, are employed in urban areas to handle both surface runoff and groundwater. Additionally, curbs and gutters are commonly used in urban and select rural areas to control surface runoff effectively (Garber & Hoel, 2015).

4.3.1.1 Transverse Slopes

The primary objective of incorporating transverse slopes is to expedite surface water removal from the pavement, ensuring efficient drainage. This is typically achieved by creating a crown or elevation at the center of the pavement, resulting in cross slopes on each side of the centerline, or by implementing a slope in one direction across the width of the pavement. In most cases, shoulders are inclined to direct water away from the pavement, except for highways featuring raised narrow medians.

However, the necessity for steep cross slopes to facilitate drainage often conflicts with the requirement for relatively flat cross slopes to enhance driver comfort. Consequently, the selection of an appropriate cross slope typically involves striking a compromise between these two competing needs. Cross slopes of 2% or less have minimal impact on driver comfort, particularly regarding the effort required for steering.

4.3.1.2 Longitudinal Slopes

In the longitudinal direction of the highway, it is essential to maintain a minimum gradient to ensure an effective slope in the longitudinal channels, especially in cut sections. For highways located in very flat terrain, it is generally advisable to maintain longitudinal channel slopes of at least 0.2%. While 0% grades may be employed on uncurbed pavements with sufficient cross slopes, it is recommended to have a minimum of 0.5% for pavements with curbing. This recommendation can be reduced to 0.3% for suitably crowned, high-type pavements constructed on solid ground.

4.3.1.3 Longitudinal Channels

Longitudinal channels or ditches are purposefully constructed alongside the highway to gather runoff from the pavement surface, subsurface drainage systems, and other parts of the highway right-of-way. In cases where the highway pavement lies at a lower elevation compared to the adjacent ground, such as in cut sections, the construction of a longitudinal drain (often referred to as an intercepting drain) at the top of the cut effectively intercepts and diverts water flow away from the pavement. Subsequently, the water accumulated in these longitudinal ditches is conveyed to a drainage channel and further directed to a natural watercourse or retention pond.

4.3.1.4 Curbs and Gutters

Curbs and gutters serve a dual purpose by managing drainage and preventing vehicles from encroaching onto adjacent areas while delineating pavement edges. They are more frequently utilized in urban settings, especially in residential areas, where they are integrated with storm sewer systems to regulate street runoff effectively. When there is a need to provide relatively lengthy, continuous stretches of curbs in urban areas, careful design of storm sewer inlets becomes essential to prevent the pooling of substantial amounts of water on the pavement surface.

4.3.1.5 Highway Drainage Structures

Drainage structures serve the dual purpose of facilitating vehicular passage over natural waterways that flow beneath the highway's right-of-way while also permitting the uninterrupted flow of water within the natural channel without significant modification or disruption of its regular course. Ensuring that these structures are adequately sized to accommodate the expected water flow is crucial. Inadequately sized structures can lead to water accumulation, potentially resulting in the submersion of adjacent highway segments for extended periods, ultimately compromising their integrity.

Drainage structures are typically classified into two categories: major and minor. Major structures encompass those with clear spans exceeding 20 feet and often comprise large bridges, though they may also include multi-span culverts. On the other hand, minor structures are characterized by clear spans of 20 feet or less, encompassing small bridges and culverts.

4.3.1.6 Major Structures

This book's focus does not extend to discussing the various types of bridges. Instead, the emphasis lies in the span and vertical clearance criteria selection for these structures. A critical criterion for elevating the bridge superstructure is that it must be positioned above the high-water mark. The elevation clearance above the high-water mark varies depending on the navigability of the waterway. In cases where the waterway is navigable, the clearance above the high-water mark should be adequate to enable the largest vessel using the channel to pass beneath the bridge without any risk of collision with the bridge deck. The clearance height and the type and spacing of piers also factor in considerations like the likelihood of ice jams and the presence of floating logs and debris during periods of high water.

Determining the high-water mark involves examining the banks on either side of the waterway, typically revealing evidence of erosion and debris deposits. Local residents who have observed the waterway during flood stages over an extended period can also provide valuable insights into the location of the high-water mark. Additionally, stream gauges installed in the waterway for many years can offer similar data.

4.3.1.7 Minor Structures

Minor structures, primarily consisting of short-span bridges and culverts, are the predominant type of drainage structures found on highways. While the openings of these structures aren't designed to handle the most extreme flood conditions, they should still be sized to accommodate flow conditions that might occur over the expected lifespan of the structure. Adequate provisions should also be made to prevent the obstruction of these structures due to floating debris and the potential displacement of large boulders from steep channel banks.

Culverts are constructed using a variety of materials and come in different shapes. Common materials used for culvert construction include concrete (both reinforced and unreinforced), corrugated steel, and corrugated aluminium. In some cases, other materials may be employed to line the interior of the culvert, either to prevent corrosion and abrasion or to reduce hydraulic resistance. For example, asphaltic concrete may be used to line corrugated metal culverts. Various shapes are typically used in culvert construction, including circular, rectangular (box), elliptical, pipe arch, metal box, and arch.

4.3.2 Sediment and Erosion Control

Soil erosion stemming from the continuous flow of surface water across shoulders, side slopes, and unlined channels adjacent to the pavement can lead to detrimental conditions for the pavement structure and surrounding facilities. Shoulder and side slope erosion may result in embankment and cut section failures, while erosion within highway channels can contaminate nearby lakes and streams. Therefore, preventing erosion is critical in highway drainage during construction and upon the highway's completion. The subsequent sections detail the methods employed to prevent erosion and control sediment.

4.3.2.1 Intercepting Drains

Strategically positioned at the crest of cut sections, intercepting drains effectively thwart erosion of cut side slopes by intercepting water before it can directly affect these slopes. These drains collect and transport the water to paved spillways thoughtfully placed on the cut's sides. The water is then conveyed through protected spillways to longitudinal ditches running alongside the highway.

4.3.2.2 Curbs and Gutters

On rural highways, curbs and gutters serve as protective measures against the erosion of unsurfaced shoulders. Placed along the pavement's edge, they prevent surface water from flowing over and eroding unpaved shoulders. When paved shoulders are employed, curbs and gutters can also safeguard embankment slopes from erosion. Positioned on the outer edge of paved shoulders, they direct surface water to strategically located paved spillways, which then transport it to the longitudinal drain at the base of the embankment.

4.3.2.3 Turf Cover

Turf cover serves as an efficient and cost-effective method for preventing erosion on unpaved shoulders, ditches, embankments, and cut slopes with slopes shallower than 3:1. Turf cover is established by seeding suitable grasses immediately after grading. However, it's worth noting that turf cover cannot withstand continuous traffic and may lose firmness following heavy rains.

4.3.2.4 Slope and Channel Linings

In cases where extensive erosion threatens a highway, more robust preventive measures are necessary. For instance, when cut and embankment side slopes are steep and located in regions prone to heavy rain or snow, a standard approach is to line the slope surface with rip-rap or hand-placed rock.

Channel linings also play a crucial role in safeguarding longitudinal channels from erosion, with linings applied to both the sides and bottom of the channel. These protective linings can be either flexible or rigid. Flexible options include dense-graded bituminous mixtures and rock rip-rap, while rigid linings encompass Portland cement concrete and soil cement. Rigid linings are more effective in preventing erosion under severe conditions but tend to be costlier and, due to their smoothness, can generate unacceptably high flow velocities. In instances where high velocities are a concern, the installation of an appropriate energy dissipater at the channel's lower end becomes necessary to prevent excessive erosion. This is not needed when water discharges into a rocky stream or a deep pool. Comprehensive design procedures for linings are provided later in this chapter.

4.3.2.5 Erosion Control During Construction

Special measures are imperative for controlling erosion and sediment accumulation during highway construction. Techniques employed include sediment basins, check dams, silt fences/filter barriers, brush barriers, diversion dikes, slope drains, and dewatering basins. Sediment basins become mandatory when runoff from drainage areas exceeding three acres flows across a disturbed area. These basins allow sediment-laden runoff to collect and settle. Check dams are employed to reduce the velocity of concentrated water flow and are typically constructed using local materials like rock, logs, or straw bales. Silt fences consist of fabric, often reinforced with wire mesh. Brush barriers are fashioned from construction spoil material, often combined with filter fabric. Diversion dikes are earthen berms that redirect water to a sediment basin, while slope drains convey water downhill, preventing erosion before permanent drainage facilities are established.

4.4 Questions

1. What are the key elements of transportation planning, and why are data collection and public participation considered crucial in this process?
2. Explain how transportation planning can influence various decisions, such as policy development and fund allocation.
3. How does transportation planning incorporate external factors like land use, air quality, and environmental policy into its comprehensive planning framework? Provide examples.
4. Describe the importance of demand analysis in transportation planning and give examples of how it can help identify specific transportation needs.
5. What is meant by "traffic optimization" in transportation planning, and what strategies can be used to alleviate congestion?
6. Discuss the objectives of transportation planning and explain how they contribute to the development of thriving and sustainable communities.
7. In the context of smart transportation, what key components and technologies should transportation planners consider integrating into their plans?
8. How can transportation infrastructure be more resilient to extreme weather events and related challenges, and why is this important?
9. How can transportation planners adapt to changing climate patterns and incorporate sustainability principles into their planning process?
10. What is the role of geometric design in shaping highway elements, and why is it crucial for ensuring road safety and efficiency?
11. Explain how traffic analysis and the behavior of drivers, pedestrians, and vehicles influence the geometric design of roads.
12. What are some of the roadway system factors that influence the standards for geometric design? Provide examples of how these factors interact with each other.
13. Why is it necessary to consider the specific characteristics of a highway when selecting geometric design standards? Give an example of how standards for a scenic mountain road might differ from those for a freeway.

14. What are some of the key elements of geometric road design, and why are they important for providing safe and efficient transportation?
15. What are sag and crest vertical curves in highway design, and what considerations are important when determining their lengths and shapes?
16. Describe the different types of horizontal alignment options used in connecting straight sections of roadway. How do these options accommodate various vehicle capabilities?
17. Why is preventing erosion and controlling sediment important in highway construction and maintenance?
18. Describe the role of intercepting drains in erosion control and where they are typically placed.
19. How do curbs and gutters contribute to preventing erosion on highways with paved shoulders, and what is their role in protecting embankment slopes?
20. Explain the purpose of turf cover in erosion prevention. Under what conditions is it effective?
21. What are slope and channel linings, and why are they used for erosion control? Discuss the differences between flexible and rigid linings.
22. During highway construction, what measures are taken to control erosion and sediment accumulation? Provide examples of erosion control techniques used in construction.

References

AASHTO (2018). *A Policy on Geometric Design of Highways and Streets*. 7th ed. American Association of State Highway and Transportation Officials (AASHTO), Washington, DC.

Austroads (1993). *Rural Road Design – Guide to the Geometric Design of Rural Roads*. Austroads National Office, Sydney.

Garber, N.J. & Hoel, L.A. (2015). *Traffic and Highway Engineering*. 5th ed. Cengage Learning, Toronto.

Mannering, F.L., Kilareski, W.P., & Washburn, S.S. (2019). *Principles of Highway Engineering and Traffic Analysis*. 7th ed. Wiley, Hoboken, NJ.

Underwood, R.T. (1991). *The Geometric Design of Roads*. Macmillan Company of Australia. 1 Market Street, Sydney, NSW, Australia.

Chapter 5

Pavement Design

5.1 Design of Flexible Pavement

Flexible pavements, named for their ability to flex under traffic loads, typically consist of multiple layers of aggregates and asphalt. Flexible pavements derive their strength from the layered structure and composition, where each layer bears the stresses transferred from the above layer, spreading them over a larger area as the load moves down.

The design of flexible pavements incorporates different methodologies, such as empirical or mechanistic-empirical approaches. Empirical methods rely on the outcomes of past road performance and experimental data, such as the 1993 American Association of State Highway and Transportation Officials (AASHTO, 1993) Pavement Design Guide method, while the mechanistic-empirical approach combines theoretical analysis with empirical data. Recent innovations, such as the Mechanistic-Empirical Pavement Design Guide (MEPDG), (AASHTO, 2007) developed under the National Cooperative Highway Research Program (NCHRP) Study 1–37A, provide a detailed procedure for flexible pavement design, taking into account traffic loadings, climatic effects, and material properties.

5.1.1 The AASHTO Pavement Design Guide (1993) Method

The AASHTO 1993 Guide is an empirical method based on the results of the AASHTO Road Test conducted during the late 1950s. The following factors need to be considered to ensure the constructed roadways are durable, cost-effective, and safe.

5.1.1.1 Pavement Performance

The structural and functional performance of the pavement are primary factors that should be considered during the design process. Structural performance is determined by the physical condition of the pavement, such as surface distresses. Functional performance is an indicator of how the pavement serves the public, such as riding comfort.

Present serviceability index (PSI) is typically used to quantify pavement performance. PSI ranges from 0 to 5, where 0 indicates the pavement has the lowest performance, and 5 is the highest. Two PSI values are used in the design of flexible pavements: the initial PSI (p_i) and the terminal PSI (p_t). p_i is the PSI value right after the construction of the pavement; p_t is the minimum acceptable PSI value before a pavement treatment is needed. AASHTO recommends a value of 4.2 for p_i, and 2.5 or 3.0 for p_t for major highways and 3.2 for roadways with a lower classification.

DOI: 10.1201/9781003197768-5

5.1.1.2 Traffic Load

In the AASHTO design method, traffic load is typically represented as the cumulative 18-kip (kilo-pound) Equivalent Single Axle Loads (ESALs) expected over the design life of the pavement. During the design process, traffic volumes need to be predicted and then converted into ESALs. The equation for calculating the ESAL for each axle load category is:

$$ESAL_i = f_d \times G_{rn} \times AADT_i \times 365 \times N_i \times F_{Ei}$$ \hfill Eq. 5.1

where

$ESAL_i$ = equivalent accumulated 18,000-lb (80 kN) single-axle load for the axle category i
f_d = design lane factor, the percent of traffic on the design lane
G_{rn} = growth factor for a given growth rate r and design period n
$AADT_i$ = first year annual average daily traffic for axle category i
N_i = number of axles on each vehicle in category i
F_{Ei} = load equivalency factor for axle category i

The growth factor (G_{rn}) for different growth rates and design periods can be calculated using the equation below:

$$G_{rn} = \left[(1+r)^n - 1 \right] / r$$ \hfill Eq. 5.2

where

r = i/100
i = growth rate
n = design period, years

Calculated growth factors G_{rn} are also included in Table 5.1 which can be used to calculate the ESAL over the design period.

The load equivalency factor (F_{Ei}) for p_t of 2.5 for single axles and for tandem axles are provided in Table 5.2.

5.1.1.3 Material Properties

Material properties include the layer coefficients for each layer (a_1 for the asphalt concrete surface, a_2 for the base, a_3 for the subbase) and the resilient modulus (M_r) of the subgrade soil. The layer coefficient of a layer is a measure of its structural contribution to the pavement strength. Figures 5.1–5.3 show layer coefficients corresponding to different material engineering properties.

For subgrade, the 1993 AASHTO guide uses the resilient modulus (M_r) to define its property. The equations that can be used to convert the CBR or R value to an equivalent M_r value (lb/in^2) are:

$$M_r = 1500 \times CBR\left(for\ fine-grade\ soils\ with\ soaked\ CBR\ of\ 10\ or\ less \right)$$ \hfill Eq. 5.3

$$M_r = 1000 + 500 \times R\left(for\ R \le 20 \right)$$ \hfill Eq. 5.4

Table 5.1 Growth Factors

Design Period, Years (n)	Annual Growth Rate, Percent (r)						
	No Growth 2	4	5	6	7	8	10
1	1.0 1.0	1.0	1.0	1.0	1.0	1.0	1.0
2	2.0 2.02	2.04	2.05	2.06	2.07	2.08	2.10
3	3.0 3.06	3.12	3.15	3.18	3.21	3.25	3.31
4	4.0 4.12	4.25	4.31	4.37	4.44	4.51	4.64
5	5.0 5.20	5.42	5.53	5.64	5.75	5.87	6.11
6	6.0 6.31	6.63	6.80	6.98	7.15	7.34	7.72
7	7.0 7.43	7.90	8.14	8.39	8.65	8.92	9.49
8	8.0 8.58	9.21	9.55	9.90	10.26	10.64	11.44
9	9.0 9.75	10.58	11.03	11.49	11.98	12.49	13.58
10	10.0 10.95	12.01	12.58	13.18	13.82	14.49	15.94
11	11.0 12.17	13.49	14.21	14.97	15.78	16.65	18.53
12	12.0 13.41	15.03	15.92	16.87	17.89	18.98	21.38
13	13.0 14.68	16.63	17.71	18.88	20.14	21.50	24.52
14	14.0 15.97	18.29	19.16	21.01	22.55	24.21	27.97
15	15.0 17.29	20.02	21.58	23.28	25.13	27.15	31.77
16	16.0 18.64	21.82	23.66	25.67	27.89	30.32	35.95
17	17.0 20.01	23.70	25.84	28.21	30.84	33.75	40.55
18	18.0 21.41	25.65	28.13	30.91	34.00	37.45	45.60
19	19.0 22.84	27.67	30.54	33.76	37.38	41.45	51.16
20	20.0 24.30	29.78	33.06	36.79	41.00	45.76	57.28
25	25.0 32.03	41.65	47.73	54.86	63.25	73.11	98.35
30	30.0 40.57	56.08	66.44	79.06	94.46	113.28	164.49
35	35.0 49.99	73.65	90.32	111.43	138.24	172.32	271.02

Drainage: The effect of drainage on the performance of flexible pavements is considered by incorporating a factor m_i for the base and subbase layer coefficients (a_1 and a_2). The m_i factors are determined by drainage quality (Table 5.3) and recommended m_i values are included in Table 5.4.

Reliability: Reliability is the probability that the pavement will perform satisfactorily over the design life under given traffic and environmental conditions. A survey of the AASHTO pavement design task force provided suggested reliability values that are included in Table 5.5.

Estimated overall standard deviation (S_0) for a given reliability accounts for the variation in the traffic forecast and the variation in actual pavement performance. Its recommended values are:

Pavement Type	Standard Deviation, S_0
Flexible pavements	0.40–0.50
Rigid pavements	0.30–0.40

5.1.1.4 Structural Design of Flexible Pavements

Using the variables determined in the previous sections, the structural number (SN) can be computed using the AASHTO design equation:

$$SN = a_1 D_1 + a_2 D_2 m_2 + a_3 D_3 m_3$$

Eq. 5.5

Table 5.2 Axle Load Equivalency Factors for Flexible Pavements, Single Axles, and p_t of 2.5

Axle Load (kips)	Pavement Structural Number (SN)					
	1	2	3	4	5	6
2	0.0001	0.0001	0.0001	0.0000	0.0000	0.0000
4	0.0005	0.0005	0.0004	0.0003	0.0003	0.0002
6	0.002	0.002	0.002	0.001	0.001	0.001
8	0.004	0.006	0.005	0.004	0.003	0.003
10	0.008	0.013	0.011	0.009	0.007	0.006
12	0.015	0.024	0.023	0.018	0.014	0.013
14	0.026	0.041	0.042	0.033	0.027	0.024
16	0.044	0.065	0.070	0.057	0.047	0.043
18	0.070	0.097	0.109	0.092	0.077	0.070
20	0.107	0.141	0.162	0.141	0.121	0.110
22	0.160	0.198	0.229	0.207	0.180	0.166
24	0.231	0.273	0.315	0.292	0.260	0.242
26	0.327	0.370	0.420	0.401	0.364	0.342
28	0.451	0.493	0.548	0.534	0.495	0.470
30	0.611	0.648	0.703	0.695	0.658	0.633
32	0.813	0.843	0.889	0.887	0.857	0.834
34	1.06	1.08	1.11	1.11	1.09	1.08
36	1.38	1.38	1.38	1.38	1.38	1.38
38	1.75	1.73	1.69	1.68	1.70	1.73
40	2.21	2.16	2.06	2.03	2.08	2.14
42	2.76	2.67	2.49	2.43	2.51	2.61
44	3.41	3.27	2.99	2.88	3.00	3.16
46	4.18	3.98	3.58	3.40	3.55	3.79
48	5.08	4.80	4.25	3.98	4.17	4.49
50	6.12	5.76	5.03	4.64	4.86	5.28
52	7.33	6.87	5.93	5.38	5.63	6.17
54	8.72	8.14	6.95	6.22	6.47	7.15
56	10.3	9.6	8.1	7.2	7.4	8.2
58	12.1	11.3	9.4	8.2	8.4	9.4
60	14.2	13.1	10.9	9.4	9.6	10.7
62	16.5	15.3	12.6	10.7	10.8	12.1
64	19.1	17.6	14.5	12.2	12.2	13.7
66	22.1	20.3	16.6	13.8	13.7	15.4
68	25.3	23.3	18.9	15.6	15.4	17.2
70	29.0	26.6	21.5	17.6	17.2	19.2
72	33.0	30.3	24.4	19.8	19.2	21.3
74	37.5	34.4	27.6	22.2	21.3	23.6
76	42.5	38.9	31.1	24.8	23.7	26.1
78	48.0	43.9	35.0	27.8	26.2	28.8
80	54.0	49.4	39.2	30.9	29.0	31.7
82	60.6	55.4	43.9	34.4	32.0	34.8
84	67.8	61.9	49.0	38.2	35.3	38.1
86	75.7	69.1	54.5	42.3	38.8	41.7
88	84.3	76.9	60.6	46.8	42.6	45.6
90	93.7	85.4	67.1	51.7	46.8	49.7

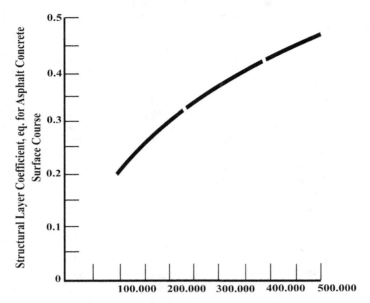

Figure 5.1 Chart for Estimating Structure Layer Coefficient of Dense-Graded Asphalt Concrete Based on the Elastic (Resilient) Modulus

where
  SN = structural number
  a_1, a_2, a_3 = layer coefficients for surface, base, and subbase layers
  D_1, D_2, D_3 = actual thickness in inches of surface, base, and subbase layers
  m_i = drainage coefficient for layer i

This equation is used for ESALs greater than 50,000 for the performance period. For ESALs less than 50,000, it is usually considered the design of low-volume roads.

The structural number represents the structural capacity of the pavement needed to carry the predicted design ESAL over the design life. It can be calculated using the equation below or obtained from the chart in Figure 5.4.

$$log_{10}W_{18} = Z_R S_0 + 9.36 log_{10}(SN+1) - 0.20 + \frac{log_{10}\left[\Delta PSI / (4.2-1.5)\right]}{0.40 + \left[1094 / (SN+1)^{5.19}\right]}$$ Eq. 5.6

where
  W_{18} = predicted number of 18,000-lb (80kN) single-axle load applications
  Z_R = standard normal deviation for a given reliability
  S_0 = overall standard deviation
  SN = structural number indicative of the total pavement thickness

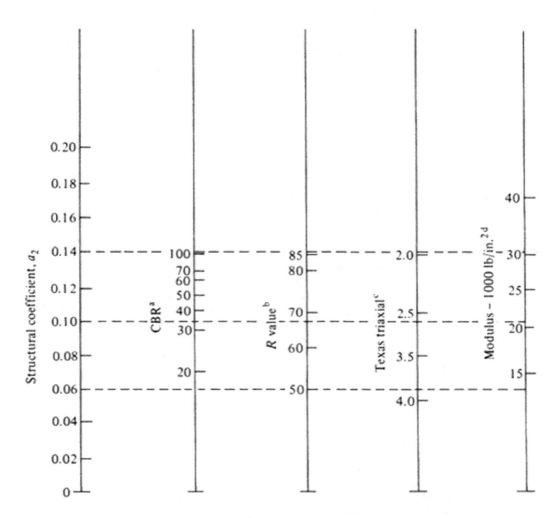

^aScale derived by averaging correlations obtained from Illinois.
^bScale derived by averaging correlations obtained from California,
New Mexico, and Wyoming.
^cScale derived by averaging correlations obtained from Texas.
^dScale derived on NCHRP project 128, 1972.

Figure 5.2 Variation in Granular Base Layer Coefficient, a_2, with Various Subbase Strength Parameters

Source: Redraw from AASHTO Guide for Design of Pavement Structures, American Association of State Highway and Transportation Officials, Washington, DC, 1993, Used with Permission

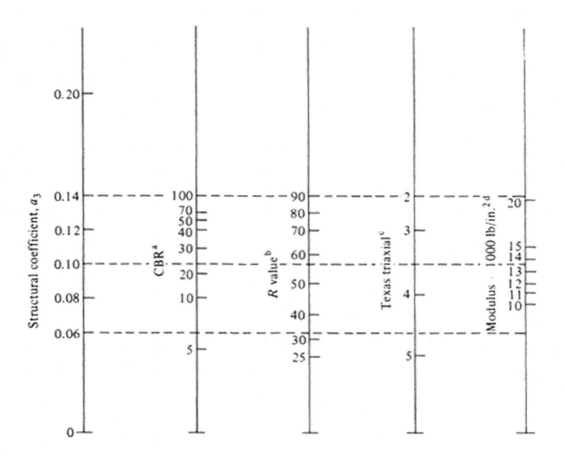

[a] Scale derived from correlations from Illinois.
[b] Scale derived from correlations obtained from The Asphalt
Institute, California, New Mexico, and Wyoming.
[c] Scale derived from correlations obtained from Texas.
[d] Scale derived on NCHRP project 128, 1972.

Figure 5.3 Variation in Granular subbase Layer Coefficient, a_3, with Various Subbase Strength Parameters

Source: Redraw from AASHTO Guide for Design of Pavement Structures, American Association of State Highway and Transportation Officials, Washington, DC, 1993, Used with Permission

$\Delta PSI = p_i - p_t$
p_i = initial serviceability index
p_t = terminal serviceability index
M_t = resilient modulus (lb/in²)

5.1.2 The Mechanistic-Empirical Pavement Design Guide (MEPDG) Method

The MEPDG method uses mechanistic principles to predict pavement responses, such as stresses, strains, and deformations, and these responses are then used in empirical or semi-empirical models

Table 5.3 Definition of Drainage Quality

Quality of Drainage	Water Removed Within*
Excellent	2 hours
Good	1 day
Fair	1 week
Poor	1 month
Very poor	(water will not drain)

Source: Adapted with Permission from AASHTO Guide for Design of Pavement Structures, American Association of State Highway and Transportation Officials, Washington, DC, 1993

*Time required to drain the base layer to saturation.

Table 5.4 Recommended m_i Values

	Percent of Time Pavement Structure Is Exposed to Moisture Levels Approaching Saturation			
Quality of Drainage	Less Than 1%	1 to 5%	5 to 25%	Greater Than 25%
Excellent	1.40–1.35	1.35–1.30	1.30–120	1.20
Good	1.35–1.25	1.25–1.15	1.15–1.00	1.00
Fair	1.25–1.15	1.15–1.05	1.00–0.80	0.80
Poor	1.15–1.05	1.05–0.80	0.80–0.60	0.60
Very poor	1.05–0.95	0.95–0.75	0.75–0.40	0.40

Source: Adapted with permission from AASHTO Guide for Design of Pavement Structures, American Association of State Highway and Transportation Officials, Washington, DC, 1993

Table 5.5 Suggested Levels of Reliability for Various Functional Classifications

Recommended Level of Reliability

Functional Classification	Urban	Rural
Interstate and other freeways	85–99.9	80–99.9
Other principal arterials	80–99	75–95
Collectors	80–95	75–95
Local	50–80	50–80

Source: Adapted with permission from AASHTO Guide for Design of Pavement Structures, American Association of State Highway and Transportation Officials, Washington, DC, 1993

Note: Results based on a survey of the AASHTO Pavement Design Task Force

to predict pavement distress and performance over time. The procedure is iterative in that a trial design is first selected and evaluated for its adequacy to carry the expected traffic load on the pavement. If the resulting performance criteria do not satisfy the input performance criteria at the specified reliability, the trial design is revised, and the evaluation repeated. A conceptual flow chart of the design process is illustrated in Figure 5.5.

The main advantage of the MEPDG is that it provides a more reliable prediction of pavement performance, taking into account the specific conditions that a pavement is expected to encounter. This leads to more efficient use of resources, as pavement designs can be optimized for specific situations, rather than following a "one-size-fits-all" approach.

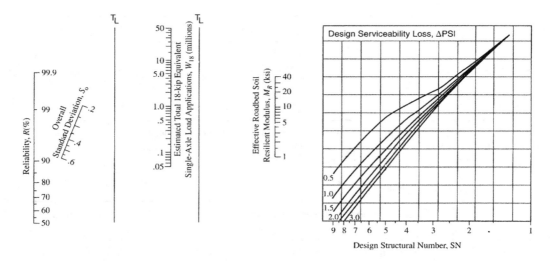

Figure 5.4 Design Chart for Flexible Pavements

Source: Guide for Design of Pavement Structures, 1993, AASHTO, Washington, DC. Used with Permission

Moreover, using the MEPDG allows for a more detailed analysis of potential pavement problems, which can help in deciding on the most effective maintenance and rehabilitation strategies. For example, the MEPDG can predict different types of distress separately, such as rutting, cracking, and roughness. This can help in prioritizing the most critical issues and addressing them effectively.

The design procedure includes the following steps.

• Step 1: Select a trial design

This trial design can either be determined by another design method, such as the 1993 AASHTO design method described in the previous section, or be specified by the agency.

• Step 2: Select the appropriate performance indicator criteria (threshold values) and design reliability level for the design

The agency's policies on rehabilitation or reconstruction can be used to identify performance indicator criteria, including rut depth, non-load-related cracking, load-related longitudinal cracking, load-related alligator cracking, and the pavement smoothness represented by the international roughness index (IRI). Suggested critical values for flexible pavements are included in Table 5.6. Suggested reliability levels for different types of roadways are presented in Table 5.7.

• Step 3: Obtain all inputs for the trial design

Based on the importance of the project, the importance of the input parameter, and the available resources, all the inputs required by the MEPDG software can be at one of the following three levels. It should be noted that it is not necessary to use the same level for all the inputs.

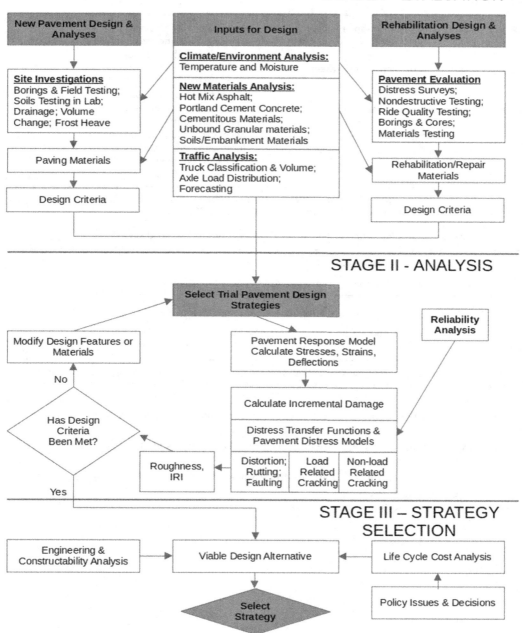

Figure 5.5 Conceptual Flow Chart of the Three-Stage Design/Analysis Process for the MEPDG

Source: Interim Mechanistic-Empirical Pavement Design Guide Manual. National Cooperative Highway Research Program, Transportation Research Board, National Research Council, Washington, DC, 2007

Table 5.6 Recommended Design Criteria or Threshold Values for Flexible Pavements

Performance Criteria	Maximum Value at End of Design Life
Alligator cracking (HMA bottom-up cracking)	Interstate: 10% lane area Primary: 20% lane area Secondary: 35% lane area
Rut depth (permanent deformation in wheel paths)	Interstate: 0.40 in Primary: 050 in Others (<45 mi/h): 0.65 in
Traverse cracking length (thermal cracks)	Interstate: 500 ft/mi Primary: 700 ft/mi Secondary: 700 ft/mi
IRI (smoothness)	Interstate: 160 in/mi Primary: 200 in/mi Secondary: 200 in/mi

Source: Interim Mechanistic-Empirical Pavement Design Guide Manual. National Cooperative Highway Research Program, Transportation Research Board, National Research Council, Washington, DC, 2007

Table 5.7 Reliability Levels for Different Functional Classifications of the Roadway

Functional Classification	Reliability Level (%)	
	Urban	Rural
Interstate/freeways	95	95
Principal arterials	90	85
Collectors	80	75
Local	75	70

Source: Interim Mechanistic-Empirical Pavement Design Guide Manual. National Cooperative Highway Research Program, Transportation Research Board, National Research Council, Washington, DC, 2007

Input Level 1: Inputs at this level are project specific; they are directly collected at the project site.

Input Level 2: Inputs at this level are derived from Level 1 inputs at other sites through correlations and regression analyses.

Input Level 3: Inputs at this level are usually estimated or default values based on the median of global or regional default values.

The inputs can be classified into six areas:

1. General project information
2. Design criteria
3. Traffic
4. Climate
5. Structural layering
6. Material properties

- Step 4: Evaluate the trial design and revise the trial design as needed

The predicted pavement distresses and the IRI values of the trial design are evaluated to determine if they are satisfactory. If the MEPDG reported liability is less than the targeted reliability, the trial design needs to be revised. f the predicted performance at the design reliability with the stipulated design criteria. If any of the computed performance indicators (the rut depth, the load-related cracking, nonload related cracking (transverse cracking) and smoothness) do not satisfy the stipulated criteria, the trail design needs to be revised and the analysis repeated for the new design.

5.2 Design of Rigid Pavement

Rigid pavements are typically made of concrete and are designed to distribute loads over a wide area of the subgrade due to their high flexural strength.

Rigid pavements can be designed using empirical or mechanistic-empirical approaches, such as the 1993 AASHTO Pavement Design Guide method and the Mechanistic-Empirical Pavement Design Guide (MEPDG) method.

5.2.1 The AASHTO Pavement Design Guide (1993) Method

The following factors should be considered in the AASHTO procedure for the design of rigid pavements.

5.2.1.1 Pavement Performance

Pavement performance considerations are the same as the ones for flexible pavement. The initial serviceability index (p_i) can be set as 4.5, and the terminal serviceability index (p_t) can be determined by the designer.

5.2.1.2 Subgrade Strength

Either graded granular materials or suitably stabilized materials can be used for the subbase layer. Recommended specifications for six types of subbase materials are included in Table 5.8. According to AASHTO, the first five types – A through E – can be used within the upper 4 in layer of the subbase, whereas Type F can be used below the uppermost 4 in layer.

5.2.1.3 Subbase Strength

The Westergaard modulus of subgrade reaction k, defined as the load in lb/in^2 on a loaded area divided by the deformation in inches, presents the strength of the subgrade. Values of k can be estimated either from experience or by correlating with other tests. Figures 5.6–5.9 and Table 5.9 can be used to calculate effective k values.

Traffic: ESALs are used to represent the traffic load for the design of rigid pavements. Tables 5.10 and 5.11 provide ESAL factors based on the slab thickness and the p_i value.

Concrete Properties: The flexural strength at 28 days of the concrete is used in the design process.

Drainage: A drainage coefficient, C_d, is used to represent the drainage quality of the rigid pavement. AASHTO-recommended C_d values are included in Table 5.12.

Reliability: Similar to the design of flexible pavements, Reliability and the overall standard deviation are included in the design charts.

5.2.1.4 Structural Design of Rigid Pavements

The pavement thickness of rigid pavements can be determined using the equation below or obtained from the chart in Figures 5.6 to 5.11.

$$log_{10}W_{18} = Z_R S_0 + 7.35log_{10}(D+1) - 0.06 + \frac{log_{10}\left[\dfrac{\Delta PSI}{4.5-1.5}\right]}{1+\left[\dfrac{1.624\times10^7}{(D+1)^{8.46}}\right]}$$

$$+(4.22-0.32P_t)log_{10}\left\{\frac{S_c'C_d}{215.63J}\left(\frac{D^{0.75}-1.132}{D^{0.75}-\left[\dfrac{18.42}{\left(\dfrac{E_c}{k}\right)^2}\right]}\right)\right\}$$

Eq. 5.7

where

W_{18} = predicted number of 18,000-lb (80kN) single-axle load applications
Z_R = standard normal deviation for a given reliability
S_0 = overall standard deviation
D = thickness of concrete pavement (to the nearest half-inch)
ΔPSI = design serviceability loss = p_i - p_t
p_i = initial serviceability index
p_t = terminal serviceability index
E_c = elastic modulus of the concrete to be used in construction (lb/in²)
Sc = modulus of rupture of the concrete to be used in construction (lb/in²)
J = load transfer coefficient = 3.2 (assumed)
C_d = drainage coefficient

5.2.2 The Mechanistic-Empirical Pavement Design Guide (MEPDG) Method

In this book, the MEPDG method is used for designing jointed plain concrete pavement (JPCP). The design procedure mirrors that of flexible pavements. Typically, a designer initiates with a trial design and then evaluates its adequacy against predetermined criteria. The entire design process is segmented into three key stages:

Stage 1: Determination of input values for the trial design
Stage 2: Analysis of the trial pavement's structural capability by predicting specific performance
   indicators
Stage 3: Evaluation of the trial design's structural viability

Table 5.8 Recommended Particle Size Distributions for Different Types of Subbase Materials

Sieve Designation	Types of Subbase					
	Type A	Type B	Type C (Cement Treated)	Type D (Lime Treated)	Type E (Bituminous Treated)	Type F (Granular)
Sieve analysis percent passing						
2 in.	100	100	-	-	-	-
1 in.	-	75-95	100	100	100	100
3/8 in.	30-65	40-75	50-85	60-100	-	-
No. 40	25-55	30-60	35-65	50-85	55-100	70-100
No. 10	15-40	20-45	25-50	40-70	40-100	55-100
No. 40	20-Aug	15-30	15-30	25-45	20-50	30-70
No. 200	8-Feb	20-May	15-May	20-May	20-Jun	25-Aug
(The minus No. 200 material should be held to a practical minimum.)						
Compressive strength lb/in² at 28 days			400-750	100		
Stability						
Hveem Stabilometer					20 min	
Hubbard field					1000 min	
Marshall stability					500 min	
Marshall flow					20 max	
Soil constants						
Liquid limit	25 max	25 max				25 max
Plasticity index(a)	N.P.	6 max	10 max (b)		6 max (b)	6 max

(a): As performed on samples prepared by AASHTO Designation T87.
(b): These values apply to the mineral aggregate before mixing with the stabilizing agent.

Source: Adapted with permission from Standard Specifications for Transportation Materials and Methods of Sampling and Testing, American Association of State Highway and Transportation Officials, Washington, DC, 2007.

The predicted distresses for a new JPCP are the mean transverse joint faulting, the transverse slab cracking (bottom-up and top-down), and the Smoothness. The design criteria or threshold values recommended in the MEPDG for these parameters are shown in Table 5.13.

If the predicted values of the trial design fail to meet the set criteria during its structural viability evaluation, the design is modified, and the evaluation process is repeated.

5.3 Design of Subdivision Roads

A subdivision road serves a parcel of land that is subdivided into two or more lots, building sites, or other divisions for sale or building development for residential purposes, providing access to the individual lots or properties within the subdivision.

Subdivision roads can be designed using simpler design options, such as design charts shown below in Tables 5.14 and 5.15. With this option, the design engineer can determine the minimum thicknesses of base and surface courses based on subgrade soil type (Group I or Group II). Heavier pavement designs than the minimum designs should be used when route classification and traffic loading impacts are higher. Tables 5.14 and 5.15 show Groups I and II design parameters (NCDOT, 2010).

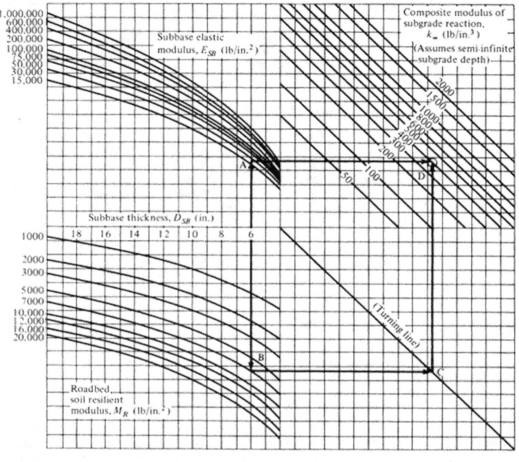

Figure 5.6 Chart for Estimating Composite Modulus of Subgrade Reaction, *k*... Assuming a Semi-Infinite
          Subgrade Depth

Source: AASHTO Guide for Design of Pavement Structures, American Association of State Highway and Transportation
Officials, Washington, DC, 1993, Used with Permission

5.4 Low Volume Road Design

For low-volume roads, the simplified table is used to design the base and asphalt layers.

This table is an easy method for determining designs typically used for low-volume pavements
based on the required structural number. However, minimum pavement depths based on road
types, if required, are still needed, regardless of Table 5.15.

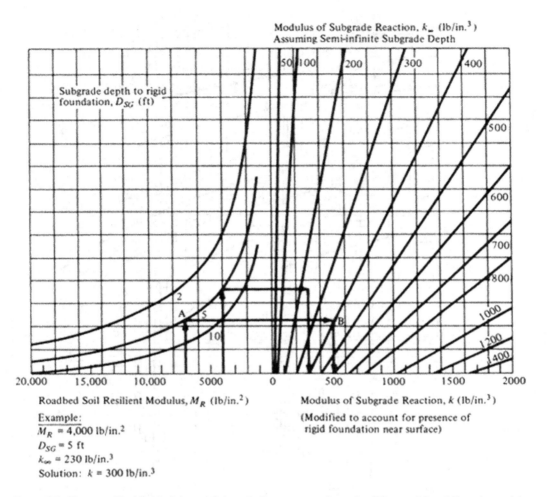

Modulus of Subgrade Reaction, k_∞ (lb/in.^{3})
Assuming Semi-infinite Subgrade Depth

Figure 5.7 Chart to Modify Modulus of Subgrade Reaction to Consider Effects of Rigid Foundation Near Surface (within 10 ft)

Source: AASHTO Guide for Design of Pavement Structures, American Association of State Highway and Transportation Officials, Washington, DC, 1993, Used with Permission

5.5 Granular Base Equivalency Method – Experience-Based Standard Sections

This method was derived from analyses of the in-service performance of historical pavement test sections and data from laboratory tests. Some sites have known historically plate-bearing tests and California Bearing Ratio (CBR) tests on various subgrade types. From the results, Over the years, these can be updated to reflect the observations and experiences. The concept of granular base equivalency (GBE) is used based on experiences, which equates the strength of various pavement materials in terms of their thicknesses. GBE thickness is the required overall structural pavement

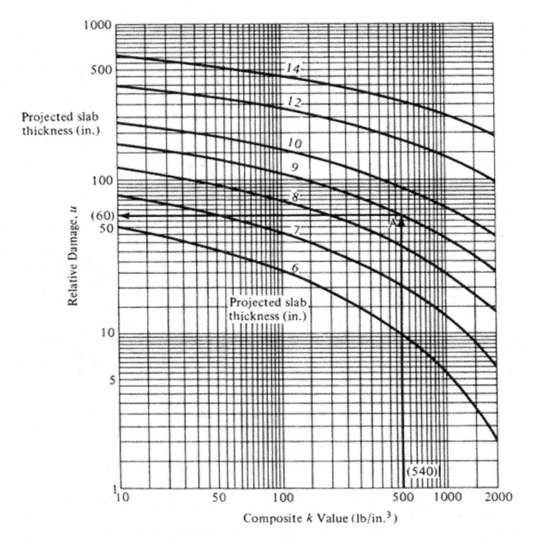

Figure 5.8 Chart for Estimating Relative Damage to Rigid Pavement Based on Slab Thickness and Underlying Support

Source: AASHTO Guide for Design of Pavement Structures, American Association of State Highway and Transportation Officials, Washington, DC, 1993, Used with Permission

thickness expressed in terms of an equivalent thickness of granular A. For example, 1 mm of hot mix is equivalent to 2 mm of granular A base, equivalent to 3 mm of granular subbase. These and other granular base equivalencies are shown on Table 5.17, which indicates the surface type and thickness and the base and subbase thicknesses for various categories of facilities and various AADT ranges and subgrade types. The table assumes 10% commercial traffic and that the foundation is a competent, non-saturated subgrade. thickness designs presented in the table assume that a

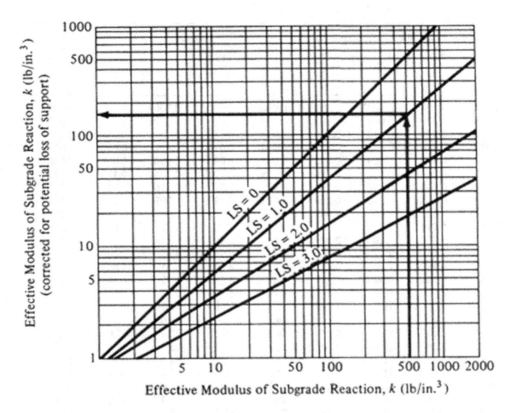

Figure 5.9 Correction of Effective Modulus of Subgrade Reaction for Potential Loss of Subbase Support

Source: AASHTO Guide for Design of Pavement Structures, American Association of State Highway and Transportation Officials, Washington, DC, 1993, Used with Permission

Table 5.9 Typical Ranges of Loss of Support Factors for Various Types of Materials

Type of Material	Loss of Support (LS)
Cement-treated granular base (E = 1,000,000 to 2,000,000 lb/in²)	0.0–1.0
Cement aggregate mixtures (E = 500,000 to 1,000,000 lb/in²)	0.0–1.0
Asphalt-treated base (E = 350,000 to 1,000,000 lb/in²)	0.0–1.0
Bituminous stabilized mixtures (E = 40,000 to 300,000 lb/in²)	0.0–1.0
Lime-stabilized mixtures (E = 20,000 to 70,000 lb/in²)	1.0–3.0
Unbound granular materials (E = 15,000 to 45,000 lb/in²)	1.0–3.0
Fine-grained or natural subgrade materials (E = 3000 to 40,000 lb/in²)	2.0–3.0

Table 5.10 ESAL Factors for Rigid Pavements, Single Axles, and P_t of 2.5

Axle Load (kip)	Slab Thickness, D (in)								
	6	7	8	9	10	11	12	13	14
2	0.0002	0.0002	0.0002	0.0002	0.0002	0.0002	0.0002	0.0002	0.0002
4	0.003	0.002	0.002	0.002	0.002	0.002	0.002	0.002	0.002
6	0.012	0.011	0.010	0.010	0.010	0.010	0.010	0.010	0.010
8	0.039	0.035	0.033	0.032	0.032	0.032	0.032	0.032	0.032
10	0.097	0.089	0.084	0.082	0.081	0.080	0.080	0.080	0.080
12	0.203	0.189	0.181	0.176	0.175	0.174	0.174	0.173	0.173
14	0.376	0.360	0.347	0.341	0.338	0.337	0.336	0.336	0.336
16	0.634	0.623	0.610	0.604	0.601	0.599	0.599	0.599	0.598
18	1.00	1.00	1.00	1.00	1.00	1.00	1.00	1.00	1.00
20	1.51	1.52	1.55	1.57	1.58	1.58	1.59	1.59	1.59
22	2.21	2.20	2.28	2.34	2.38	2.40	2.41	2.41	2.41
24	3.16	3.10	3.22	3.36	3.45	3.50	3.53	3.54	3.55
26	4.41	4.26	4.42	4.67	4.85	4.95	5.01	5.04	5.05
28	6.05	5.76	5.92	6.29	6.61	6.81	6.92	6.98	7.01
30	8.16	7.67	7.79	8.28	8.79	9.14	9.35	9.46	9.52
32	10.8	10.1	10.1	10.7	11.4	12.0	12.3	12.6	12.7
34	14.1	13.0	12.9	13.6	14.6	15.4	16.0	16.4	16.5
36	18.2	16.7	16.4	17.1	18.3	19.5	20.4	21.0	21.3
38	23.1	21.1	20.6	21.3	22.7	24.3	25.6	26.4	27.0
40	29.1	26.5	25.7	26.3	27.9	29.9	31.6	32.9	33.7
42	36.2	32.9	31.7	32.2	34.0	36.3	38.7	40.4	41.6
44	44.6	40.4	38.8	39.2	41.0	43.8	46.7	49.1	50.8
46	54.5	49.3	47.1	47.3	49.2	52.3	55.9	59.0	61.4
48	66.1	59.7	56.9	56.8	58.7	62.1	66.3	70.3	73.4
50	79.4	71.7	68.2	67.8	69.6	73.3	78.1	83.0	87.1

Source: Adapted from AASHTO Guide for Design of Pavement Structures, American Association of State Highway and Transportation Officials, Washington, DC, 1993

granular base is constructed across the full width of the cross-section, the shoulder is constructed of granular materials with or without a paved surface, and the drainage of the roadbed is adequate. Based on regional experience, modifications may be applied to this table.

5.6 Group Index Method

The group index or GI method for constructing flexible pavement comprises measuring the several levels of pavement, such as the surface course, base course, subbase, and subgrade. This work proposes the use of the GI method to build flexible pavements, with the practical approach of defining liquid limit, plasticity index, and water content. Certain materials in the form of admixtures are added to improve the index qualities. As a result, the inclusion of admixtures will increase the flexibility of the pavement.

Group index is a number assigned to the soil based on its physical properties like particle size, liquid limit and plastic limit. It varies from a value of 0 to 20, lower the value higher is the quality

Table 5.11 ESAL Factors for Rigid Pavements, Tandem Axles, and P_t of 2.5

Axle Load (kip)	Slab Thickness, D (in)								
	6	7	8	9	10	11	12	13	14
26	0.644	0.637	0.627	0.622	0.62	0.619	0.618	0.618	0.618
28	0.855	0.854	0.852	0.85	0.85	0.85	0.849	0.849	0.849
30	1.11	1.12	1.13	1.14	1.14	1.14	1.14	1.14	1.14
32	1.43	1.44	1.47	1.49	1.50	1.51	1.51	1.51	1.51
34	1.82	1.82	1.87	1.92	1.95	1.96	1.97	1.97	1.97
36	2.29	2.27	2.35	2.43	2.48	2.51	2.52	2.52	2.53
38	2.85	2.80	2.91	3.03	3.12	3.16	3.18	3.20	3.20
40	3.52	3.42	3.55	3.74	3.87	3.94	3.98	4.00	4.01
42	4.32	4.16	4.3	4.55	4.74	4.86	4.91	4.95	4.96
44	5.26	5.01	5.16	5.48	5.75	5.92	6.01	6.06	6.09
46	6.36	6.01	6.14	6.53	6.90	7.14	7.28	7.36	7.40
48	7.64	7.16	7.27	7.73	8.21	8.55	8.75	8.86	8.92
50	9.11	8.50	8.55	9.07	9.68	10.14	10.42	10.58	10.66
52	10.8	10.0	10.0	10.6	11.3	11.9	12.3	12.5	12.7
54	12.8	11.8	11.7	12.3	13.2	13.9	14.5	14.8	14.9
56	15.0	13.8	13.6	14.2	15.2	16.2	16.8	17.3	17.5
58	17.5	16.0	15.7	16.3	17.5	18.6	19.5	20.1	20.4
60	20.3	18.5	18.1	18.7	20.0	21.4	22.5	23.2	23.6
62	23.5	21.4	20.8	21.4	22.8	24.4	25.7	26.7	27.3
64	27.0	24.6	23.8	24.4	25.8	27.7	29.3	30.5	31.3
66	31.0	28.1	27.1	27.6	29.2	31.3	33.2	34.7	35.7
68	35.4	32.1	30.9	31.3	32.9	35.2	37.5	39.3	40.5
70	40.3	36.5	35.0	35.3	37.0	39.5	42.1	44.3	45.9
72	45.7	41.4	39.6	39.8	41.5	44.2	47.2	49.8	51.7
74	51.7	46.7	44.6	44.7	46.4	49.3	52.7	55.7	58.0
76	58.3	52.6	50.2	50.1	51.8	54.9	58.6	62.1	64.8
78	65.5	59.1	56.3	56.1	57.7	60.9	65.0	69.0	72.3
80	73.4	66.2	62.9	62.5	64.2	67.5	71.9	76.4	80.2
82	82.0	73.9	70.2	69.6	71.2	74.7	79.4	84.4	88.8
84	91.4	82.4	78.1	77.3	78.9	82.4	87.4	93.0	98.1
86	102.0	92.0	87.0	86.0	87.0	91.0	96.0	102.0	108.0
88	113.0	102.0	96.0	95.0	96.0	100.0	105.0	112.0	119.0
90	125.0	112.0	106.0	105.0	106.0	110.0	115.0	123.0	130.0

Source: Adapted from AASHTO Guide for Design of Pavement Structures, American Association of State Highway and Transportation Officials, Washington, DC, 1993

of the sub-grade and greater the value, poor is the sub-grade. Using the sieve analysis test, we can determine the group index value of soil subgrade from the equation:

$$GI = (F - 35) [0.2 + 0.005 (W_L - 40)] + 0.01 (F - 15)(I_p - 10)$$

where
F = percentage of soil passing 0.074 mm sieve
W_l = liquid limit of the subgrade soil
I_p = plasticity index in excess of 10

Table 5.12 Recommended Values for Drainage Coefficient, Cd, for Rigid Pavements

	Percent of Time Pavement Structure is Exposed to Moisture Levels Approaching Saturation			
Quality of Drainage	Less Than 1%	1–5%	5–25%	Greater Than 25%
Excellent	1.2–1.20	1.20–1.15	1.15–1.10	1.10
Good	1.20–1.15	1.15–1.10	1.10–1.00	1.00
Fair	1.15–1.10	1.10–1.00	1.00–0.90	0.90
Poor	1.10–1.00	1.00–0.90	0.90–0.80	0.80
Very poor	1.00–0.90	0.90–0.80	0.80–0.70	0.70

Source: Adapted from AASHTO Guide for Design of Pavement Structures, American Association of State Highway and Transportation Officials, Washington, DC, 1993

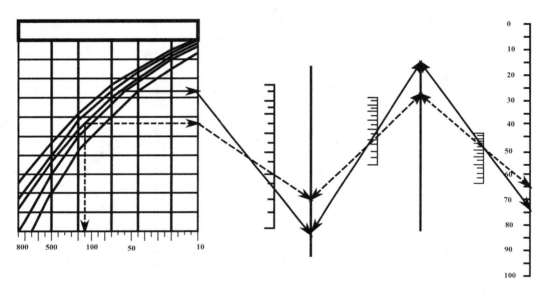

Figure 5.10 Design Chart for Rigid Pavements Based on Using Values for Each Input Variable (Segment 1)

Source: AASHTO Guide for Design of Pavement Structures, American Association of State Highway and Transportation Officials, Washington, DC, 1993, Used with Permission

By using Figure 5.12, the thickness of each layer of the pavement can be found.

5.7 Questions

1. How do the design approaches for flexible and rigid pavements differ according to this chapter?
2. What is the Mechanistic-Empirical Pavement Design Guide (MEPDG), and how does it improve pavement design?
3. Describe the importance of the present serviceability index (PSI) in the design of flexible pavements.
4. How is traffic load considered in the AASHTO 1993 pavement design method?
5. Explain the significance of the growth factor and how it is calculated in pavement design.

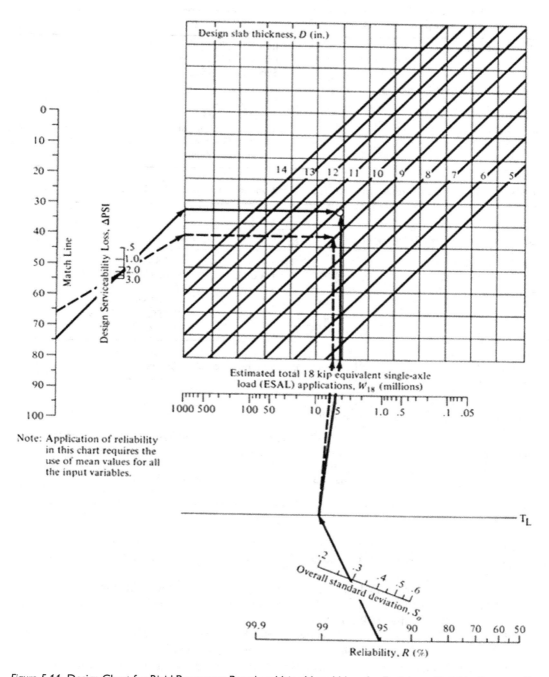

Figure 5.11 Design Chart for Rigid Pavements Based on Using Mean Values for Each Input Variable (Segment 2)

Source: AASHTO Guide for Design of Pavement Structures, American Association of State Highway and Transportation Officials, Washington, DC, 1993, Used with Permission

Table 5.13 Design Criteria or Threshold Values Recommended for Use in Judging the Acceptability of a New JCP, CPR, and Overlay Trial Design

Mean Joint Faulting	Interstate: 0.15 in Primary: 0.20 in Secondary: 0.25 in
Percent transverse slab cracking	Interstate: 10% Primary: 15% Secondary: 20%
IRI (smoothness)	Interstate: 160 in/mi Primary: 120 in/mi Secondary: 200 in/mi

Table 5.14 Subdivision Road Design Parameters Based on Good to Excellent Subgrade

Group I

Good to Excellent Subgrade Soil Types	Base Course	Pavement Surface
A-1-a, A-l-b, A-3 A-2-4, A-2-5, A-2-6 A-2-7	7" STBC, Type A or C 9" STBC, Type A or C 8" ABC or STBC, Type B 6" ABC or STBC, Type B 3" B25.0B 4" ABC	2" SF9.5A (1 layer) or S9.5B 1 ½" SF9.5A AST 1 ½" SF9.5A or S9.5B 1 ½" SF9.5A or S9.5B 5" Jointed Concrete

Table 5.15 Subdivision Road Design Parameters Based on Poor to Fair Subgrade

Group II

Poor to Fair Subgrade Soil Types	Base Course	Pavement Surface
A-44, A-5, A-6 A-7-5, A-7-6	9" STBC, Type A or C 8" ABC or STBC, Type B 10" ABC or STBC, Type B 4" B25.0B 4" ABC	2" SF9.5A (1 layer) or S9.5B 1 ½" SF9.5A or S9.5B AST 1 ½" SF9.5A or S9.5B 6" Jointed Concrete

6. What are the different levels of input data considered in the MEPDG method?
7. What criteria are used to determine the adequacy of a trial pavement design using the MEPDG method?
8. Discuss how environmental conditions impact pavement design and performance.
9. What are the recommended reliability levels for different functional classifications of roadways according to the MEPDG?
10. How does the design of subdivision roads differ from that of major highways?
11. What are the key factors to consider when designing a low-volume road?
12. Explain how the granular base equivalency (GBE) method contributes to pavement design.
13. Discuss the role of the group index method in constructing flexible pavements.
14. Calculate the group index (GI) for a soil with the following properties: 61% passing the #200 sieve, a liquid limit (W_l) of 35, and a plasticity index (I_p) of 22.

Table 5.16 Low-Volume Road Design Without Curb and Gutter Designs

1 Recommended Pavement Design – (Full Depth)

Required Structural Number	Surface	Inter.	Base	Structural Number Provided
	when 20 year ESALS are <0.3 Million			
<2.37	2.5" S9.5B		4.0" B25.0C	2.30
2.38–2.52	2.5" S9.5B		4.5" B25.0C	2.45
2.53–2.67	2.5" S9.5B		5.0"" B25.0C	2.60
2.68–2.82	2.5" S9.5B		5.5" B25.0C	2.75
2.83–3.47	2.5" S9.5B	2.5" I19.0C	4.0" B25.0C	3.40
	when 20 year ESALS are >0.3 Million			
<2.59	3.0" S9.5X		4.5" B25.0C	2.52
2.60–2.74	3.0"" S9.5X		4.5" B25.0C	2.67
2.75–2.89	3.0" S9.5X		5.0" B25.0C	2.82
2.90–3.04	3.0" S9.5X		5.5" B25.0C	2.97
3.05–3.69	3.0" S9.5X	2.5" I19.0C	4.0" B25.0C	3.62
3.70–3.84	3.0" S9.5X	2.5"""" I19.0C	4.5" B25.0C	3.77
3.85–3.99	3.0" S9.5X	2.5" I19.0C	5.0" B25.0C	3.92
4.00–4.14	3.0" S9.5X	2.5" I19.0C	5.5" B25.0C	4.07

2 Recommended Pavement Design (ABC)

Required Structural Number	Surface	Inter.	Base	Structural Number Provided
	when 20 year ESALS are <0.3 Million			
<2.01	2.5" S9.5B*		6" ABC	1.94
2.02–2.29	2.5" S9.5B*		8" ABC	2.22
2.30–2.57	2.5" S9.5B*		10"" ABC	2.50
2.58–3.11	2.5" S9.5B	2.5" I19.0C	6" ABC	3.04
3.12–3.39	2.5" S9.5B	2.5" I19.0C	8" ABC	3.32
3.40–3.67	2.5" S9.5B	2.5" I19.0C	10" ABC	3.60
	when 20 year ESALS are >0.3 Million			
<2.23	3.0" S9.5X		6" ABC	2.16
2.24–2.51	3.0" S9.5X		8" ABC	2.44
2.52–2.79	3.0" S9.5X		10" ABC	2.72
2.80–3.33	3.0" S9.5X	2.5" I19.0C	6" ABC	3.26
3.34–3.61	3.0" S9.5X	2.5" I19.0C	8" ABC	3.54
3.62–3.89	3.0" S9.5X	2.5" I19.0C	10" ABC	3.82

Curb and Gutter Designs

1 Recommended Pavement Design – (Full Depth)

Required Structural Number	Surface	Inter.	Base	Structural Number
<4.35	3.0" S9.5X	4.0" I19.0C	4.0" B25.0C	4.28
4.36–4.50	3.0" S9.5X	4.0" I19.0C	4.5" B25.0C	4.43

2 Recommended Pavement Design (ABC)

Required Structural Number	Surface	Inter.	Base	Structural Number
<3.99	3.0" S9.5X	4.0" I19.0C	6" ABC	3.92
4.00–4.27	3.0" S9.5X	4.0" I19.0C	8" ABC	4.20
4.28–4.55	3.0" S9.5X	4.0" I19.0C	10" ABC	4.48

* Prime coat required

Table 5.17 Structural Design Guidelines for Flexible Pavements (Thickness in mm)

AADT	Pavement Structure Elements	Subgrade Materials					
		Gravels and sands suitable as granular-borrow	Sands and Silts			Lacustrinc Clays	Varved and Leda Clays
			5–75 µm	5–75 µm	5–75 µm		
>4000 AADT	HMA	130	130	130	130	130	130
	B	150-250	150	150	150	150	150
	SB	---	300-450	450-600	600-800	450	450-1100
	GBE	410-510	610-710	710-810	810-945	710	710-1145
3000-4000 AADT	HMA	120-130	120-130	120-130	120-130	120-130	120-130
	B	150-250	150	150	150	150	150
	SB	---	300-450	450-600	600-800	450	450-1100
	GBE	390-510	590-710	690-810	790-945	690-710	690-1145
2000–3000 AADT	HMA	90	90	90	90	90	90
	B	150	150	150	150	150	150
	SB	---	300	450	600	450	800
	GBE	330	530	630	730	630	865
1000–2000 AADT	HMA	50	50	50	50	50	50
	B	150	150	150	150	150	150
	SB	---	250	300	450	300	450 (300-600)
	GBE	250	415	450	550	450	550 (415-550)
200–1000 AADT	HMA	50	50	50	50	50	50
	B	150	150	150	150	150	150
	SB	---	150	250	300	250	300 (250-450)
	GBE	250	350	415	450	415	250 (415-550)

B: Base thickness, SB: Subbase thickness, and GBE: Granular base equivalency thickness (1 mm HMA = 2 mm B = 3 mm SB)

Note: All AADT volumes refer to present traffic.

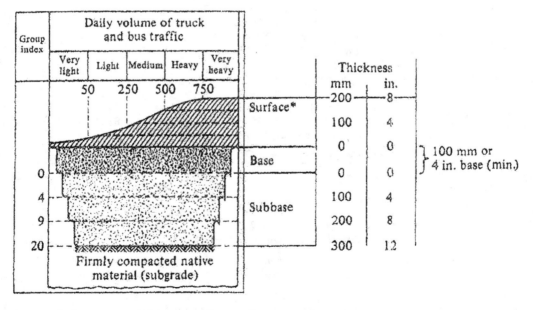

Figure 5.12 Group Index (GI) Pavement Design Chart

References

AASHTO (1993). *AASHTO Guide for Design of Pavement Structures (1993)*. American Association of State Highway and Transportation Officials (AASHTO), Washington, DC.

AASHTO (2007). *Standard Specifications for Transportation Materials and Methods of Sampling and Testing (2007)*. American Association of State Highway and Transportation Officials (AASHTO), Washington, DC.

NCDOT (2010). *Subdivision Roads Minimum Construction Standards (2010)*. North Carolina Department of Transportation (NCDOT), Raleigh, NC.

Chapter 6

Hot-Mix Asphalt Composing Materials

6.1 Introduction

The United States has 2.6 million miles of paved road, over 94% of which are paved with asphalt. From October 1987 through March 1993, the Strategic Highway Research Program (SHRP) conducted a $50 million research effort to develop new ways to specify, test, and design asphalt materials. Near the end of SHRP, the Federal Highway Administration assumed leadership in implementing SHRP research, out of which Superpave emerged.

Superpave stands for Superior Performing Asphalt Pavement System (Superpave), which was introduced in 1993 when the SHRP was completed. This new mix design method replaces the Hveem and Marshall methods. The Superpave system ties asphalt binder and aggregate selection into the mix design process and considers traffic and climate. The compaction device is a Superpave gyratory compactor (SGC), and the compaction effort in mix design is tied to expected traffic (Brown, Kandhal, Roberts, Kim, & Kennedy, 2009).

Superpave mix design has seven basic steps: (i) Asphalt binder selection; (ii) Aggregate selection; (iii) Sample preparation (including compaction); (iv) Performance tests; (v) Density and voids calculations; (vi) Optimum asphalt binder content selection; and (vii) Moisture susceptibility evaluation.

6.2 Asphalt Binder Selection

For asphalt paving projects, asphalt binders are selected based on the climate and traffic conditions in which they are intended to serve. The Superpave is a unique, improved system for testing, specifying, and selecting asphalt binders. The procedure to select the correct performance grade (PG) of asphalt binder for a particular application is as follows.

The physical property requirements within the Superpave binder specification are constant among all PG grades. What differentiates the various binder grades is the temperature at which the requirements must be met. For example, a binder classified as a PG 64-22 means that the binder must meet high-temperature physical property requirements at least up to a temperature of 64°C, and low-temperature physical properties must be met at least down to −22°C. These physical properties are directly related to field performance, so the greater the first (high) temperature value is, the more resistant the binder should be to high-temperature distress, such as rutting or shoving. Likewise, the lower the second (low) PG temperature value is, the more resistant the binder should be to low-temperature cracking. The high and low-temperature designations extend in both directions as far as necessary in six-degree increments, almost unlimited in the number of possible

DOI: 10.1201/9781003197768-6

grades. The more common paving grades used in the United States are PG 64–22, PG 70–22, PG 76–22, PG 58–22, PG 64–28, PG 58–28, and PG 52–34.

Specifiers select a binder grade based on the environment (historical temperature data), traffic conditions, and the desired reliability factor. Most state highway agencies (SHA) specify the binder grade to be used. A database of temperature information from over 7,000 weather stations in the United States and Canada allows users to select binder grades for the climate at a particular project location. For each year the weather stations have been in operation, the hottest seven-day period was identified, and the average maximum air temperature for this seven-day period was calculated. The mean and standard deviation of the seven-day average maximum air temperature were calculated for all the years of operation. Similarly, each year's one-day minimum air temperature was identified, and the mean and standard deviation were calculated (AI, 2001).

However, the design temperatures to be used for selecting asphalt binder grades are the pavement temperatures, not the air temperatures. Therefore, air temperatures from the weather station database must be converted into pavement temperatures. For surface layers, Superpave defines the location for the high pavement design temperature at a depth of 20 mm below the pavement surface and the low pavement design temperature at the pavement surface.

Using theoretical analyses of actual conditions performed with models for net heat flow and energy balance and assuming typical values for solar absorption (0.90), radiation transmission through the air (0.81), atmospheric radiation (0.70), and wind speed (4.5 m/sec), this equation was developed to convert the seven-day high air temperature to the high pavement design temperature:

$$T_{20mm} = (T_{air} - 0.00618 \text{ Lat}^2 + 0.2289 \text{ Lat} + 42.2)(0.9545) - 17.78 \qquad \text{Eq. 6.1}$$

where

T_{20mm} = high pavement design temperature at a depth of 20 mm
T_{air} = seven-day average high air temperature, °C
Lat = the geographical latitude of the project in degrees

The preferred method of determining the low pavement design temperature in Superpave is to utilize the LTPP Bind software. Before the availability of this software, there were two possible ways to establish this temperature. The first way was to assume that the low pavement design temperature was the same as the low air temperature. This method was initially recommended by SHRP researchers. This is a very conservative assumption because pavement temperature is almost always warmer than the air temperature in cold weather. The second method uses an equation developed by Canadian SHRP researchers:

$$T_{min} = 0.859 T_{air} + 1.7°C \qquad \text{Eq. 6.2}$$

where

T = minimum pavement design temperature in °C
T_{air} = minimum air temperature in an average year in °C

The Superpave system allows designers to use reliability measurements to assign a degree of design risk to the high and low pavement temperatures used in selecting the binder grade. Reliability is the percent probability in a single year that the actual temperature (one-day low or seven-day high) will not exceed the design temperature. Higher reliability means lower risk. For example, consider summer air temperatures in Cleveland, Ohio, which has a mean seven-day maximum of 32°C and a standard deviation of 2°C. In an average year, there is a 50% chance that the seven-day maximum air temperature will exceed 32°C. However, assuming a normal statistical frequency distribution, there is only a 2% chance that the seven-day maximum will exceed 36°C (mean plus two standard deviations); therefore, a design air temperature of 36°C will provide 98% reliability.

6.3 Air Temperature Selection

Continuing the example, assume that an asphalt mixture will be designed for Cleveland. In a typical summer, the average seven-day maximum air temperature is 32°C, and in a "very hot" summer, this average may reach 36°C. Using the same approach for winter conditions, Cleveland has a one-day minimum air temperature of −21°C with a standard deviation of 4°C. Consequently, in an average winter, the coldest temperature is −21°C.

For a "very cold" winter, the temperature may reach −29°C. The standard deviations show there is more variation in the one-day low temperatures than in the seven-day average high temperatures.

6.4 Pavement Temperature Selection

Continuing the example, for a surface course in Cleveland, the design pavement temperatures are about 52°C and −16°C for 50% reliability and about 56°C and −23°C for 98% reliability (mean plus two standard deviations).

6.4.1 Binder Grade Selection Based on Pavement Temperatures

To achieve reliability of at least 50% and provide an average maximum pavement temperature of at least 52°C, the standard high-temperature grade, PG 52, matches the design temperature of 52°C. Using the same reasoning, the standard low-temperature grade to attain 50% reliability is a PG −16, which again happens to match the design reliability. To obtain at least 98% reliability, select a standard high-temperature grade of PG 58 to protect above 56°C and a standard low-temperature grade of PG −28 to protect below −23°C. In both the high and low-temperature cases of the PG 58–28 binder grade, the reliability exceeds 99% because of the "rounding up" caused by the six-degree difference between standard grades. This "rounding up" introduces conservatism into the binder selection process.

Another source of conservatism occurs when considering the actual asphalt binder's physical properties. The specific binder may possess the actual properties of a PG 60–24, but it will be classified as a standard PG 58–22 grade. The net result is that a significant factor of safety is included in the binder selection scheme. For example, it is possible that the PG 52–16 binder, selected previously for a minimum of 50% reliability for Cleveland, may actually have been graded as a PG 56–20 had such a grade existed. Users of the PG grading system for binder selection should recognize that considerable safeguards are already included. Because of these factors, it may not be necessary or cost-effective to require indiscriminately high values of reliability or abnormally conservative high or low-temperature grades.

The Superpave computer programs perform all of these calculations based on minimal user input. The user can enter minimum reliability for any location, and the software will calculate the required asphalt binder grade. Alternatively, the user can specify a desired asphalt binder grade, and Superpave will calculate the reliability. Agencies face engineering management decisions when deciding on the asphalt binder PG grades to specify for their climatic and loading conditions. They will have to decide on the level of reliability to be used. Depending on the policy established by each agency, the selected reliability may be a function of road classification, traffic level, binder cost and availability, and other factors.

6.5 Adjusting Binder Grade Selection for Traffic Speed and Loading (Grade Bumping)

The asphalt binder selection procedure described is the basic procedure for typical highway loading conditions. Under these conditions, the pavement is assumed to be subjected to a number of fast, transient loads. For the high-temperature design situation, controlled by specified properties relating to permanent deformation, the loading speed has an additional effect on performance.

Superpave requires an additional shift in the selected high-temperature binder grade for slow and standing load applications. For slow-moving design loads, the binder would be selected one high-temperature grade higher, such as a PG 64 instead of a PG 58. For standing design loads, the binder would be selected two high-temperature grades higher, such as a PG 70 instead of a PG 58.

Also, an additional shift is needed for extraordinarily high numbers of heavy traffic loads. If the design traffic is expected to be between 10,000,000 and 30,000,000 equivalent single axle loads (ESAL), selecting one high-temperature binder grade higher than the selection based on climate is encouraged. If the design traffic is expected to exceed 30,000,000 ESAL, then the binder is required to be selected with one higher temperature grade than the selection based on climate. This practice of adjusting high-temperature grades for traffic loading and speed is sometimes called "grade-bumping." Table 6.1 summarizes AASHTO's grade-bumping policy as it is presented in MP2 of the AASHTO Provisional Standards (AASHTO, 2000, 2001).

It should be emphasized that proper or conservative binder selection does not guarantee total pavement performance. Fatigue cracking performance is greatly affected by the pavement structure

Table 6.1 Binder Selection on the Basis of Traffic Speed and Traffic Level (Data from AI, 2001)

Design ESALs[1] (Million)	Adjustment to Binder PG Grade[5] Traffic Load Rate		
	Standing[2]	Slow[3]	Standard[4]
<0.3	–		
0.3 to <3	2	1	
3 to <10	2	1	
10 to <30	2	1	–
≥30	2	1	1

[1]Design ESALs are the anticipated project traffic level expected on the design lane over a 20-year period. Regardless of the actual design life of the roadway, determine the design ESALs for 20 years and choose the appropriate N_{design} level.
[2]Standing Traffic – where the average traffic speed is less than 20 km/h.
[3]Slow Traffic – where the average traffic speed ranges from 20 to 70 km/h.
[4]Standard Traffic – where the average traffic speed is greater than 70 km/h.
[5]Increase the high-temperature grade by the number of grade equivalents indicated (one-grade equivalent to 6°C). Do not adjust the low-temperature grade.
Consideration should be given to increasing the high-temperature grade by one grade equivalent.
Practically, performance-graded binders stiffer than PG 82-XX should be avoided. In cases where the required adjustment to the high-temperature binder grade would result in a grade higher than a PG 82, consideration should be given to specifying a PG 82-XX and increasing the design ESALs by one level (e.g., 10 to <30 million increased to 30 million).

and traffic. Permanent deformation or rutting is directly a function of the shear strength of the mix, which is greatly influenced by aggregate properties. Pavement low-temperature cracking correlates most significantly to the binder properties. Engineers should try to achieve a balance among the many factors when selecting binders (Table 6.2).

6.6 Aggregate Selection

SHRP researchers found that there was general agreement that aggregate properties played an integral role in overcoming permanent deformation. Fatigue cracking and low-temperature cracking were less affected by aggregate characteristics. These survey results were used to identify two categories of aggregate properties needed in the Superpave system: consensus properties and source properties. Additionally, a new way of specifying the design aggregate gradation was developed.

6.7 Consensus Aggregate Properties

The pavement experts agreed that certain aggregate characteristics were critical to well-performing HMA. These characteristics were called "consensus properties" because there was wide agreement on their use and specified values. The consensus properties are (i) coarse aggregate angularity; (ii) fine aggregate angularity; (iii) flat and elongated particles; and (iv) clay content.

The criteria for these consensus aggregate properties are based on traffic levels and position within the pavement structure. Materials near the pavement surface subjected to high traffic levels require more stringent consensus properties. The criteria are intended to be applied to a proposed aggregate blend rather than individual components. However, many agencies currently apply such requirements to individual aggregates so that undesirable components can be identified. The consensus properties are detailed in Table 6.1.

6.7.1 Coarse Aggregate Angularity

This property ensures a high degree of aggregate internal friction and rutting resistance. It is defined as the percentage (by mass) of aggregates larger than 4.75 mm with one or more fractured faces. The test method specified by Superpave is ASTM D5821, the Test Method for Determining the Percentage of Fractured Faces in Coarse Aggregate. Table 6.1 gives the required minimum values for coarse aggregate angularity, which is a function of traffic level and position within the pavement.

6.7.2 Fine Aggregate Angularity

This property ensures a high degree of fine aggregate internal friction and rutting resistance. It is defined as the percent of air voids in loosely compacted aggregates smaller than 2.36 mm. The test method specified by Superpave is ASTM C1252 and AASHTO T304, Uncompacted Void Content of Fine Aggregate. This property is influenced by particle shape, surface texture, and grading. Higher void contents mean more fractured faces.

In the test procedure, a sample of fine, washed, and dried aggregate is poured into a small, calibrated cylinder through a standard funnel. By measuring the mass of fine aggregate (W) in the filled cylinder of known volume (V), the void content can be calculated as the difference between the cylinder volume and fine aggregate volume collected in the cylinder. The fine aggregate bulk

Table 6.2 Performance-Graded Asphalt Binder System

Performance Grade (°C)	PG 46-			PG 52-							PG 58-					PG 64-						PG 70-						PG 76-					PG 82-				
	34	40	46	10	16	22	28	34	40	46	16	22	28	34	40	10	16	22	28	34	40	10	16	22	28	34	40	10	16	22	28	34	10	16	22	28	34
Average 7-day max. pavement design temperature	<46			<52							<58					<64						<70						<76					<82				
Min. pavement design temperature	>−34	>−40	>−46	>−10	>−16	>−22	>−28	>−34	>−40	>−46	>−16	>−22	>−28	>−34	>−40	>−10	>−16	>−22	>−28	>−34	>−40	>−10	>−16	>−22	>−28	>−34	>−40	>−10	>−16	>−22	>−28	>−34	>−10	>−16	>−22	>−28	>−34
Original Binder																																					
Flashpoint temp, T48: min.																			230																		
Viscosity, ASTM D 4402: Max 3 Pa.s, test temp.																			135																		
Dynamic shear, TP5: G*/sinδ, min. 1.00 kPa, test temp. @10 rad/s	46			52							58					64						70						76					82				
Rolling thin film oven residue (T240)																																					
Mass loss, max. %																			1.0																		
Dynamic shear, TP5: G*/sinδ, min. 2.20 kPa, test temp. @10 rad/s	46			52							58					64						70						76					82				
Pressure aging vessel residue (PP1)																																					
PAV aging temp.	90			90							100					100						100(110)						100(110)					100(110)				
Dynamic shear, TP5: G*/sinδ, min. 5,000 kPa, test temp. @10 rad/s	10	7	4	25	22	19	16	13	10	7	25	22	19	16	13	31	28	25	22	19	16	37	34	31	28	25	22	40	37	34	31	28	40	37	34	31	28
Physical hardening																			Report																		
Creep stiffness TP1: S, max. 300 kPa, m-value, min. 0.3000, test temp. @60 S	−24	−30	−36	0	−6	−12	−18	−24	−30	−36	−6	−12	−18	−24	−30	0	−6	−12	−18	−24	−30	0	−6	−12	−18	−24	−30	0	−6	−12	−18	−24	0	−6	−12	−18	−24
Direct tension, TP3: failure strain, min. 1.0%, test temp. @10 mm/min	−24	−30	−36	0	−6	−12	−18	−24	−30		−6	−12	−18	−24	−30	0	−6	−12	−18	−24	−30	0	−6	−12	−18	−24	−30	0	−6	−12	−18	−24	0	−6	−12	−18	−24

specific gravity (G_{sb}) is used to compute the fine aggregate volume. Figure 6.1 shows the sand uncompacted voids measuring.

Table 6.3 gives the required minimum values for fine aggregate angularity (Uncompacted Void Content of Fine Aggregate) as a function of traffic level and position within the pavement.

6.7.3 Flat and Elongated Particles

This characteristic is the percentage by mass of coarse aggregates that have a maximum-to-minimum dimension ratio greater than five. Flat and elongated particles are undesirable because they have a tendency to break during construction and under traffic. The test procedure used is ASTM D4791, Flat or Elongated Particles in Coarse Aggregate, and it is performed on coarse aggregates larger than 4.75 mm.

The procedure uses a proportional caliper device, as shown in Figure 6.2, to measure the dimensional ratio of a representative sample of aggregate particles. The aggregate particle is first placed with its largest dimension between the swinging arm and the fixed post at position A. The swinging arm remains stationary while the aggregate is placed between the swinging arm and the fixed post at position B. If the aggregate does not fill this gap, it is counted as a flat or elongated particle.

The required maximum values for flat and elongated particles in the coarse aggregate are given in Table 6.3.

Figure 6.1 Measuring Fine Aggregate Uncompacted Voids

Table 6.3 Superpave Aggregate Consensus Property Requirements

Design ESALs[1] (Million)	Coarse Aggregate Angularity (Percent), Minimum		Uncompacted Void Content of Fine Aggregate (Percent), Minimum		Sand Equivalent (Percent), Minimum	Flat and Elongated[3] (Percent), Maximum
	≤100 mm	>100 mm	≤100 mm	>100 mm		
<0.3	55/–	–/–	–	–	40	–
0.3 to <3	75/–	50/–	40	40	40	10
3 to <10	85/80[2]	60/–	45	40	45	10
10 to <30	95/90	80/75	45	40	45	10
≥30	100/100	100/100	45	45	50	10

[1]Design ESALs are the anticipated project traffic level expected on the design lane over a 20-year period. Regardless of the actual design life of the roadway, determine the design ESALs for 20 years and choose the appropriate N_{design} level.
[2]85/80 denotes that 85% of the coarse aggregate has one fractured face, and 80% has 2 or more fractured faces.
[3]Criterion based upon a 5:1 maximum-to-minimum ratio.
(If less than 25% of a layer is within 100 mm of the surface, the layer may be considered to be below 100 mm for mixture design purposes.)

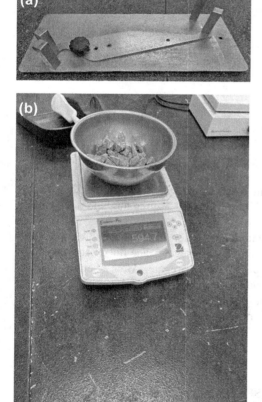

Figure 6.2 Measuring Flat and Elongated Particles of Coarse Aggregates

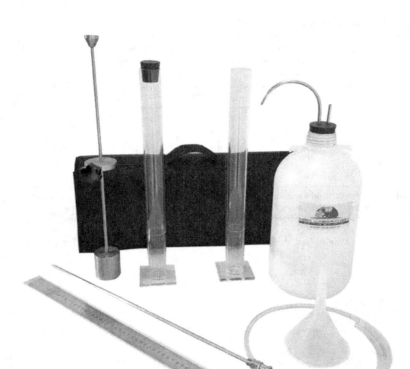

Figure 6.3 Fine Aggregate Sand Equivalent Test Apparatus

6.7.4 Clay Content (Sand Equivalent)

Clay content is the percentage of clay material contained in the aggregate fraction finer than a 4.75-mm sieve. It is measured by AASHTO T176, Plastic Fines in Graded Aggregates, and Soils by Use of the Sand Equivalent Test (ASTM D2419) (Figure 6.3).

A sample of fine aggregate is mixed with a flocculating solution in a graduated cylinder and agitated to loosen the clayey fines present in and coating the aggregate. The flocculating solution forces the clayey material into suspension above the granular aggregate. After a settling period, the cylinder height of suspended clay and settled sand is measured. The sand equivalent value is computed as the ratio of the sand to clay height readings, expressed as a percentage. The allowable clay content values for fine aggregate, expressed as a minimum percentage of sand equivalent, are given in Table 6.3.

6.8 Source Aggregate Properties

In addition to the consensus aggregate properties, SHRP researchers believed that certain other aggregate characteristics were critical. However, critical values of these properties could not be reached by consensus because the needed values were source-specific. Consequently, a set of source properties was recommended. Specified values are established by local agencies. While these properties are relevant during the mix design process, they may also be used for source acceptance control. Those properties are (i) toughness; (ii) soundness; and (iii) deleterious materials.

6.8.1 Toughness

Toughness is the percent loss of material from an aggregate blend during the Los Angeles abrasion test (AASHTO T96 or ASTM C131 or C535). This test estimates the resistance of coarse aggregate to abrasion and mechanical degradation during handling, construction, and in-service. It is performed by subjecting the coarse aggregate, usually larger than 2.36 mm, to impact and grinding by steel spheres. The result is the mass percentage of coarse material lost during the test due to mechanical degradation. Maximum loss values typically range from 35 to 45%.

6.8.2 Soundness

Soundness is the percent loss of material from an aggregate blend during the sodium or magnesium sulfate soundness test (AASHTO T104 or ASTM C88). This test estimates the resistance of an aggregate to in-service weathering. It can be performed on both coarse and fine aggregates. The test is performed by exposing an aggregate sample to repeated immersions in saturated sodium or magnesium sulfate solutions, followed by oven drying. One immersion and drying is considered one soundness cycle. During the drying phase, salts precipitate in the permeable void space of the aggregate. Upon re-immersion, the salt rehydrates and exerts internal expansive forces that simulate the expansive forces of freezing water.

The test result is the total percent loss over various sieve intervals for a required number of cycles. Maximum loss values typically range from 10–20% for 5 cycles.

6.8.3 Deleterious Materials

Deleterious materials are defined as the mass percentage of contaminants such as clay lumps, shale, wood, mica, and coal in the blended aggregate (AASHTO T112 or ASTM C142). The analysis can be performed on both coarse and fine aggregates. The test is performed by wet sieving aggregate size fractions over specified sieves. The mass percentage of material lost as a result of wet sieving is reported as the percent of clay lumps and friable particles.

A wide range of criteria for the maximum allowable percentage of deleterious particles exists. Values range from as little as 0.2% to as high as 10%, depending on the exact composition of the contaminant.

6.9 Gradation

To specify graduation, Superpave modifies an approach already used by some agencies. The 0.45-power gradation chart is used to define a permissible gradation. The "point 45 power" chart uses a unique graphing technique to show an aggregate blend's cumulative particle size distribution. The ordinate (vertical axis) of the chart is percent passing. The abscissa (horizontal axis) is an arithmetic scale of sieve size in millimeters, raised to the 0.45 power. Figure 6.4 illustrates how the abscissa is scaled.

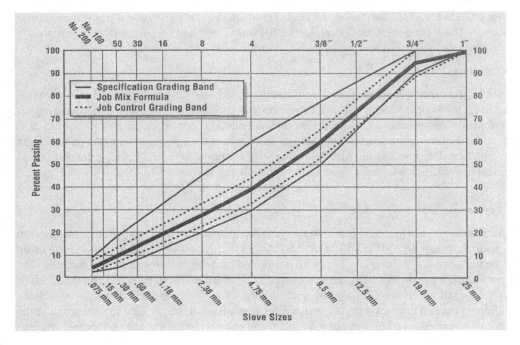

Figure 6.4 Typical 0.45 Power Chart and Example of Grading Band

In this example, the 4.75-mm sieve is plotted at 2.02, which is the sieve size of 4.75 mm raised to the 0.45 power. Traditionally, 0.45-power charts do not show arithmetic abscissa labels such as those in Figure 3.8. Instead, the scale is marked with the actual sieve size.

An important feature of the 0.45 power chart is the maximum density gradation. The maximum density line plots as a straight line from the maximum aggregate size to the origin. Below are the Superpave definitions for maximum and nominal maximum size:

Maximum Size: One sieve size larger than the nominal maximum size.

Nominal Maximum Size: One sieve size larger than the first sieve to retain more than 10%.

The maximum density line in Figure 6.4 represents a gradation where the aggregate particles fit together in their densest possible arrangement. Figure 6.4 shows a 0.45-power gradation chart with a maximum density line for a 19 mm maximum size aggregate (12.5 mm nominal maximum aggregate size).

Two additional features have been added to the 0.45-power chart – control points and a restricted zone to specify aggregate gradation.

6.9.1 Control Points

Control points function as master ranges through which gradations must pass. Control points are placed at the nominal maximum size, an intermediate size (2.36 mm), and the smallest size (0.075 mm). The control point limits vary depending on the nominal maximum aggregate size of the design mixture.

6.9.2 Restricted Zone

The restricted zone prevents a gradation from following the maximum density line in the fine aggregate sieves. Gradations that follow the maximum density line often have inadequate VMA to allow room for sufficient asphalt for durability. These gradations are typically very sensitive to asphalt content and can easily become plastic with even minor variations in asphalt content. The restricted zone is no longer a Superpave requirement (Figure 6.5).

While Superpave originally recommended that gradations pass below the restricted zone, it is not a requirement. Several highway agencies have successfully used gradings for passing above the restricted zone. Experience has shown that some gradations passing through the restricted zone perform satisfactorily. Before using such gradations, it is recommended that experience or testing be evaluated to determine if the particular aggregate structure performs satisfactorily (adequate VMA, non-tender mix behavior, etc.).

In most cases, a humped gradation indicates an over-sanded mixture and/or a mixture that possesses too much fine sand in relation to total sand. This gradation often results in tender mix behavior, which is manifested by compaction problems during construction. During their performance life, these mixtures may also offer reduced resistance to permanent deformation (rutting).

The restricted zone prevents a gradation from following the maximum density line in the fine aggregate sieves. Gradations that follow the maximum density line often have inadequate VMA

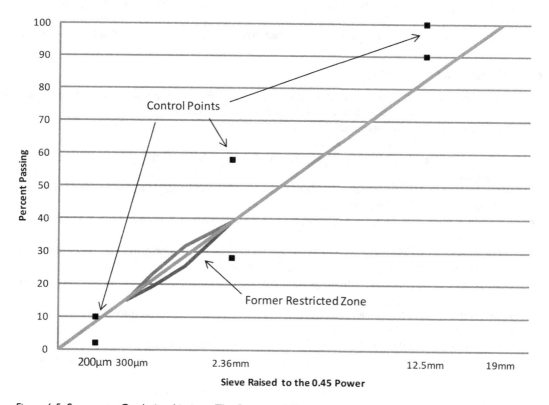

Figure 6.5 Superpave Gradation Limits – The Restricted Zone is no Longer a Superpave Requirement

Table 6.4 Superpave Mixture Gradations

Superpave Designation	Nominal Maximum Size, mm	Maximum Size, mm
37.5 mm	37.5	50.0
25.0 mm	25.0	37.5
19.0 mm	19.0	25.0
12.5 mm	12.5	19.0
9.5 mm	9.5	12.5

to allow room for sufficient asphalt for durability. These gradations are typically very sensitive to asphalt content and can easily become plastic with even minor variations in asphalt content.

While Superpave originally recommended that gradations pass below the restricted zone, it is not a requirement. Several highway agencies have successfully used gradings for passing above the restricted zone. Experience has shown that some gradations passing through the restricted zone perform satisfactorily. Before using such gradations, it is recommended that experience or testing be evaluated to determine if the particular aggregate structure performs satisfactorily (adequate VMA, non-tender mix behavior, etc.).

The design aggregate structure is the term used to describe the distribution of aggregate particle sizes. A design aggregate structure that lies between the control points meets the Superpave gradation requirements.

Superpave defines five mixture gradations by their nominal maximum aggregate size (Table 6.4).

6.10 Questions

1. How does Superpave tie asphalt binder and aggregate selection into the mix design process?
2. What are the seven basic steps involved in Superpave mix design?
3. How are asphalt binders selected for paving projects in the Superpave system, and what is the significance of performance grade (PG)?
4. How do specifiers determine the appropriate binder grade for a specific project, and what factors do they consider?
5. What factors and equations are involved in converting seven-day high air temperatures to high pavement design temperatures in Superpave?
6. What does reliability mean in the context of the Superpave system, and how is it related to design risk?
7. What factors other than binder selection influence pavement performance, particularly in terms of fatigue cracking and permanent deformation?
8. How do aggregate properties impact permanent deformation in asphalt mixtures, according to SHRP research?
9. How do the criteria for consensus aggregate properties vary based on traffic levels and position within the pavement structure?
10. Explain fine aggregate angularity and the test method used to measure it in Superpave.
11. What is the significance of flat and elongated particles in aggregate selection, and how are they tested?
12. What are source properties in Superpave, and why are they not universally standardized like consensus properties?

13. Describe the significance of toughness, soundness, and deleterious materials as source properties in aggregate selection.

14. How is toughness measured, and what does it indicate about an aggregate's resistance to abrasion?

15. Explain the concept of soundness and the test used to determine it in Superpave.

16. What are deleterious materials in aggregates, and how are they quantified in the Superpave system?

17. How does the Superpave system specify aggregate gradation, and what is the purpose of the 0.45-power gradation chart?

18. The Superpave specification is based on the high and low pavement temperatures expected at the highway's geographic location. High temperature grades vary from 46°C to 82°C, by 6°C increments. Low temperature ranges are from −10°C to −46°C. An asphalt binder of performance grade PG58-28 would be suitable for pavements designed for maximum temperatures of 58°C (the next lower grade is PG 52) and minimum temperatures down to −28°C.

19. Describe general requirements for HMA aggregates: (i) Maximum size and gradation; (ii) Cleanliness; (iii) Toughness; (iv) Particle shape; (v) Surface texture; (vi) Absorptive capacity; (vii) Affinity for asphalt cement; (viii) Frictional properties (surface course).

20. Describe consensus aggregate properties – 'Consensus' opinion of an expert panel determined that the following aggregate properties are critical to asphalt pavement performance: (i) Coarse aggregate angularity; (ii) Fine aggregate angularity, or Fine Aggregate Angularity (FAA) Uncompacted Voids Test; (iii) Flat or elongated particles; (iv) Clay content.

21. Describe source aggregate properties. Source-specific properties where 'consensus' of critical values could not be reached: (i) Toughness (Los Angeles abrasion); (ii) Soundness (magnesium or sodium soundness test); (iii) Deleterious materials (clay lumps, shale, wood, mica, coal, etc.).

References

AASHTO (2000; 2001). *AASHTO Provisional Standards*. American Association of State Highway and Transportation Officials (AASHTO). Washington, DC.

AI (2001). *Superpave Mix Design*. Superpave Series No. 2 (SP-2). 3rd ed. Asphalt Institute (AI), Lexington, KY.

(2001). *The Asphalt Binder Handbook*. Asphalt Institute (AI). Manual Series, No. 26 (MS-26). 1st ed. Asphalt Institute (AI), Lexington, KY.

Brown, E.R., Kandhal, P.S., Roberts, F.L., Kim, Y.R., Lee, D., & Kennedy, T.W. (2009). *Hot-Mix Asphalt Materials, Mixture Design and Construction*. NAPA Research and Education Foundation, Lanham, MD.

Chapter 7

Superpave Hot-Mix Asphalt Mix Design

7.1 Introduction

Superpave is a new rational approach, after Marshall and Hveem methods, to the design of asphalt pavement that can extend the life of a typical asphalt pavement by up to 30%. The objective of a mix design is to determine the combination of asphalt binder and aggregate that will give long-lasting performance as part of the pavement structure. HMA mix design involves laboratory procedures to establish the necessary proportions of materials for use in the asphalt mixture. In Superpave methods, when a mix design is conducted in the laboratory, the mix is analyzed to determine its probable performance in a pavement structure. Volumetric analysis is typically conducted on all mixtures regardless of the mix design methodology employed. The volumetric analysis focuses on the following five characteristics of the mixture and the influence those characteristics are likely to have on mix behavior: mix density, design air voids, voids in mineral aggregate, voids filled with asphalt, and asphalt content. This chapter describes the Superpave mix design procedures.

7.2 HMA Mixture Volumetrics

The volumetric proportions of asphalt binder and aggregate components must be considered when considering asphalt mixture behavior. The developers of Superpave felt that the volumetric properties of asphalt mixtures were so important that a volumetric mixture design protocol was developed. Volumetric analysis, which plays a significant role in most mixture design procedures of HMA, is described below.

Figure 7.1 visually represents the different components involved in creating hot-mix asphalt. This illustration is divided into several sections, each highlighting a specific phase of the composition.

Asphalt Binder: This area of the illustration would show a viscous, dark liquid that acts as the glue holding the asphalt mixture together. It might be represented by a flowing, sticky substance that binds the aggregates.

Coarse Aggregates: These are depicted as large, irregularly shaped stones or rocks varying in size. They are illustrated to emphasize their role in providing strength and structure to the asphalt. The aggregates are shown in reddish colors. The actual colors of the aggregates indicate different types of materials like granite or limestone.

Fine Aggregates: Smaller than the coarse aggregates, these could be represented by grains of sand or small stones. The fine aggregates fill in the gaps left by the coarse aggregates, and their detailed illustration shows a finer texture.

Air Voids: An essential component for durability and water drainage, air voids are illustrated as spaces between the aggregates and asphalt binder. They might be depicted as white or light-colored

DOI: 10.1201/9781003197768-7

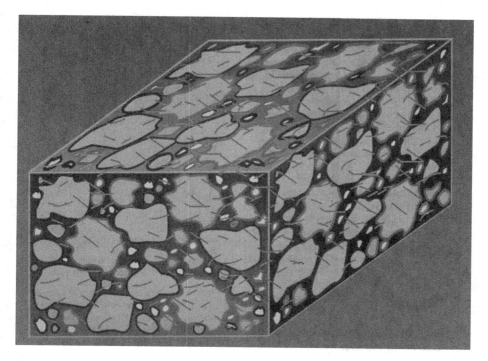

Figure 7.1 Composition Elements Hot-Mix Asphalt Concrete, Asphalt Binder, Coarse and Fine Aggregates, and Air Voids

empty spaces that intersperse the other materials, indicating areas where air is trapped within the asphalt mix (Brown, Kandhal, Roberts, Kim, Lee, & Kennedy, 2009).

The illustration as a whole shows how these components are blended together in precise proportions to create hot-mix asphalt used for paving roads. The different textures and colors used for each element visually explain their distinct roles and how they combine to form a cohesive and durable paving material.

The volumetric properties of a compacted paving mixture, including air voids (V_a), voids in the mineral aggregate (VMA), voids filled with asphalt (VFA), and effective asphalt content (P_{be}) provide some indication of the mixture's probable pavement service performance. It is necessary to understand the definitions and analytical procedures described hereafter to be able to make informed decisions concerning the selection of the design asphalt mixture. The information here applies to both paving mixtures that have been compacted in the laboratory and to undisturbed samples that have been cut from a pavement in the field.

7.2.1 Definitions

Mineral aggregate is porous and can absorb water and asphalt to a variable degree. Furthermore, the ratio of water to asphalt absorption varies with each aggregate. The three methods of measuring aggregate specific gravity consider these variations. These methods are bulk, apparent, and effective specific gravity:

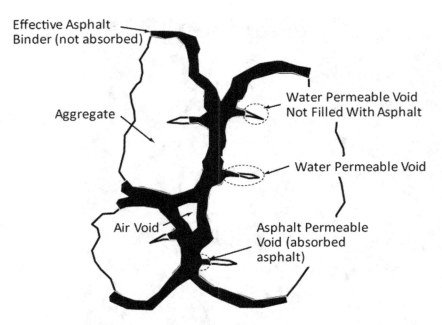

Figure 7.2 Illustration of Bulk, Effective and Apparent Specific Gravity, and Effective Asphalt Binder Content in Compacted Paving Mixtures

Bulk Specific Gravity, G_{sb}: The ratio of the mass in air of a unit volume of a porous material (including both permeable and impermeable voids regular to the material) at a stated temperature to the mass in the air of equal density of an equal volume of gas-free distilled water at a stated temperature.

Apparent Specific Gravity, G_{sa}: The ratio of the mass in air of a unit volume of an *impermeable* material at a stated temperature to the mass in the air of equal density of an equal volume of gas-free distilled water at a stated temperature.

Effective Specific Gravity, G_{se}: The ratio of the mass in air of a unit volume of a porous material (excluding voids permeable to asphalt) at a stated temperature to the mass in the air of equal density of an equal volume of gas-free distilled water at a stated temperature.

The definitions for VMA, P_{be}, V_{a}, and VFA are:

Voids in the Mineral Aggregate, VMA: The volume of intergranular void space between the aggregate particles of a compacted paving mixture, which includes the V_{a} and the P_{be}, expressed as a percentage of the total volume of the sample.

Effective Asphalt Content, P_{be}: The paving mixture's total asphalt content minus the asphalt portion absorbed by the aggregate particles.

Air Voids, V_{a}: The total volume of the small pockets of air between the coated aggregate particles throughout a compacted paving mixture, expressed as a percentage of the bulk volume of the compacted paving mixture.

Voids Filled with Asphalt, VFA: The percentage portion of the volume of intergranular void space between the aggregate particles that is occupied by the effective asphalt. It is expressed as the ratio of (VMA:V_{a}) to VMA (AI, 2001).

The Superpave mix design procedures require calculating VMA values for compacted paving mixtures in terms of the aggregate's bulk specific gravity, G_{sb}. The use of other aggregate-specific gravities to compute VMA means that the VMA criteria no longer apply. The effective specific gravity, G_{se}, should be the basis for calculating the V_a in a compacted asphalt paving mixture.

VMA and V_a are expressed as a percent by volume of the paving mixture. VFA is the percentage of VMA filled by the effective asphalt. Depending on the specified asphalt content, the P_{be} may be expressed as either a percent, by mass, of the *total mass* of the paving mixture or as a percent, by mass, of the *aggregate* in the paving mixture.

Because V_a, VMA, and VFA are volume quantities and, therefore, cannot be weighed, a paving mixture must first be designed or analyzed on a volume basis. This volume approach can easily be converted to a mass basis for design purposes to provide a job-mix formula (Figure 7.3).

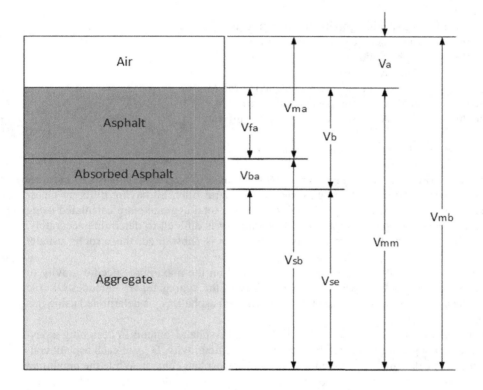

V_{ma} = Volume of voids in mineral aggregate
V_{mb} = Bulk volume of compacted mix
V_{mm} = Voidless volume of paving mix
V_{fa} = Volume of voids filled with asphalt
V_a = Volume of air voids
V_b = Volume of asphalt
V_{ba} = Volume of absorbed asphalt
V_{sb} = Volume of mineral aggregate (by bulk specific gravity)
V_{se} = Volume of mineral aggregate (by effective specific gravity)

Figure 7.3 HMA Superpave Volumetrics

7.2.2 Analyzing a Compacted Paving Mixture

The measurements and calculations needed for a void analysis are

- Measure the bulk specific gravity of the coarse aggregate (AASHTO T85 or ASTM C127) and the fine aggregate (AASHTO T84 or ASTM C128)
- Measure the specific gravity of the asphalt cement (AASHTO T228 or ASTM D70) and the mineral filler (AASHTO T100 or ASTM D854)
- Calculate the bulk specific gravity of the aggregate combination in the paving mixture
- Measure the maximum specific gravity of the loose paving mixture (ASTM D2041 or AASHTO T209)
- Measure the bulk specific gravity of the compacted paving mixture (ASTM D1188 / D2726 or AASHTO T166)
- Calculate the effective specific gravity of the aggregate
- Calculate the maximum specific gravity at other asphalt contents
- Calculate the asphalt absorption of the aggregate
- Calculate the P_{be} of the paving mixture
- Calculate the percent VMA in the compacted paving mixture
- Calculate the percent V_a in the compacted paving mixture
- Calculate the percent VFA in the compacted paving mixture

Equations for these calculations are found in Table 7.1:

1. Bulk Specific Gravity of Aggregate: When the total mass of aggregate consists of separate fractions of coarse aggregate, fine aggregate, and mineral filler, all having different measured specific gravities, the bulk specific gravity for the total aggregate are calculated using this equation. The bulk specific gravity of mineral filler is difficult to determine accurately. However, if the apparent specific gravity of the filler is substituted, the error is usually negligible.

2. Effective Specific Gravity of Aggregate: When based on the maximum specific gravity of a paving mixture, G_{mm}, the effective specific gravity of the aggregate, G_{se}, includes all void spaces in the aggregate particles except those that absorb asphalt. G_{se} is determined using this equation.

3. Maximum Specific Gravity of Mixtures with Different Asphalt Content: In designing a paving mixture with a given aggregate, the maximum specific gravity, G_{mm}, at each asphalt content is needed to calculate the percentage of V_a for each asphalt content. While the maximum specific gravity can be determined for each asphalt content, the precision of the test is best when the mixture is close to the design asphalt content. After calculating the effective specific gravity of the aggregate from the measured maximum specific gravity and averaging the G_{se} results, the maximum specific gravity for any other asphalt content can be obtained using the equation shown in the Table. The equation assumes the effective specific gravity of the aggregate is constant, which is valid since asphalt absorption does not vary appreciably with changes in asphalt content.

4. Asphalt Absorption: Absorption is expressed as a percentage by mass of aggregate rather than as a percentage by total mass of the mixture. Asphalt absorption, P_{ba}, is determined using this equation.

Table 7.1 Eight Equations for Calculation of Volumetric Properties

Properties	Equation	Eq.	Where
1 Bulk Specific Gravity of Aggregate	$G_{sb} = \dfrac{P_1 + P_2 + \ldots + P_N}{\dfrac{P_1}{G_1} + \dfrac{P_2}{G_2} + \ldots + \dfrac{P_N}{G_N}}$	7.1	G_{sb} = bulk specific gravity for the total aggregate; P_1, P_2, P_N = individual percentages by mass of aggregate; G_1, G_2, G_N = individual (e.g. coarse, fine) bulk specific gravity of aggregate
2 Effective Specific Gravity of Aggregate	$G_{se} = \dfrac{P_{mm} - P_b}{\dfrac{P_{mm}}{G_{mm}} - \dfrac{P_b}{G_b}}$	7.2	G_{se} = effective specific gravity of aggregate; G_{mm} = maximum specific gravity of paving mixture (no V_a); P_{mm} = percent by mass of total loose mixture = 100; P_b = asphalt content, percent by total mass of the mixture; G_b = specific gravity of asphalt
3 Maximum Specific Gravity of Mixtures with Different Asphalt Content	$G_{mm} = \dfrac{P_{mm}}{\dfrac{P_s}{G_{se}} + \dfrac{P_b}{G_b}}$	7.3	P_s = aggregate content, percent by total mass of the mixture
4 Asphalt Absorption	$P_{ba} = 100 \times \dfrac{G_{se} - G_{sb}}{G_{sb} G_{se}} \times G_b$	7.4	P_{ba} = absorbed asphalt, percent by mass of aggregate
5 P_{be} of a Paving Mixture	$P_{be} = P_b - \dfrac{P_{ba}}{100} \times P_s$	7.5	P_{be} = effective asphalt content, percent by total mass of the mixture
6 Percent VMA in Compacted Paving Mixture	$VMA = 100 - \dfrac{G_{mb} \times P_s}{Gsb}$	7.6	VMA = voids in the mineral aggregate, percent of bulk volume; G_{mb} = bulk specific gravity of compacted mixture
7 Percent V_as in Compacted Mixture	$V_a = 100 \times \dfrac{G_{mm} - G_{mb}}{G_{mm}}$	7.7	V_a = air voids in compacted mixture, percent of total volume
8 Percent VFA in Compacted Mixture	$VFA = 100 \times \dfrac{VMA - Va}{VMA}$	7.8	VFA = voids filled with asphalt, percent of VMA

5. Effective Asphalt Content of a Paving Mixture: The effective asphalt content, P_{be}, of a paving mixture is the total asphalt content minus the quantity of asphalt lost by absorption into the aggregate particles. It is the portion of the total asphalt content that remains as a coating on the outside of the aggregate particles, and it is the asphalt content that governs the performance of an asphalt mixture.

6. Percent VMA in Compacted Paving Mixture: The VMA are defined as the intergranular void space between the aggregate particles in a compacted paving mixture, which includes the V_a and the P_{be}, expressed as a percentage of the total volume. The VMA is calculated based on

the bulk specific gravity of the aggregate and is expressed as a percentage of the bulk volume of the compacted paving mixture. Therefore, the VMA can be calculated by subtracting the volume of the aggregate, determined by its bulk specific gravity, from the bulk volume of the compacted paving mixture. The calculation is illustrated for each type of mixture percentage content.

7. Percent V_a in Compacted Mixture: The air voids, V_a, in the total compacted paving mixture consist of the small air spaces between the coated aggregate particles. The volume percentage of V_a in a compacted mixture can be determined using this equation.

8. Percent VFA in Compacted Mixture: The percentage of the VMA that are filled with asphalt, VFA, not including the absorbed asphalt, is determined using this equation.

7.2.2.1 Calculation Example

Table 7.2 provides example data for a sample of paving mixture. These design data are used for the sample calculations in the following practice.

1. Calculate Bulk Specific Gravity of Aggregate by using Equation 7.1 in Table 7.1:

$$G_{sb} = \frac{50.0 + 50.0}{\dfrac{50.0}{2.716} + \dfrac{50.0}{2.689}} = \frac{100}{18.41 + 18.59} = 2.703 \qquad \text{Eq. 7.9}$$

2. Calculate Effective Specific Gravity of Aggregate by using Equation 7.22 in Table 7.1:

$$G_{se} = \frac{100 - 5.3}{\dfrac{100}{2.535} - \dfrac{5.3}{1.030}} = \frac{94.7}{39.45 - 5.15} = 2.761 \qquad \text{Eq. 7.10}$$

Table 7.2 Basic Data for a Sample of Paving Mixture

Mixture Components

Material	Specific Gravity		Test Methods		Mix Composition	
	Bulk		AASHTO	ASTM	By Mass of Total Mix	By Mass of Total Aggr.
Asphalt Cement	1.030 (G_b)		T228	D70	5.3 (P_b)	5.6 (P_b)
Coarse Aggregate		2.716 (G_1)	T85	C127	47.4 (P_1)	50.0 (P_1)
Fine Aggregate		2.689 (G_2)	T84	C128	47.3 (P_2)	50.0 (P_2)
Mineral Filler	–		T100	D854	–	–

Paving Mixture

Bulk specific gravity of compacted paving mixture specimen, $G_{mb} = 2.442$
Maximum specific gravity of paving mixture specimen, $G_{mm} = 2.535$

3. Calculate the Maximum Specific Gravity of Mixtures with Different Asphalt Content by using Equation 7.3 in Table 7.1:

$$G_{mm} = \frac{100}{\dfrac{96.0_s}{2.761} + \dfrac{4.0}{1.030}} = \frac{100}{34.77 + 388} = 2.587 \qquad \text{Eq. 7.11}$$

4. Calculate Asphalt Absorption by using Equation 7.4 in Table 7.1:

$$P_{ba} = 100 \times \frac{2.761 - 2.703}{2.703 \times 2.761} \times 1.030 = 100 \times \frac{0.058}{7.463} \times 1.030 = 0.8 \qquad \text{Eq. 7.12}$$

5. Calculate P_{be} of a Paving Mixture by using Eq. 7.5 in Table 7.1:

$$P_{be} = 5.3 - \frac{0.8}{100} \times 94.7 = 4.5 \qquad \text{Eq. 7.13}$$

6. Calculate Percent VMA in Compacted Paving Mixture by using Equation 7.6 in Table 7.1:

$$VMA = 100 - \frac{2.442 \times 94.7}{2.703} = 100 - 85.6 = 14.4 \qquad \text{Eq. 7.14}$$

7. Calculate Percent V_a in Compacted Mixture by using Equation 7.7 in Table 7.1:

$$V_a = 100 \times \frac{2.535 - 2.442}{2.535} = 3.7 \qquad \text{Eq. 7.15}$$

8. Calculate Percent VFA in Compacted Mixture by using Equation 7.8 in Table 7.1:

$$VFA = 100 \times \frac{14.4 - 3.7}{14.4} = 74.3 \qquad \text{Eq. 7.16}$$

Note: The volume of asphalt binder absorbed by an aggregate is almost invariably less than the volume of water absorbed. Consequently, the value for the effective specific gravity of an aggregate should be between its bulk and apparent specific gravities. When the effective specific gravity falls outside these limits, its value must be assumed to be incorrect. The calculations, the maximum specific gravity of the total mix by ASTM D2041/AASHTO T209, and the composition of the mix in terms of aggregate and total asphalt content should then be rechecked to find the source of the error.

7.3 Superpave Mix Design Procedures

Superpave mix design procedures involve:

- Selecting asphalt and aggregate materials that meet their respective criteria
- Developing several aggregate trial blends to meet Superpave gradation requirements
- Blending asphalt with the trial blends and short-term oven aging the mixtures
- Compacting specimens and analyzing the volumetrics of the trial blends
- Selecting "the best" trial blend as the design aggregate structure and compacting samples of the design aggregate structure at several asphalt contents to determine the design asphalt content

The testing procedures presented here are paraphrased from the AASHTO standards. The primary device used in Superpave mix design is the Superpave gyratory compactor (SGC). The SGC is used to produce specimens for volumetric analysis, and it also records data to provide a measure of specimen density throughout the compaction procedure.

7.3.1 Superpave Gyratory Compactor

SHRP researchers had several goals in developing a laboratory compaction method. Most importantly, they wanted to realistically compact mixture test specimens to densities achieved under actual pavement climate and loading conditions. The compaction device needed to be capable of accommodating large aggregates and be able to measure compactability, so that potential tender mix behavior and similar compaction problems could be identified. A high priority for SHRP researchers was a device portable enough for use in the quality control operations of a mixing facility. Since no existing compactor achieved all these goals, the SGC was developed.

7.3.2 Test Equipment

The basis for the SGC was a Texas gyratory compactor modified to use the compaction principles of a French gyratory compactor. The modified Texas gyratory accomplished the goals of realistic specimen densification, and it was reasonably portable. Its 6-inch sample diameter (ultimately 150 mm on an SGC) could accommodate mixtures containing aggregate up to 50 mm maximum size (37.5 mm nominal maximum size). SHRP researchers modified the Texas gyratory compactor by lowering its angle and speed of gyration and adding real-time specimen height-recording capabilities.

The SGC consists of these components:

- Reaction frame, rotating base, and motor
- Loading system, loading ram, and pressure gauge
- Height measuring and recordation system
- Mold and base plate
- Specimen extruding device (Figure 7.4)

A loading mechanism presses against the reaction frame and applies a load to the loading ram to produce a 600-kPa compaction pressure on the specimen. A pressure gauge measures the ram loading to maintain constant pressure during compaction. The SGC mold has an inside diameter of

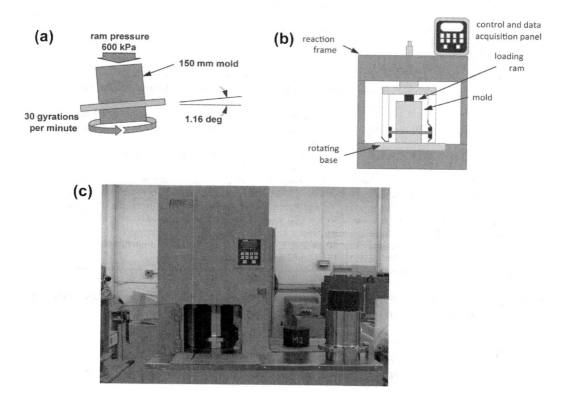

Figure 7.4 Superpave Gyratory Compactor (SGC)

150 mm, and a base plate at the bottom of the mold provides confinement during compaction. The SGC base rotates at a constant rate of 30 revolutions per minute during compaction, with the mold positioned at a compaction angle of 1.25 degrees.

Specimen height measurement is an important function of the SGC. Specimen density can be estimated during compaction by knowing the mass of material placed in the mold, the inside diameter of the mold, and the specimen height. Height is measured by recording the position of the ram throughout the test. Using these measurements, a specimen's compaction characteristics are developed.

The density of the asphalt mixture increases with increasing gyrations. As with other mix design procedures, asphalt mixtures are designed at a specific level of compactive effort. In Superpave, this is a function of the design number of gyrations, N_{des}. N_{des} is used to vary the compactive effort of the design mixture, and it is a function of traffic level. Traffic is represented by the design ESALs. The range of values for N_{des} is shown in Table 7.3.

Two other gyration levels are also of interest: the initial number of gyrations (N_{ini}), and the maximum number of gyrations (N_{max}). Test specimens is compacted using N_{des} gyrations, and an estimation of the compactability of the mixture is determined using N_{ini}. N_{max} is determined (using additional SGC specimens) after the mixture properties are established as a check to help guard

Table 7.3 Superpave Gyratory Compactive (SGC) Effort

Design ESALs (Millions)	Compaction Parameters			Typical Roadway Applications
	N_{ini}	N_{des}	N_{max}	
<0.3	6	50	75	Very light traffic (local/county; city streets where truck traffic is prohibited)
0.3 to <3	7	75	115	Medium traffic (collector roads; most county roadways)
3 to <30	8	100	160	Med. to high traffic (city street; state route; US highways; some rural interstates)
≥30	9	125	205	High traffic (most of the interstate system; climbing lanes; truck weighing stations)

When specified by the agency and the top of the design layer is ≥100 mm from the pavement surface and the estimated design traffic level is ≥0.3 million ESALs, decrease the estimated design traffic level by one, unless the mixture will be exposed to significant mainline and construction traffic prior to being overlaid. If less than 25% of the layer is within 100 mm of the surface, the layer may be considered to be below 100 mm for mixture design purposes.

When the design ESALs are between 3 to 10 million ESALs, the agency may, at their discretion, specify $N_{initial}$ at 7, N_{design} at 75, and N_{max} at 115, based on local experience.

against plastic failure caused by traffic in excess of the design level. N_{max} and N_{ini} are calculated from N_{des}, using the following relationships:

$$\text{Log } N_{max} = 1.10 \text{ Log } N_{des}$$
$$\text{Log } N_{ini} = 0.45 \text{ Log } N_{des}$$

The values of N_{ini}, N_{des} and N_{max} are shown for Superpave-defined traffic levels in Table 7.3.

7.3.3 Additional Test Equipment

Ancillary test equipment required in the preparation of Superpave asphalt mixtures include

1. Ovens, thermostatically controlled, for heating aggregates, asphalt, and equipment
2. Commercial bread dough mechanical mixer with a 10-liter (10-qt.) capacity or larger, equipped with metal mixing bowls and wire whips
3. Flat-bottom metal pans for heating aggregates and aging mixtures
4. Round metal pans, approximately 10-liter (10-qt.) capacity, for mixing asphalt and aggregate
5. Scoops for batching aggregates
6. Containers such as gill-type tins, beakers, or pouring pots, for heating asphalt
7. Thermometers, either armored, glass, or dial-type with metal stem, 10°C to 235°C, for determining the temperature of aggregates, asphalt, and asphalt mixtures
8. Balances with 10-kg capacity, sensitive to 1 g, for weighing aggregates and asphalt; 10-kg capacity, sensitive to 0.1 g, for weighing compacted specimens
9. Large mixing spoon or small trowel
10. Large spatula
11. Welders' gloves (or similar) for handling hot equipment
12. Paint, markers, or crayons, for identifying test specimens
13. Paper disks, 150 mm, for compaction
14. Fans for cooling compacted specimens
15. Computer/printer for data collection and recording

7.3.4 Select Design Aggregate Structure

Prior to selecting a design aggregate structure, the individual asphalt and aggregate materials must be selected and approved.

To select the design aggregate structure, trial blends are established by mathematically combining the gradations of the individual aggregates into a single gradation. The blend gradation is then compared to the specification control requirements for the appropriate sieves.

Trial blending consists of varying the stockpile percentages of each aggregate to obtain blend gradations that meet the gradation requirements for that particular mixture. There is no set number of trial blends that should initially be attempted. Three blends that cover a range of gradations are often sufficient for a starting point. Note that while Superpave recommends that gradations pass below the restricted zone, this is not a requirement. A trial gradation can plot above or even through the restricted zone. As an agency or contractor begins testing materials using the Superpave system, it would be beneficial to conduct analyses on many trial blends to determine the mixture behavior of the local materials.

Once the trial blends are selected, a preliminary evaluation of the blended aggregate properties is necessary. This includes the four consensus properties, the bulk and apparent specific gravity of the aggregate, and any source aggregate properties. These values can initially be mathematically estimated from the individual aggregate properties. Actual tests should be performed on the aggregate blend for final approval.

After the aggregate properties have been evaluated, the next step is to compact specimens and determine the volumetric properties of each trial blend. The trial asphalt binder content for each trial blend can be calculated by following the procedure detailed in AASHTO Provisional Standard, PP-28, Appendix XI.

Rather than calculate the trial asphalt binder contents, many designers prefer to estimate the trial values based on experience or other information. The following values are typical for aggregate blends having combined aggregate bulk specific gravity of approximately 2.65. Aggregate combinations having significantly higher G_{sb}, may need less asphalt, and those with a lower G_{sb} may need more asphalt (Table 7.5).

A minimum of two specimens for each trial blend is compacted using the Superpave gyratory compactor. Two samples are also prepared for determining the mixture's maximum theoretical specific gravity. An aggregate mass of 4,700 g is usually sufficient for the compacted specimens. An aggregate mass of 2,000 g is usually sufficient for the specimens used to determine maximum theoretical specific gravity (G_{mm}), although AASHTO T209 (ASTM D2041) should be consulted to determine the minimum sample size required for various mixtures.

7.3.5 Specimen Preparation and Compaction

This procedure outlines the preparation of HMA test specimens using the SGC. It includes guidelines for mixing and compacting test specimens.

7.3.5.1 Preparation of Aggregates

Prepare a batching sheet showing the batch weights of each aggregate component and the asphalt binder. Weigh into a pan the proper weights of each aggregate component. Three sample sizes are used depending on their final use. For compacted specimens that will be used in Superpave mix design, the specimen size is 150 mm (diameter) by 115 mm (height) and requires approximately

4,700 g of aggregate. Samples to be used for the determination of maximum theoretical specific gravity by AASHTO T209/ASTM D2041 remain uncompacted, and their sizes vary by aggregate size and range from 1,000 to 2,500 grams. Moisture damage testing using AASHTO T283 requires a specimen height of 95 mm and approximately 3,700 g of aggregate.

7.3.5.2 Mixing and Compaction Temperatures

Determine the laboratory mixing and compaction temperatures using a plot of viscosity versus temperature. Select mixing and compaction temperatures corresponding with binder viscosity values of 0.17 ± 0.02 Pa·s and 0.28 ± 0.03 Pa·s, respectively.

Note that these viscosity ranges are not valid for modified asphalt binders. The designer should consider the manufacturer's recommendations when establishing mixing and compaction temperatures for modified binders. Practically, the mixing temperature should not exceed 165°C, and the compaction temperature should not be lower than 115°C.

Note: The mixing and compaction temperatures described above are intended for laboratory mix design purposes only. Field mixing and compaction temperatures must be determined from trial applications and experience.

Place the pan containing the aggregate in an oven set approximately 15°C higher than the mixing temperature. Two to four hours are required for the aggregate to reach the mixing temperature. While the aggregate is heating, heat all mixing implements such as spatulas, the mixing bowl, and other tools. Heat the asphalt binder to the desired mixing temperature. The time required for this step varies depending on the amount of asphalt and the heating method (NCDOT, 2024).

7.3.5.3 Preparation of Mixtures

- Place the hot mixing bowl on a balance and zero the balance.
- Charge the mixing bowl with the heated aggregates and mix thoroughly.
- Form a crater in the blended aggregate and weigh the required asphalt into the mixture to achieve the desired batch weight.
- Remove the mixing bowl from the scale and mix the asphalt and aggregate using a mechanical mixer.
- Mix the specimen until the aggregate is thoroughly coated.
- Place the mix in a flat, shallow pan at an even thickness ranging between 25 mm and 50 mm.
- Place the mix and pan in the conditioning oven for 2 hours ± 5 minutes at a temperature equal to the mixture's specified compaction temperature ± 3°C.
- Short-term age the specimen for 2 hours.
- Repeat this procedure until the desired number of specimens is produced.
- Proper timing of the gyratory compaction steps can be achieved by spacing approximately 20 minutes between mixing each specimen.
- At the end of the short-term aging period, proceed to AASHTO T209/ASTM D2041 if the mixture is to be used to determine the maximum theoretical specific gravity. Otherwise, proceed with compaction.

Note: Experience has shown that aggregates having more than 2% (water) absorption should be aged for four hours to allow additional asphalt absorption to occur. This adjustment should more closely align laboratory values with volumetric calculations and measurements made on plant-produced mixture determined in production. For evaluation of properties of specimens compacted from mixing-plant produced asphalt mixtures containing absorptive aggregates, consideration should be given to adding a short-term aging (STOA) period. STOA periods of up to two hours may be necessary to allow asphalt binder absorption to occur.

7.3.5.4 Compaction of Volumetric Specimens

Prepare the compactor while the mixture specimens are short-term aging. This includes verifying that the compaction pressure, the compaction angle, and speed of gyration are set to their proper values, and setting the desired number of gyrations, N_{des}. Also, ensure that the data acquisition device is functioning.

Approximately 45–60 minutes before compaction of the first specimen, place the compaction molds and base/top plates in an oven set at the compaction temperature. Remove the mold and base plate from the oven, place the base plate in the mold, and a paper disk on top of the base plate.

Place the short-term aged mixture in the mold , level the mixture, and place a paper disk on top of the leveled mixture. Place the mold containing the specimen into the compactor. Center the mold under the loading ram and start the system so that the ram extends down into the mold cylinder and contacts the specimen. The ram will stop when the pressure reaches 600 kPa.

Apply the 1.25° angle of gyration and start the gyratory compaction. Compaction will proceed until N_{des} has been completed. During compaction, the ram loading system will maintain a constant pressure of 600 kPa. The specimen height is continually monitored during compaction, and a height measurement is recorded after each revolution.

The compactor will cease compacting after reaching N_{des}, and the angle of gyration will be released and the loading ram raised. Remove the mold containing the compacted specimen from the compactor and slowly extrude the specimen from the mold. A 5-minute cooling period will facilitate specimen removal without undue distortion.

Remove the paper disk from the top and bottom of the specimen and allow the specimen to cool undisturbed. Place the mold and base plate back in the oven to reach compaction temperature for the next specimen. Additional molds will avoid the delay caused by this step. Repeat the compaction procedure for each specimen. Identify each specimen with a suitable marker.

7.3.6 Data Analysis and Presentation

Superpave gyratory compaction data are analyzed by computing the estimated bulk specific gravity, corrected bulk specific gravity, and corrected percentage of maximum theoretical specific gravity for each desired gyration. During compaction, the height is measured and recorded after each gyration. G_{mb} of the compacted specimen and G_{mm} of the loose mixture are measured, and an estimate of G_{mb} at any value of gyration is made by dividing the mass of the mixture by the volume of the compaction mold:

$$G_{mb}\left(estimated\right) = \frac{W_m \quad V_{mx}}{\gamma_m}$$

Eq. 7.17

where

G_{mb} (estimated) = estimated bulk specific gravity of specimen during compaction
W_m = mass of specimen, grams
γ_w = density of water = 1 g/cm^3
V_{mx} = volume of compaction mold (cm^3), calculated using the equation:

$V_{mx} = \pi d^2/4 \times h_x$

where

d = diameter of mold (150 mm)

h_x = height of specimen in mold during compaction (mm)

π = 3.1416

This calculation assumes that the specimen is a smooth-sided cylinder, which it is not. Surface irregularities cause the volume of the specimen to be slightly less than the volume of a smooth-sided cylinder. Therefore, the final estimated G_{mb} at N_{des} is different from the measured G_{mb} at N_{des}. Consequently, the estimated G_{mb} is corrected by a ratio of the measured to the estimated bulk specific gravity:

$$C = \frac{G_{mb}\left(measured\right)}{G_{mb}\left(estimated\right)}$$

Eq. 7.18

where

C = correction factor

G_{mb} (measured) = measured bulk specific gravity after N_{des}

G_{mb} (estimated) = estimated bulk specific gravity at N_{des}

The estimated G_{mb} at any other gyration level is then determined using:

$$G_{mb} \text{ (corrected)} = C \times G_{mb}(\text{estimated})$$

Eq. 7.19

where

G_{mb} (corrected) = corrected bulk specific gravity of the specimen at any gyration

C = correction factor

G_{mb} (estimated) = estimated bulk specific gravity at any gyration.

Percent G_{mm} at any gyration level is then calculated as the ratio of G_{mb} (corrected) to G_{mm} (measured), and the average percent G_{mm} values for the two companion specimens are calculated.

Using the N_{max}, N_{des}, and N_{ini} gyration levels previously determined from the design traffic level, Superpave volumetric mix design criteria (VMA, VFA, and dust ratio) are established on a four-percent air void content at N_{des}. Superpave mix design also specifies criteria for the mixture density at N_{ini}, N_{des}, and N_{max}.

The percent air voids at N_{des} are determined from the equation:

$$V_a = 100 - \%G_{mm} @ N_{des}$$

Eq. 7.20

where

V_a = air voids @ N_{des}, percent of total volume

$\%G_{mm}@N_{des}$ = maximum theoretical specific gravity @ N_{des}, percent

The percent voids in the mineral aggregate are calculated using:

$$\%VMA = 100 - \left(\frac{\%G_{mm} @ N_{des} \times G_{mm} \times P_s}{G_{sb}} \right)$$

Eq. 7.21

where

VMA = voids in the mineral aggregate, percent of bulk volume

$\%G_{mm}@N_{des}$ = maximum theoretical specific gravity @ N_{des}, percent

G_{mm} = maximum theoretical specific gravity

G_{sb} = bulk specific gravity of total aggregate

P_s = aggregate content, cm³/cm³, by total mass of the mixture

If the percentage of air voids is equal to four percent, then this data is compared to the volumetric criteria and an analysis of this blend is completed. However, if the air void content at N_{des} varies from four percent (and this will typically be the case), an estimated design asphalt content to achieve four% air voids at N_{des} are determined; and the estimated design properties at this estimated design asphalt content are calculated.

The estimated asphalt content for N_{des} = four percent air voids is calculated using this equation:

$$P_{b\,estimated} = P_{bi} - \left(0.4 \times \left(4 - V_a\right)\right)$$

Eq. 7.22

where

P_b = estimated asphalt content, percent by mass of mixture

P_{bi} = initial (trial) asphalt content, percent by mass mixture

V_a = percent air voids at N_{des} (trial)

The volumetrics (VMA and VFA) at N_{des} and mixture density at N_{ini} are then estimated at this asphalt binder content using the equations that follow.

For VMA:

$$VMA_{estimated} = \%VMA_{initial} + C \times \left(4 - V_a\right)$$

Eq. 7.23

where

$\%VMA_{initial}$ = %VMA from trial asphalt binder content

C = constant = 0.1 if V_a is less than 4%; = 0.2 if V_a is greater than 4%

Specified minimum values for VMA at the design air void content of four percent are a function of nominal maximum aggregate size and are given in Table 7.4.

For VFA:

$$\%VFA_{estimated} = 100 \times \frac{\left(\%VMA_{estimated} - 4.0\right)}{\%VMA_{estimated}}$$

Eq. 7.24

The acceptable range of design VFA at four percent air voids as a function of traffic level is shown in Table 7.4.

For $\%G_{mm}$ at N_{ini}:

$$\%G_{mm\,estimated} @ N_{ini} = \%G_{mm\,trial} @ N_{ini} - \left(4.0 - V_a\right)$$

Eq. 7.25

The maximum allowable mixture density at Nin for the various traffic levels is shown in Table 7.4.

Finally, there is a requirement for the dust proportion. It is calculated as the percent by mass of the material passing the 0.075-mm sieve (by wet sieve analysis) divided by the effective asphalt binder content (expressed as a percent by mass of the mix). The effective asphalt binder content is calculated using:

$$P_{be} = -\left(P_s \times G_b\right) \times \left(\frac{G_{se} - G_{sb}}{G_{se} \times G_{sb}}\right) + P_{b\,estimated}$$

Eq. 7.26

Table 7.4 Superpave Volumetric Mixture Design Requirements

Design ESALs (Million)	Requirement Density (% of Theoretical Maximum Specify Gravity)			Voids-in-the-Mineral Aggregate (Percent), Minimum — Nominal Maximum Aggregate Size, mm					Voids Filled with Asphalt (Percent)	Dust-to-Binder Ratio
	$N_{initial}$	N_{design}	N_{max}	37.5	25.0	19.0	12.5	9.5		
<0.3	≤91.5	96.0	≤98.0	11.0	12.0	13.0	14.0	15.0	70–80	0.6–1.2
0.3 to <3	≤90.5								65–78	
3 to <10	≤89.0								65–75	
10 to <30										
≥30										

Design ESALs are the anticipated project traffic level expected on the design lane over a 20-year period. Regardless of the actual design life of the roadway, determine the design ESALs for 20 years, and choose the appropriate N_{design} level.

For 9.5-mm nominal maximum size mixtures, the specified VFA range shall be 73% to 76% for design traffic levels of 3 million ESALs.

For 25.0-mm nominal maximum size mixtures, the specified lower limit of the VFA shall be 67% for design traffic levels of <0.3 million ESALs.

For 37.5-mm nominal maximum size mixtures, the specified lower limit of the VFA shall be 64% for all design traffic levels.

If the aggregate gradation passes beneath the boundaries of the aggregate restricted zone,, consideration should be given to increasing the dust-to-binder ratio criteria from 0.6–1.2 to 0.8–1.6

Table 7.5 Trial Asphalt Binder Content

Nominal Maximum Aggregate Size (mm)	Trial Asphalt Binder Content (%)
37.5	3.5
25.0	4.0
19.0	4.5
12.5	5.0
9.5	5.5

where

P_{be} = effective asphalt content, percent by total mass of the mixture
P_s = aggregate content, percent by total mass of the mixture
G_b = specific gravity of asphalt
G_{se} = effective specific gravity of aggregate
G_{sb} = bulk specific gravity of aggregate
P_b = asphalt content, percent by total mass of the mixture
Dust proportion is calculated using:

$$DP = \frac{P_{0.075}}{P_{be}}$$

Eq. 7.27

where

$P_{0.075}$ = aggregate content passing the 0.075-mm sieve, percent by mass of aggregate
P_{be} = effective asphalt content, percent by total mass of the mixture
An acceptable dust proportion ranges from 0.6 to 1.2. However, consideration may be given to increasing the ratio from 0.8 to 1.6 if the aggregate gradation falls below the restricted zone.

After establishing all the estimated mixture properties, the designer can look at the trial blends and decide if one or more are acceptable, or if further trial blends need evaluation.

7.3.7 Design Asphalt Binder Content

Once the design aggregate structure is selected from the trial blends, specimens is compacted at varying asphalt binder contents. The mixture properties are then evaluated to determine the design asphalt binder content.

A minimum of two specimens is compacted at the trial blend's estimated asphalt content, at ±0.5% of the estimated asphalt content, and at +1.0% of the estimated asphalt content. The four asphalt contents are the minimum required for Superpave mix design.

A minimum of two specimens is also prepared for determination of maximum theoretical specific gravity at the estimated binder content. Specimens are prepared and tested in the same manner as the specimens from Section 7.3.4.

Mixture properties are evaluated for the selected blend at different asphalt binder contents by using the densification data at N_{ini} and N_{des}. The volumetric properties are calculated at N_{des} for each asphalt content. From these data points, the designer can generate graphs of air voids, VMA, and VFA versus asphalt content. The design asphalt binder content is established at 4.0 percent air voids. All other mixture properties are checked at the design asphalt binder content to verify that they meet the criteria.

After verifying that all other mixture properties meet criteria at the design asphalt binder content, two additional SGC specimens is compacted to N_{max} (from Table 7.3) to ensure that N_{max} does not exceed 98% G_{mm}.

7.3.8 Moisture Sensitivity

The final step in the Superpave mix design process is to evaluate the moisture sensitivity of the design mixture. This step is accomplished by performing the AASHTO T283 test, "Resistance of Compacted Bituminous Mixtures to Moisture Induced Damage," on the design aggregate blend at the design asphalt binder content. Specimens for this test are compacted to approximately 7% air voids. One subset, consisting of three specimens, is considered the control set. The other subset of three specimens is conditioned.

The conditioned specimens are subjected to partial vacuum saturation, followed by an optional freeze cycle, followed by a 24-hour thaw cycle at 60°C. All specimens are tested to determine their indirect tensile strengths. The moisture sensitivity is determined as the ratio of the average tensile strengths of the conditioned subset divided by the average tensile strengths of the control subset. The Superpave criterion for the tensile strength ratio is 80%, minimum.

7.4 Summary of Superpave Mix Design

7.4.1 Selection of Materials

7.4.1.1 Selection of Asphalt Binder

1. Determine project weather conditions using a weather database
2. Select reliability
3. Determine design temperatures
4. Verify asphalt binder grade
5. Temperature-viscosity relationship for lab mixing and compaction

7.4.1.2 Selection of Aggregates

1. Consensus properties
   a. Combined gradation
   b. Coarse aggregate angularity
   c. Fine aggregate angularity
   d. Flat and elongated particles
   e. Clay content
2. Agency and other properties
   a. Specific gravity
   b. Toughness
   c. Soundness
   d. Deleterious materials
   e. Other

7.4.1.3 Selection of Modifiers

7.4.2 Selection of Design Aggerate Structure

7.4.2.1 Establish Trial Blends

1. Develop three blends
2. Evaluate combined aggregate properties

7.4.2.2 Compact Trial Blend Specimens

1. Establish trial asphalt binder content:
   a. Superpave method
   b. Engineering judgment method
2. Establish trial blend specimen size
3. Determine $N_{initial}$ and N_{design}
4. Batch trial blend specimens
5. Compact specimens and generate densification tables
6. Determine mixture properties (G_{mm} and G_{mb})

7.4.2.3 Evaluate Trial Blends

1. Determine $\%G_{mm}$ @ $N_{initial}$ and N_{design}
2. Determine $\%V_a$s and %VMA
3. Estimate asphalt binder content to achieve 4% V_a
4. Estimate mix properties @ estimated asphalt binder content
5. Determine dust–asphalt ratio
6. Compare mixture properties to criteria

7.4.2.4 Select Most Promising Design Aggregate Structure for Further Analysis

7.4.3 Selection of Design Asphalt Binder Content

7.4.3.1 Compact Design Aggregate Structure Specimens at Multiple Binder Contents

1. Batch design aggregate structure specimens
2. Compact specimens and generate densification tables

7.4.3.2 Determine Mixture Properties versus Asphalt Binder Content

1. Determine $\%G_{mm}$ @ $N_{initial}$ and N_{design} and $N_{maximum}$
2. Determine volumetric properties
3. Determine dust-asphalt ratio
4. Graph mixture properties versus asphalt binder content

7.5 Questions

1. What is the primary objective of Superpave HMA mix design in the context of asphalt pavement construction?
2. Describe the five characteristics of the mixture analyzed during Superpave mix design.
3. Why is volumetric analysis important in the design of asphalt mixtures?
4. What are the definitions of voids in the mineral aggregate (VMA), effective asphalt content (P_{be}), air voids (V_a), and voids filled with asphalt (VFA)?
5. Explain the importance of calculating effective specific gravity (G_{se}) in the context of asphalt mixture design.
6. What role does asphalt absorption play in determining the effective asphalt content of a paving mixture?
7. How are percent voids in the mineral aggregate (VMA), percent air voids (V_a), and percent voids filled with asphalt (VFA) calculated?
8. What are the primary steps involved in Superpave mix design procedures?
9. How does Superpave address the selection of asphalt and aggregate materials for mix design?
10. What is the significance of short-term oven aging in Superpave mix design?
11. What were the main goals of developing the Superpave gyratory compactor (SGC)?
12. How does the Superpave gyratory compactor (SGC) achieve realistic specimen densification?
13. What components make up the Superpave gyratory compactor (SGC), and what are their functions?
14. How is specimen height measured during compaction in the Superpave gyratory compactor (SGC)?
15. What is the significance of the design number of gyrations (Ndes) in Superpave mix design, and how is it determined?
16. What are the relationships between the initial number of gyrations (N_{ini}), design number of gyrations (N_{des}), and maximum number of gyrations (N_{max}) in Superpave mix design?
17. What are some ancillary test equipment required in the preparation of Superpave asphalt mixtures?
18. How is the design aggregate structure selected in the Superpave mix design process?
19. What is trial blending, and why is it done during the Superpave mix design process?
20. What are the key components of the Superpave mix design process?
21. Why does the Superpave mix design procedure control various volumetric parameters, and which parameter is to be controlled eventually?

References

AI (2001). *Superpave Mix Design*. 3rd ed. Superpave Series No. 2 (SP-2). Asphalt Institute (AI), Lexington, KY.

Brown, E.R., Kandhal, P.S., Roberts, F.L., Kim, Y.R., Lee, D. & Kennedy, T.W. (2009). *Hot-Mix Asphalt Materials, Mixture Design and Construction*. NAPA Research and Education Foundation, Lanham, MD.

NCDOT (2024). *Asphalt Quality Management System*. Materials and Tests Unit, North Carolina Department of Transportation (NCDOT), Raleigh, NC.

Chapter 8

Hot-Mix Asphalt Production and Plant Operation

8.1 Asphalt Mixing Plants

Hot-mix asphalt (HMA) materials are prepared and combined at an asphalt mixing plant. Here, aggregates are blended, heated, dried, and mixed with asphalt binder to create a hot paving mixture. The complexity and size of the mixing plant can vary, ranging from small and simple to large and complex, depending on the type and volume of asphalt mixture being produced. These plants may be either stationary (permanent) or portable. The two most common types of asphalt plants are batch plants and drum mix plants. Additionally, plants used to produce asphalt mix must be certified by state highway agencies and professional organizations to meet equipment qualification standards and environmental requirements.

8.1.1 Batch Plant

During operation, the batch plant produces asphalt mix in discrete batches, sequentially processing one batch at a time. The size of each batch depends on the capacity of the plant's pugmill – the mixing chamber where the aggregate and binder are combined – ranging from a minimum capacity of 3,000 lbs to a maximum of 12,000 lbs.

The essential operations for batch plants include

- Storage and cold feeding of aggregates
- Drying and heating of aggregates
- Screening and storage of hot aggregates
- Storage and heating of the asphalt binder
- Measuring and mixing the asphalt binder and aggregate
- Loading of the finished asphalt mix

Figure 8.1 shows the process of a typical asphalt batch plant. It helps to understand the functions and relationships of the various plant components.

Cold aggregates stored in the cold bins are precisely metered by cold-feed gates and transferred via a belt conveyor or bucket elevator to the dryer, where they are dried and heated. The baghouse efficiently removes dust from the dryer exhaust, with the remaining gases expelled through the exhaust stack. Once dried and heated, the aggregates are conveyed by a hot elevator to a screening unit equipped with a scalping screen, which removes any oversized material, depositing it into a reject chute for disposal (NCDOT, 2024).

The aggregates are then sorted into different sizes and temporarily stored in hot bins. These aggregates are accurately measured and sent to the weigh box when required. They are then

DOI: 10.1201/9781003197768-8

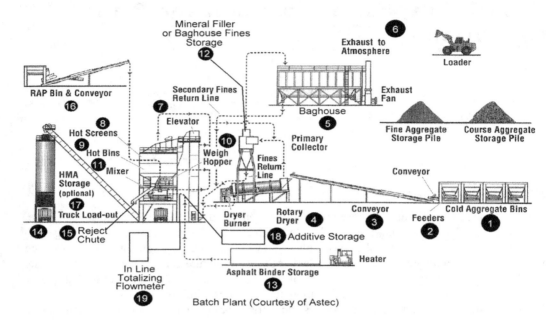

Figure 8.1 Major Batch Plant Components

transferred to the mixing chamber or pugmill, where they are combined with the necessary amount of mineral filler from storage or baghouse fines, depending on the mix requirements.

A recycled asphalt pavement (RAP) bin and conveyor are essential for plants producing HMA using recycled pavement. The heated asphalt binder is drawn from the storage tank, measured in the weigh bucket, and mixed thoroughly with the aggregates and any fines or mineral filler in the pugmill. The resulting asphalt mix is then loaded into trucks or stored in silos or surge bins.

When anti-strip additives are used, an additive storage tank with a non-resettable totalizing flow-meter is installed in the additive feed line to ensure precise introduction into the binder feed line.

8.1.2 Drum Mix Plant

Drum mixing is a straightforward method for producing asphalt mix, distinguished from batch mixing primarily by how the components are combined. In drum-mix plants, aggregates are dried, heated, and mixed with the asphalt binder within the drum. However, some modern drum-mix plants introduce the asphalt binder outside the drum. These include designs like the coater box, a pugmill-type device located at the drum's discharge end, where the asphalt binder is added, and the double barrel-type drum-mix plants that add the binder between an inner and outer drum. Despite these variations, the underlying principle remains a continuous mixing process, unlike the batch-by-batch mixing at batch plants. Drum-mix plants have no gradation screens, hot bins, or weigh hoppers. Aggregate gradation is managed at the cold feed and through the gradations of the individual aggregates used.

Drum mix plants come in various sizes and rated capacities, ranging from about 60 tons to several hundred tons per hour. Many construction contracts require them to have a minimum production capacity of 90 tons per hour.

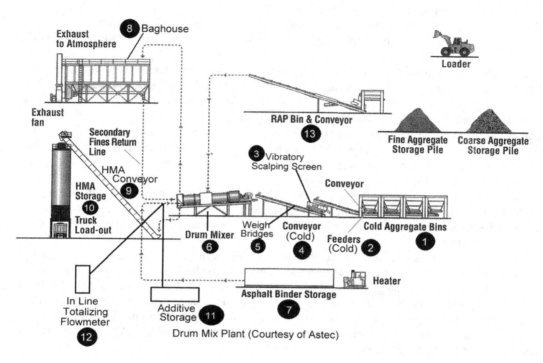

Figure 8.2 Basic Drum-Mix Plant

Figure 8.2 outlines the sequence of processes in a typical drum-mix plant operation: Aggregates are deposited into cold feed bins, from which they are fed in exact proportions by cold feeders through a vibratory scalping screen onto a cold-feed conveyor. An automatic aggregate weighing system or weigh bridge monitors the amount of aggregate flowing into the drum mixer. This weighing system is interlocked with the controls on the asphalt binder storage pump, which draws asphalt binder from a storage tank and introduces it into the drum, coater box, or between an inner and outer drum, where asphalt and aggregate are thoroughly blended through a mixing action. A dust collection system baghouse captures excess dust escaping from the drum. From the drum, the asphalt concrete is transported by a mix conveyor to a surge bin or silo, from which it is loaded into trucks and hauled to the paving site. All plant operations are monitored and controlled from instruments in the control room. When anti-strip additives are introduced at the plant site, an additive storage tank with a non-resettable totalizing flowmeter is required, mounted in the additive feed line just before introduction into the binder feed line.

8.2 Operational Standards for Asphalt Production Facilities

8.2.1 Asphalt Binder Storage and Handling

For effective asphalt mix production, it is crucial to use the correct grade of asphalt binder as specified in the job mix formula (JMF). The asphalt binder must be certified or tested and sourced according to the specifications in the JMF.

The storage capacity for asphalt binders at the plant must be ample to support smooth operations. If a project requires multiple grades of asphalt binder, separate tanks should be used for each grade. Alternatively, a tank must be emptied before switching to a different grade to prevent the mixing of various grades.

Asphalt binder storage tanks must be equipped with measurement devices to monitor the quantity of material at any time. The tanks also require heating to maintain the asphalt binder in a fluid state suitable for movement through delivery and return lines. Heating can be performed electrically or by circulating hot oil through coils inside the tank. It is imperative that an open flame never comes in direct contact with the tank or its contents. The temperature of the asphalt binder should not exceed the supplier's recommended limits during storage or use in mix production. If hot oil is used for heating, it is essential to regularly check the oil level in the reservoir to identify any potential leaks that could contaminate the asphalt binder.

All transfer lines, pumps, and weigh buckets should have heating coils or jackets to keep the asphalt binder fluid enough for pumping. Temperature control is crucial. Hence, thermometers must be installed in the asphalt binder feed line to monitor the temperature as the binder is introduced into the mixer or drum.

Asphalt binder tanks should feature a circulation system designed to disperse and mix any additives evenly throughout the asphalt binder. Additives should be introduced well before production starts to ensure complete integration. Adequate pumps must be provided to unload asphalt binders from tankers and maintain plant operations. A valve or spigot should be installed in the circulating system or directly on the tank to facilitate sampling. Extreme caution must be exercised when sampling from the circulating system due to the high pressure in the lines, which can cause the hot asphalt binder to splatter (AI, 2001).

8.2.2 Asphalt Mix Temperature Requirements

The binder and aggregates' temperature is critical to ensure proper adhesion to the aggregates and adequate mixing to achieve a consistent asphalt mixture. The final asphalt mixture must fall within a specific temperature range to ensure it can be placed appropriately and compacted without damaging the binder. The target mixing temperature at the asphalt plant will be set according to the JMF.

When checking the temperature of mixes in the truck at the asphalt plant, it should be within 25°F of the specified JMF temperature. Additionally, the temperature of the asphalt mixture at the point of discharge from the mixer must not exceed 350°F to prevent degradation of mix quality (AI, 2017A; 2017B).

8.2.3 Aggregate Storage and Cold Feed System

8.2.3.1 Aggregate Storage

Effective stockpiling and storage techniques are essential for producing high-quality asphalt mix. Properly stockpiled aggregates maintain their gradation, while poorly managed stockpiles can lead to segregation (separation by size) and variability in gradation throughout the stockpile. Handling can degrade individual aggregate particles and cause segregation, particularly with differently sized particles, so it should be minimized to maintain the integrity of the aggregates. Technicians should be vigilant about the impact of stockpiling and handling on aggregate gradation and consistently promote best practices.

Aggregates must be handled and stored to minimize segregation and prevent contamination. They should be stockpiled near the plant on well-drained, firm ground, cleared of vegetation, and prepared to shield the aggregates from contaminants. Stockpiles should be distinctly separated to avoid intermixing, achieved through clear demarcation using sturdy bulkheads, silos, or other effective means. Bulkheads, if utilized, must reach the full depth of the stockpile and withstand operational pressures.

Stockpiles should be built in layers rather than cone-shaped mounds to ensure even coverage. Aggregates should be closely placed to maintain consistent layer thickness across the stockpile. Care must be taken during the construction, maintenance, or dismantling of stockpiles to minimize aggregate degradation by equipment, with minimal operation of rubber-tired or tracked vehicles on top of the stockpiles.

8.2.3.2 Aggregate Cold Feed System

In both batch and drum mix plants, the journey of aggregates through the plant starts at the cold feed bins, consisting of several cold bins (typically three to five) with gates and feeders mounted beneath, arranged above a collector conveyor belt. Each aggregate size used in a specific mix requires a dedicated cold feed bin, commonly set up with four bins. When incorporating RAP or recycled asphalt shingles (RAS), separate feeder bins must be used.

Each cold bin should be sized sufficiently to hold an ample supply of aggregates. Baffles between adjacent bins prevent overflow and ensure consistent mix properties by preventing inter-bin material mixing. Each bin features an adjustable gate and a feeder belt calibrated to deliver aggregates at a controlled rate, synchronizing the volume of material to meet the JMF requirements. Aggregates from the feeders are deposited onto a collector conveyor and then moved into the dryer, ensuring a steady and precise flow essential for quality mix production (Figure 8.3).

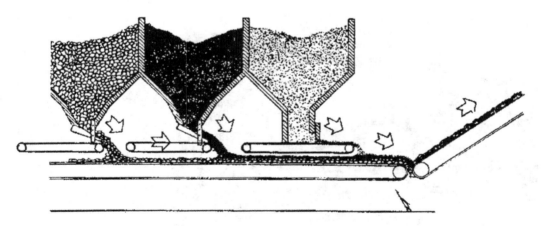

Figure 8.3 Three-Bin Cold Feeder and Belt

8.4 Batch Plant Operations

8.4.1 The Dryer

The dryer receives aggregates from the cold feeder and performs two primary functions: (i) it removes moisture from the aggregates; and (ii) it elevates the temperature of the aggregates to the required level. The main components of the dryer are: (i) a rotating cylinder, typically ranging from 3 to 10 feet in diameter and 20 to 40 feet in length; (ii) a burner, which can be gas or oil-fired; and (iii) a fan, primarily part of the dust collection system but also essential for providing combustion air in the dryer (Figure 8.4). The dryer features longitudinal troughs or channels known as lifting flights. These flights lift the aggregate, allowing it to fall through the burner flame and hot gases, enhancing heat transfer.

The drum's slope, rotational speed, diameter, length, and the layout and quantity of the lifting flights govern the time it takes for the aggregate to move through the dryer. The processed aggregate exits the dryer through a discharge chute near the burner end and is transported to the hot elevator.

An automatic burner control system is integrated into the dryer, equipped with a thermometric device approved for the purpose, positioned in the aggregate discharge chute. This system regulates the burner to maintain a consistent mix temperature and prevent the aggregate from overheating, which could damage the asphalt during the mixing process. Stability in mix temperature is crucial for achieving consistent laydown and compaction quality. Variations in mix temperature from one batch to another can hinder achieving the desired uniform density and degree of compaction.

8.4.2 Screening Unit

The screening unit depicted in Figure 8.5 comprises multiple vibrating screens of varying sizes. At the forefront is a scalping screen that discards oversized aggregates. This is succeeded by one or

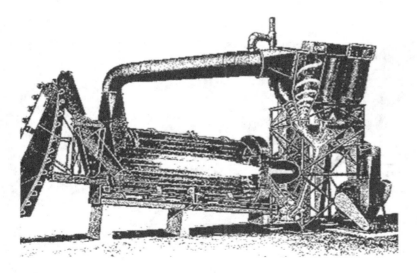

Figure 8.4 A Counter-Flow Dryer

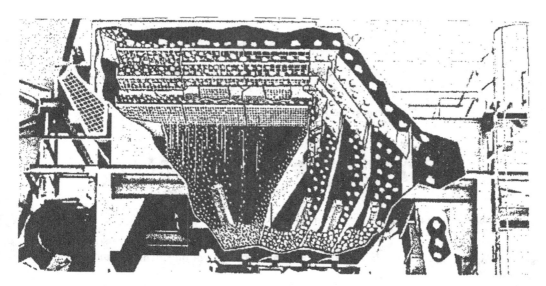

Figure 8.5 Screening Unit

two intermediate screens, arranged in descending size order. The series concludes with a fine screen at the bottom, often called a "sand" screen. These screens categorize the aggregates into designated sizes. For optimal performance, the total screen area should be sufficient to manage the volume of material input. It is crucial that the screens remain clean and well-maintained. Additionally, the screening capacity should be synchronized with the capacities of both the dryer and the pugmill.

If screens are overfed or their openings become clogged, many particles that should be filtered through might bypass the screens and fall into bins reserved for larger aggregates. Worn or damaged screens, with enlarged openings or holes, may also misdirect oversized particles into bins meant for smaller aggregates. This misplacement of finer aggregates into bins for larger particles is termed "carry-over." Excessive carry-over increases the acceptable aggregate content in the mix, thereby expanding the surface area that needs asphalt binder coverage. If unchecked, especially if carry-over in the No. 2 bin is inconsistent or unknown, it can adversely affect the mix design, impacting both gradation and asphalt binder content. Immediate corrective actions, such as cleaning the screens or adjusting the feed rate from the cold feed, are necessary if excessive carry-over is detected through sieve analysis of the hot bins' contents. While some carry-over is normal and acceptable during screening processes, it should remain relatively consistent to ensure quality.

8.4.3 Hot Bins

Hot bins are utilized to temporarily store heated and screened aggregates in required sizes. Each bin either constitutes an individual compartment or forms a segment within a larger compartment, separated by partitions (Figure 8.6). The design of a properly sized hot bin installation should accommodate sufficient quantities of each aggregate size, ensuring capacity even when the mixer operates at maximum. Bin partitions should be secure and devoid of holes, extending high enough to prevent mixing of different aggregate sizes. Each hot bin is also equipped with an overflow pipe,

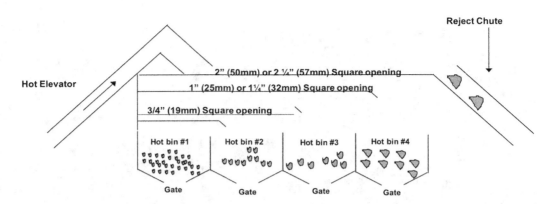

Figure 8.6 Typical Batch Plant Setup for Screens and Hot Bins

essential for avoiding aggregates from overflowing into adjacent bins, which could result in size contamination.

These overflow pipes must be inspected regularly to ensure they are functioning correctly. Additionally, the bottom of each bin features a discharge gate that should close securely to prevent any material leakage into the weigh hopper.

All hot bins must have a mechanism allowing aggregate sampling for quality control purposes. Samples can be collected through "gates" or "windows" on the bin sides or by diverting aggregate flow into a sampling container. It is critical that the sampling methods or devices used are capable of providing representative samples from the contents of the hot bins to ensure an accurate assessment of the aggregate mix.

8.4.4 Aggregate Weigh Hopper

Aggregates are dispensed from the hot bins into the weigh hopper, typically starting with the largest aggregate and proceeding to the smallest, often with mineral filler interspersed among the larger aggregates. The aggregate quantity from each bin is based on the batch size and the specific proportions or percentages needed for the blend. The methods for determining the percentages from the hot bins and the corresponding pull weights will be further explored later in this section.

The weigh hopper is mounted on a scale beam, allowing for the cumulative weighing of aggregates. It is essential to ensure enough materials are in the hot bins to complete a batch before beginning the withdrawal process. If any bin is close to being depleted or is overflowing, it may indicate that adjustments are necessary in the rates of cold feed or in the quantities being pulled from the hot bins.

8.4.5 Asphalt Binder Measurement

Asphalt binder can be measured using a designated bucket or a meter for each batch. When using a bucket, the asphalt binder is pumped into a pre-weighed bucket and weighed on a scale. Alternatively, when using metering devices, the binder is measured volumetrically. Since the volume of asphalt binder varies with temperature, some meters include temperature-compensating devices that adjust the binder flow to account for temperature fluctuations. To calibrate the meter, the volume of asphalt binder transferred between two meter readings can be weighed. This ensures accuracy in the measurement and consistency in the mix.

8.4.6 The Mixer Unit (Pugmill)

In the mixing phase, aggregates and binders are combined in the pugmill, a crucial component of batch plants. This mixer features twin shafts outfitted with paddles that blend the materials into a uniform mixture (Figure 8.7). Key elements include paddle tips, shanks, a spray bar, liners, shafts, a discharge gate, and a heated jacket. Adequate mixing relies on various factors: (i) the number, shape, and condition of the paddle tips; (ii) the rotational speed of the shafts; (iii) the mixing duration; (iv) the temperature and quantity of the materials; and (v) the clearance between the paddle tips and the liner plates. It is mandatory for all batch plants to have a mixer with a minimum rated capacity of 3,000 lbs.

The batch size is determined by the mixer's rated capacity, specified in the plant certification. It is essential to avoid underfilling or overfilling the mixer. Although the mixers in batch mix and continuous mix plants share similar designs, they differ slightly in the arrangement of the paddle tips. Materials are introduced into the center in batch-mix mixers, and the paddles are arranged to create an end-to-center or "figure-eight" mixing pattern. The materials are retained in the mixer for the necessary duration before being discharged into transport vehicles. The mixer also has an

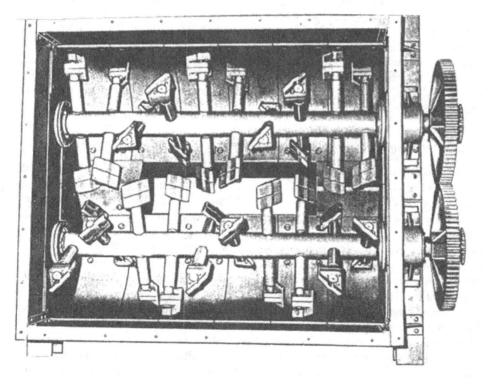

Figure 8.7 Pugmill Mixer for a Batch Plant

automatic timing device that controls the dry-mixing and wet-mixing phases, ensuring precise timing for optimal mixture quality.

8.5 Drum Mix Plant Operations

Drum mix plants streamline asphalt production by integrating drying, heating, and mixing within a single drum, hence the name. This drum closely resembles the dryer drum of a batch plant but with a key distinction: in drum mix plants, aggregates are dried, heated, and mixed with the asphalt binder directly within the drum. Recent designs of drum mix plants may introduce the asphalt binder outside of the drum. For instance, a coater box – a pugmill-type apparatus situated at the drum's discharge end – enables the addition of asphalt binder into the box rather than the drum. In "double barrel" drum mix plants, the asphalt binder is introduced between an inner and an outer drum. Despite these variations, the core principle remains the same: all these types facilitate a continuous mixing process, unlike the batch-wise mixing seen in batch plants. Drum mix plants do not utilize gradation screens, hot bins, weigh hoppers, or pugmills. Instead, aggregate gradation is managed at the cold feed stage.

8.5.1 Cold Feed System in Drum Mix Plant

In drum mix plants, where no gradation screening unit is used, precise aggregate proportioning before it enters the mixing drum is crucial for achieving the correct mix gradation and uniformity. The plant must include facilities that allow easy collection of representative samples from each cold feed and the total cold feed stream for calibration purposes. The calibration process for these feeds, ensuring compliance with the JMF, mirrors that used in batch plants. Each feeder has an automatic system that triggers a warning alarm and/or a flashing light when a bin runs empty or if the aggregate flow is restricted. This system is integrated with the plant's control system, enabling it to halt production automatically if normal aggregate flow is not restored within 60 seconds. Additionally, the design of each feeding system ensures that samples can be quickly and effectively collected.

8.5.2 Continuous Weighing of Aggregate and RAP/RAS

Drum mix plants incorporate a continuous weighing system on the cold feed conveyor belts, which include belts for RAP and RAS. These plants utilize in-line belt weighing devices, known as weighbridges, depicted in Figure 8.8. These devices continuously weigh the combined aggregates or RAP/RAS as they pass over the conveyor, and the control room displays a real-time readout of the flow weight.

Figure 8.8 illustrates one of the conveyor's weigh idlers, which is part of the belt weighbridge system mounted on a pivoted scale carriage. As the loaded conveyor belt moves over this idler, the weight is recorded in tons or tons per hour, and the reading is displayed at the control console within the control trailer. This measurement is typically adjusted to account for the moisture content in the aggregate, which is crucial since the calculations for the required percentage of asphalt binder are based on dry-aggregate data.

The weighbridge is strategically placed midway between the conveyor's head and tail pulleys to minimize reading variations due to impact loading, aggregate rollback, or changes in belt tension.

Some plants may feature a digital readout that displays the rate in tons per hour. This allows for continuous monitoring and adjustments during operations to ensure accuracy. Testing involves

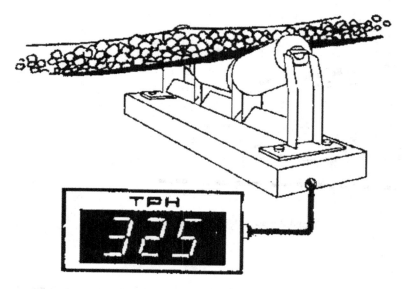

Figure 8.8 Weighing Bridge

running approximately 15 tons of clean coarse aggregate or 10 tons of RAP or RAS through a diversion chute over the weighbridge into a container, where it is weighed to determine the net weight. The material is timed from its first crossing of the weighbridge until it clears, facilitating a calculation of tons per hour using a specific formula.

In drum mix plants, weighing the aggregate before it undergoes drying is vital since undried material may contain moisture that significantly affects its weight. Accurate measurement of this moisture content is essential. These measurements allow for precise adjustments to the automatic asphalt binder metering system, ensuring the correct amount of binder is added to the aggregate, accounting for its moisture content.

8.5.3 Asphalt Binder Measuring System

Most drum mix plants incorporate a system to introduce asphalt binder into the aggregate mixture. Typically, this binder is added directly within the drum mixer, as illustrated in Figure 8.9. However, newer models may feature a setup where the asphalt binder is added outside the drum into a coater box, a pugmill-type device designed to mix binder with the aggregate before it enters the drum. Another variant, the "double barrel" drum plant, introduces asphalt binder between an inner and outer drum.

These systems utilize an asphalt binder metering and delivery system that works continuously and is mechanically proportioned. It is interlocked with the aggregate weighing system to ensure precise asphalt binder content in the mix. The quantity of asphalt binder delivered into the drum or coater box is calculated based on the weight of aggregate measured by the weight belt.

Most metering systems are volumetric, measuring the volume of asphalt binder in gallons through a volumetric flowmeter. This volume is then converted to a weight rate of flow, a necessary adjustment because both the volume and weight of the asphalt binder are temperature dependent.

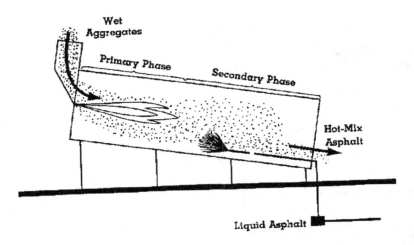

Figure 8.9 Zones in Drum Mixer

The plant control system includes adjustments for temperature and specific gravity variations of the asphalt binder and a temperature indicator within the binder feed line. Some plants also employ mass flowmeters, which measure the mass instead of the volume of the asphalt binder and provide readouts in weight, independent of temperature and specific gravity.

The proportioning of asphalt binder is determined by setting a delivery rate that matches the dry aggregate delivery rate, expressed in tons per hour. This rate adjusts automatically in proportion to the corrected dry weight of the aggregate passing over the belt scale, with the rate typically displayed on the control panel.

In addition, most drum-mixer plants are equipped with a bypass valve system. This system allows the asphalt binder to be directed through the flowmeter into a container rather than the drum mixer, facilitating calibration and testing. Typically, a minimum of 500 gallons – or an amount recommended by the manufacturer – is diverted into an asphalt distributor or supply tanker. This container is weighed empty, and then the binder is added to determine the net weight of the asphalt binder. These measurements are performed on a certified scale with sufficient capacity.

The net weight of the asphalt binder is then compared to the flowmeter reading: gallons pumped for volumetric flowmeters or the weight for mass flowmeters. A conversion between pounds and gallons is necessary for volumetric measurements to align the readings for accurate calibration.

8.5.4 Drum-Mixer Dryer Operations

The drum dryer is central to the functionality of drum mix plants (Figure 8.10). Its design and operation differ significantly from the conventional rotary dryers found in batch plants. Most drum dryers in drum-mix plants operate on the parallel flow principle, where both the air and material flow in the same direction, contrary to the counter flow principle typical in batch plants. The burner is positioned at the upper end of the drum, introducing heat to the cold, proportioned aggregates at the point of entry. This configuration ensures that the hottest gases and the flame contact the aggregates first, while the asphalt binder introduced further along the drum is shielded from the direct, potentially damaging heat by the evaporating moisture from the aggregates.

Figure 8.10 Proper 3-Dump Loading of Truck

However, newer models, such as the double barrel drum mix plants, invert this arrangement. These feature the burner at the lower end of the drum, with aggregates moving towards the flame. In these models, the inner drum acts solely as a drying chamber, not for mixing.

Advancements in drum mix plant design have led to methods that further protect the asphalt binder from excessive heat. For example, including a coater box at the drum's discharge end or implementing double-barrel drum setups, where the binder is added between the inner and outer drums, away from direct flame exposure, are such innovations. In all types of drum plants, mix temperature is continuously monitored using a thermometric device located in the dryer discharge chute. This device automatically adjusts the burner controls to maintain the desired mix temperature.

Drum mixer dryers are designed with a minimum rated capacity of 90 tons per hour, producing a finished mixture at 300°F and removing 5% moisture from the combined aggregate. This capacity ensures efficient production while maintaining the quality and integrity of the asphalt mixture.

8.5.5 Surge-Storage Silos in Drum-Mix Operations

In drum-mix plants, which produce a continuous flow of fresh asphalt mix, surge silos are essential for temporary storage and controlled loading of trucks. These silos may be equipped with a weighing system that monitors and records the amount of material loaded into each truck via a control panel typically located in the control van or trailer.

Surge silos are designed to maintain the heat and quality of the asphalt mix for up to twelve hours if insulated, with capacities that can reach several hundred tons. Conversely, non-insulated silos are generally smaller and can only store the mix for brief periods due to quicker heat loss. While silos are effective storage solutions, improper use can lead to mix segregation. To mitigate this, it is advisable to install baffle plates, a batching hopper, or a rotating chute at the conveyor's discharge end that loads the silo. These devices help distribute the mix evenly, preventing coning and segregation as it enters the silo.

Maintaining the silo at least one-third full is also recommended to minimize segregation during depletion and to preserve the mix's temperature. The asphalt mix is introduced at the top of the silo and descends vertically, with the design aiming to minimize mix segregation. When loading trucks, a method involving a minimum of three dumps is preferred to further reduce segregation risks, particularly with coarser mixes that are more prone to this issue. This three-dump loading technique is depicted in Figure 8.10 and is critical for maintaining homogeneous mix quality.

8.6 Understanding Segregation of Asphalt Mixtures

Segregation in asphalt mixtures refers to the uneven distribution of various aggregate sizes within the mix, causing deviations from the specified JMF. This phenomenon usually occurs when larger particles separate from smaller ones under conditions influenced by mixing, storing, transporting, and handling the asphalt. Although segregation is often observed in drum mix plants, no definitive evidence suggests that these plants are more susceptible to segregation than batch plants. However, segregation issues are frequently linked to surge-storage systems commonly employed in drum mix plants.

Coarse-graded mixes, such as 25.0 mm base mixes, are particularly vulnerable to segregation due to their larger stone content, lower asphalt binder content, and potential gap grading. In contrast, finer-graded mixes, like 9.5 mm surface mixes, generally exhibit fewer segregation issues due to the opposite characteristics.

Segregation can initiate at almost any stage in the asphalt production process, whether during mix design, in aggregate stockpiles, cold-feed bins, hot bins in batch plants, drum mixers, drag-slat conveyors, or surge-storage bins. Sometimes, segregation may not become apparent until the material is loaded into trucks. The earlier segregation begins, the greater the challenge, as the mix undergoes more handling, exacerbating the issue.

Addressing segregation typically involves interventions at multiple points where problems may arise. These may include alterations to the mix design, improved handling practices for aggregates, and modifications to plant operations, conveyors, and storage systems. Ensuring controlled movement of the mix, particularly minimizing downslope movement, is crucial since such movement significantly worsens segregation. For detailed troubleshooting and specific corrective actions, referring to segregation diagnostic charts available in the Appendix is recommended. These resources provide targeted solutions to various segregation issues encountered in asphalt production.

8.7 Hauling of Asphalt Mixtures

Inspecting the truck bed before loading is essential to ensure it has been lightly coated with an approved release agent. This coating prevents the asphalt mixture from sticking to the truck bed.

The temperature of the asphalt mix must be closely monitored to maintain quality during transport. At the asphalt plant, the mix temperature within the truck should be within 25°F of the JMF temperature. Similarly, just before discharging the mix at the construction site, the temperature should also be within 25°F of the JMF temperature.

Frequent observations of the mix and regular temperature checks should be conducted and recorded. The contractor must provide a platform near the truck loading area. This platform should allow for easy observation of the mix, facilitate sampling, and enable accurate temperature measurements. Additionally, each load should be covered with a solid, waterproof tarp made from canvas, vinyl, or another suitable material. The tarps must be free from rips or holes and should cover the dump box entirely to prevent moisture ingress and excessive temperature loss during transport (Brown, Kandhal, Robert, Kim, & Lee, 2023).

8.8 Questions

1. What are the primary components of hot-mix asphalt (HMA)?
2. What is the primary function of an HMA plant?
3. How do batch and drum plants differ in asphalt production processes?
4. What role do mineral fillers play in the production of HMA?
5. What are the temperature requirements for the asphalt binder and aggregates during mixing?
6. Describe the certification requirements for asphalt mixing plants.
7. What is the purpose of anti-strip additives in asphalt production?
8. How does a batch plant ensure the correct proportions of aggregate and binder?
9. What is the function of a baghouse in an asphalt plant?
10. How are different aggregate sizes managed in a drum mix plant?
11. What are the advantages of using a drum mix plant over a batch plant?
12. Discuss the environmental considerations in HMA plant operations.
13. What measures are taken to prevent segregation of the asphalt mix?
14. Explain how the asphalt binder's storage conditions affect mix production.
15. What procedures ensure the accurate measurement of asphalt binder in mix production?
16. Describe the role of recycled asphalt pavement (RAP) in asphalt mix production.
17. What are the critical considerations when handling and storing aggregates at an asphalt plant?
18. How do surge-storage silos maintain the quality of asphalt mix?
19. What operational standards must be met for effective asphalt binder handling and storage?
20. Discuss the significance of maintaining specific temperatures within asphalt mixes during transport.

References

(2001). *HMA Construction, (MS-22)*. Asphalt Institute (AI), Lexington, KY.

(2017A). *2017 Asphalt Plant Technician Instruction Manual, (MS-4)*. Asphalt Institute (AI), Lexington, KY.

(2017B). *2017 Asphalt Plant Technician Instruction Manual, (MS-20)*. Asphalt Institute (AI), Lexington, KY.

Brown, E.R., Kandhal, P.S., Roberts, F.L., Kim, Y.R., & Lee, D.Y. (2023). *Hot-Mix Asphalt Materials, Mixture Design and Construction.* 3rd ed. NCAT, Lanham, MD.

NCDOT (2024). *Asphalt QMS Manual*. Materials and Test Unit, North Carolina Department of Transportation (NCDOT), Raleigh, NC.

Chapter 9

Asphalt Paving

9.1 Introduction

Asphalt paving operations encompass a detailed process that includes material selection, mix design, HMA production, transportation, and paving. Figure 9.1 illustrates the progression from materials to the completed pavement, detailing steps such as HMA delivery, surface preparation, mix placement, and compaction. Successful adherence to these steps by both contractors and highway agencies ensures that the public receives a pavement that will perform well for a reasonable period. A haul vehicle is responsible for transporting HMA from the asphalt plant to the paving site (USACE, 2000). This transport must occur promptly and maintain the mix's characteristics with minimal alteration and no segregation during delivery. The performance of HMA under traffic heavily depends on the condition of the surface upon which the pavement layers are applied. For mixes placed on existing asphalt layers, this surface must be adequately prepared, utilizing a tack coat to bond the new asphalt overlay with the existing pavement surface. When HMA is laid on a granular base course, a prime coat is necessary to enhance adhesion.

9.2 Managing Asphalt Transportation and Haul Truck Operations

Trucks are responsible for delivering the asphalt mix to the job site. It is essential that these trucks have clean, smooth, and tightly sealed beds. Each truck should be numbered for identification purposes and equipped with a tarp. Tarps must be made of durable, waterproof materials such as canvas or vinyl to protect the asphalt mix from the elements. Additionally, all trucks are required to meet established safety standards.

Prior to loading, the truck bed must be cleared of debris and hardened asphalt. It should then be lightly sprayed with an approved release agent to prevent the fresh asphalt mix from adhering to the bed. Diesel fuel, kerosene, or fuel oil are strictly prohibited as release agents. Any excess release agent should be allowed to drain off the bed before loading.

If the asphalt plant operates with platform scales, each truck must be weighed before loading to determine its tare weight. This weight is subtracted from the gross weight after loading to calculate the net weight of the transported asphalt mix.

During delivery, the driver must position the truck squarely against the paver and halt just inches away before the truck tires touch the paver's push roller bar. This prevents the truck from pushing against the paver, which could displace the screed and create a bump in the pavement that remains even after compaction.

DOI: 10.1201/9781003197768-9

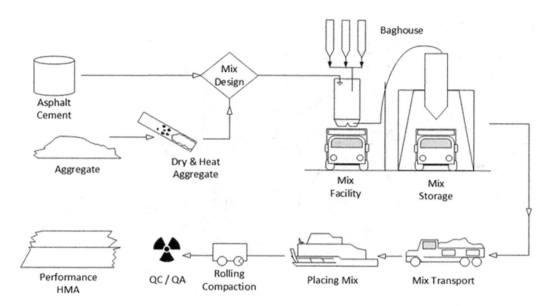

Figure 9.1 Entire Process of Asphalt Pavement Construction

To minimize segregation, the truck bed should be partially raised to allow the load to "break" before opening the tailgate. This controlled release of the asphalt into the paver hopper helps ensure a uniform mix flow, reducing potential segregation between loads.

9.3 Prime Coat and Tack Coat

Liquid asphalt materials, such as prime coat and tack coat, are applied using an asphalt distributor specifically designed for even and precise distribution of liquid asphalt on road surfaces (refer to Figure 9.2).

The asphalt distributor consists of a truck- or trailer-mounted insulated tank with an integrated heating system, typically oil-fired, to keep the asphalt at the correct application temperature. A visible thermometer must be mounted on the distributor for constant temperature monitoring. The equipment is equipped with a spray bar system designed to apply the liquid asphalt uniformly and includes a handheld spray attachment for reaching areas that the spray bars cannot access (Figure 9.3).

At the rear of the tank, a series of spray bars and nozzles disperse the asphalt under pressure onto the road surface. The spray bar maintains uniform pressure along its length to ensure consistent output from all nozzles. An essential component of the distributor, the spray bar's effectiveness depends on selecting the correct nozzle sizes for the specific conditions of the job. Additionally, the angle of the nozzle openings must be adjusted to prevent overlapping spray patterns, with typical adjustments ranging from 15° to 30° depending on the distributor model.

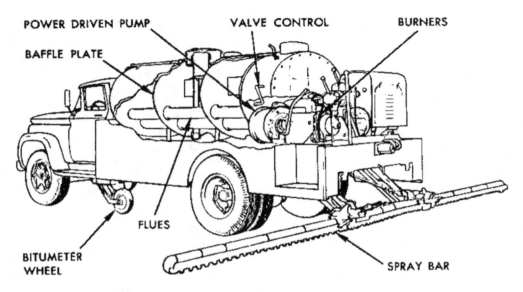

Figure 9.2 Asphalt Distributor

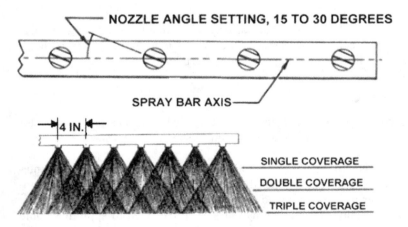

Figure 9.3 Proper Nozzle Angle Setting

9.3.1 Prime Coat

A prime coat is a spray application of low-viscosity liquid asphalt applied to a base course of untreated material. It serves as the initial layer of asphalt on a non-asphalt base, setting the stage for any additional treatments or construction. The primary purposes of a prime coat include:

(i)   Waterproofing the surface to prevent moisture penetration into the base course or subgrade, thus enhancing durability

(ii) Sealing voids, coating, and bonding loose mineral particles to stabilize the primed surface

(iii) Temporarily protect the pavement when construction is delayed or conducted in stages
(iv) Enhancing the adhesion of subsequent asphalt layers or surface treatments to granular bases, such as Aggregate Base Course (ABC) and various soil types

Prime coats are typically necessary when an asphalt pavement layer is directly placed on an aggregate or soil base. All approved prime coat materials currently in use are emulsified asphalts chosen for their cohesive properties. The application rate for prime coats should range from 0.20 to 0.50 gallons per square yard, adhering to specifications listed on approved material lists.

9.3.2 Tack Coat

A tack coat is a spray of liquid asphalt applied to existing asphalt or concrete surfaces to promote adhesion between old pavement layers and new asphalt overlays. The key requirements for a practical tack coat application include (i) the existing pavement surface must be thoroughly cleaned to remove any contaminants that could impair bonding; (ii) uniform coverage must be achieved across the entire area slated for paving; (iii) the tack coat must be allowed to fully break (cure) before the new asphalt layer is placed.

A tack coat is essential under each new layer of asphalt plant mix. It should only be applied when the surface is dry, and the air temperature is 35°F or higher, measured away from artificial heating sources. Before application, the existing asphalt or concrete surface must be free of dust, dirt, clay, fuel oil, grass, or any other foreign matter that might prevent the tack coat from adhering correctly. Failure to do so can result in a poor bond between layers. Suitable materials for tack coat include asphalt emulsions such as Grade RS-1H, CRS-1H, CRS-1, HFMS-1, or CRS-2.

Achieving uniform application and adhering to the correct application rate is critical for the success of the tack coat. A uniform application rate of 0.04 gallons per square yard is typically required for new asphalt layers. For resurfacing projects, a minimum of 0.06 gallons per square yard is recommended to compensate for the oxidized condition of the old asphalt surface. Similarly, milled asphalt surfaces should be meticulously cleaned and prepared to ensure the tack coat is applied uniformly at 0.06 gallons per square yard (NCDOT, 2024).

9.4 Asphalt Mix Equipment, Placement, and Compaction

The placement and compaction of the asphalt mixture are the culmination of all preceding processes. This involves carefully selecting and combining aggregates, precise mix design, and meticulous setup, calibration, and inspection of the plant and its auxiliary equipment. Once mixed, the materials are delivered to the paving site.

Asphalt mix can be transferred to the paving site via trucks, deposited directly into the paver, laid in windrows in front of the paver, or transferred using a material transfer vehicle. The paver then distributes the mix according to the required grades, cross-sectional thicknesses, and widths specified in the plans. As it moves forward, the paver not only spreads the mix but also initiates partial compaction, providing a preliminary smooth and uniform surface.

Subsequently, while the mix is still hot, it undergoes further compaction through the use of steel-wheeled, vibratory, or rubber-tired rollers, or a combination of these. The rollers continue over the freshly paved mat until the mix achieves the specified density or until the temperature drops to a point where further compaction could be harmful. After the pavement layer has been properly compacted and allowed to cool, it is either prepared for additional paving courses or is ready to bear traffic loads.

9.4.1 Incidental Tools for Paving Operations

For effective paving operations, it is essential to have both adequate hand tools and the proper equipment for cleaning and heating these tools on site. The list of incidental tools required includes

- Rakes, shovels, and lutes for manual adjustment and smoothing of the asphalt
- A tool heating torch and cleaning equipment to ensure tools are effective and free of debris
- Hand tampers and small mechanical vibrating compactors to achieve initial compaction in areas inaccessible to larger rollers
- Blocks and shims for properly positioning and supporting the screed of the paver at the start of operations
- Heavy paper or timbers designed for constructing joints at the ends of paving runs
- Joint cutting and tacking tools to ensure seamless transitions between asphalt batches
- A 10-foot straightedge for checking surface evenness
- A 6-inch (150 mm) core drill equipped with a 6-inch (150 mm) diameter cut core bit for sampling the asphalt
- A 4-foot level and a depth-checking device to verify uniformity and the proper depth of the asphalt layer
- An infrared thermometer to monitor the temperature of the asphalt
- A string line for accurate paver alignment to maintain straight paving lines

These tools play a critical role in ensuring the quality and precision of the paving process, from initial application to final compaction and finishing touches.

9.4.2 Asphalt Paving Equipment

The asphalt paver is a self-contained, power-propelled machine that distributes the asphalt mix in a uniform layer at the desired thickness, shape, elevation, and cross-section, setting it up for subsequent compaction. Modern asphalt pavers operate on either crawler tracks or wheels and can lay layers ranging from less than one inch to about eight inches thick across widths from 6 to 32 feet. These machines typically operate at speeds between 10 to 70 feet per minute.

An asphalt paver is composed of two main units: the tractor unit and the screed unit. Used primarily in highway construction, these machines are large and complex, featuring many intricate parts and adjustments. Despite differences in specific features, all pavers operate under the same fundamental principles, utilizing a self-leveling, floating screed to achieve a level and compacted layer of asphalt.

The operational flow of an asphalt paver is illustrated in plan and side views shown in Figure 9.4. Asphalt mix is transferred from a truck into the paver's front receiving hopper. The paver pushes the truck using rollers mounted at the front that contact the truck's rear tires, allowing continuous feeding of the mix into the hopper.

Inside the paver, the asphalt mix is moved from the hopper to the rear via two independently controlled slat conveyors, passing through control gates to the spreading screws (augers). Each auger and its corresponding conveyor is automatically controlled to ensure the mix is uniformly distributed just in front of the screed unit.

The screed unit, attached to the tractor by two long pull arms that pivot from the front, is designed to float. These arms do not support the screed vertically; as the tractor moves forward, the screed levels itself against the asphalt mix, following a path parallel to the direction of the pull, thus laying the asphalt evenly before compaction.

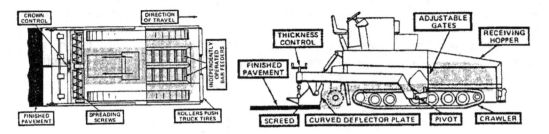

Figure 9.4 The Tractor Unit

9.4.2.1 The Tractor Unit

The tractor unit is the powerhouse of the asphalt paver, propelling the machine either on crawler tracks or pneumatic tires. This unit houses the propulsion system, push rollers, paver hopper, slat conveyors, flow gates, spreading screws (augers), material feed systems, and controls. Many pavers feature dual controls, allowing operators to manage the machine from either side.

The paver's design incorporates a receiving hopper and an automated distribution system, ensuring a consistent material supply across the entire screed length, including any screed extensions. The hopper contains two slat conveyors that transport the asphalt mix through the paver to the augers at the back. Each slat conveyor and auger operates independently, yet both sides are synchronized to ensure even mix distribution. This setup allows the operator or the automatic system to adjust the material feed to one side of the paver, accommodating specific needs such as paving ramps, mailbox turnouts, tapers, and areas with variable widths or depths.

Individual flow control gates located at the hopper's rear above the slat conveyors adjust the flow rate of material to the augers, regulating the volume delivered by the conveyors. Sensors near the augers' ends monitor the material volume in front of the screed and trigger the automatic controls to maintain a constant material depth, including for screed extensions.

The material from the slat conveyors is evenly spread across the width of the paver screed by the augers. At the center, where the augers meet near the gearbox, a specially designed auger (reverse auger or paddle) ensures consistent mix placement beneath the gearbox. If this reverse auger paddle malfunctions or wears out, it could result in a visibly segregated streak in the center of the asphalt mat. It is crucial for the augers to maintain a uniform mix volume at the front of the screed to ensure constant pressure and material head.

For extended screeds, auger extensions are necessary to maintain the proper head of mix in front of the full screed length. When the screed extends beyond one foot, the corresponding side's auger should also be extended by an equivalent amount to ensure consistency across the newly widened area.

9.4.2.2 The Screed Unit

The screed of a paver is a crucial component that operates as a free-floating unit, responsible for striking off, partially compacting, and smoothing the surface of the asphalt mat as it is towed forward by the tractor unit. It connects to the tractor at only one point on each side, known as the pull point or tow point, allowing it to "float" and adjust based on the forces exerted as it moves into the asphalt mix. This free-floating screed principle is a standard feature in all modern asphalt pavers. For specific details on a particular brand or model, consulting the service manuals and manufacturer literature is recommended. See Figure 9.5 for an illustration of various screed unit components.

The screed should be equipped with a fully functional screed plate, preheated to the necessary width and featuring operational vibrators and a heating source. The vibrators enhance the uniformity of the mixture feeding under the screed and provide initial compaction to the mat, improving

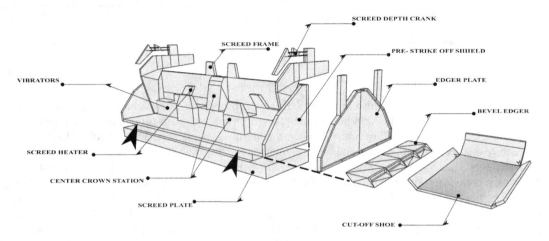

Figure 9.5 The Screed Unit

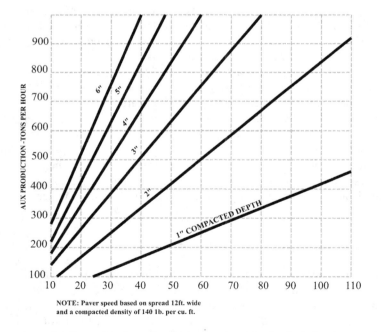

NOTE: Paver speed based on spread 12ft. wide
and a compacted density of 140 1b. per cu. ft.

Figure 9.6 Paving Machine Speeds Required to Handle Plant Production

mat thickness, density, smoothness, and surface texture. This initial compaction also minimizes screed settling when the paver stops. Vibrations are typically produced by electrically operated mechanical vibrators or eccentrically loaded turning shafts, with adjustable frequency to maximize compaction and smoothing effects.

Both the leading and trailing edges of the screed have adjustable crowns. The leading edge usually features slightly more crown (about ⅛ in) than the trailing edge to ensure smooth material flow under the screed. However, excessive crown at the lead can cause open textures along the mat edges, while insufficient crown may cause open textures at the center. Crown adjustments are often determined through trial and error and can be adjusted independently or simultaneously during paving to achieve a uniform surface texture across the mat's width.

Heaters on the screed are crucial to prevent asphalt from sticking. It is especially necessary to preheat the screed to nearly the asphalt's temperature at the paving's start. They are particularly useful during cool, windy conditions but should not be used to heat a cool mix being delivered to the paver.

Before paving, the screed must be raised to inspect the bottom surface for smoothness, holes, or excessive wear. Wear typically first appears about 4 to 6 inches in from the trailing edge. Extensions should be flush and aligned with the screed bottom. Tampers need regular checks for wear, adjustment, and function, as excessive wear can cause a pitted surface, and improper adjustment can scuff the mat. The tamper bar's bottom stroke should extend 1/64 in below the screed plate bottom.

While the screed is raised, also check the condition and adjustment of the strike-off device according to the manufacturer's recommendations. An uneven, damaged, or improperly adjusted strike-off can significantly impact the mat's smoothness, texture, and uniformity.

9.4.3 Material Transfer Vehicle

A material transfer vehicle (MTV) functions as a mobile surge bin, facilitating the transfer and remixing of asphalt mix from the haul vehicle to the paver hopper. This process maintains a uniform and continuous flow, allowing the paver to operate almost uninterrupted during truck exchanges, provided there is a steady supply of material from the plant. The key role of the MTV is to remix the asphalt mix before it enters the paver conveyor system. This step helps to minimize aggregate segregation and temperature variations that may have occurred during production, storage, and transport. Utilizing an MTV significantly enhances the uniformity and overall ride quality of the pavement.

It is advisable to employ an MTV for laying all asphalt concrete plant mix pavements that require the use of asphalt binder grade PG 76–22 and for all open-graded friction course (OGFC) types, unless specified otherwise. An MTV should be used for all surface mixes, regardless of binder grade, on primary routes like Interstate, US, and NC routes that have four or more lanes and are median divided. For these routes, the MTV is essential for placing all full-width travel lanes and collector lanes. Additionally, MTV usage is recommended for all ramps, loops, and Y-lines with four or more lanes and median divisions, as well as full-width acceleration lanes, deceleration lanes, and turn lanes exceeding 1,000 feet in length.

9.4.4 Coordinating Plant Production and Paver Speed

Achieving uniformity in asphalt paving operations is crucial. A consistent and continuous operation, including a steady forward speed of the paver, results in the highest quality pavement characterized by smoothness, uniform density, and a consistent surface texture.

To ensure a uniform and continuous laydown operation, pavers should operate at a speed that matches the plant production and material delivery, and enables the effective laying of the mixture. It's important to synchronize the paving and loading operations to maintain an adequate amount of asphalt mix in the paver hopper between truck exchanges and to prevent the hopper from becoming empty.

If issues such as uneven texture, tearing, segregation, or shoving occur during the paving process due to unsatisfactory methods or equipment, corrective actions must be taken immediately. Operating the paver faster than the plant can produce or deliver the mix leads to frequent stops, waiting for additional mix. This not only disrupts the paving process but also affects the smoothness and surface texture of the finished pavement, particularly if the paver has to restart after cooling, causing the screed to adjust unevenly.

Moreover, when the paver starts up again, if the mix has cooled, this can cause the screed to rise and fall, disrupting the evenness of the layer. If using automatic screed controls, the system may overcorrect, leading to a rough ride quality and possibly a segregated and less dense mix.

The operation should never exceed a speed that compromises the quality of the pavement. If the paver cannot keep pace with the plant's output, the production rate may need to be adjusted, or other corrective measures should be considered. Using a chart, as shown in Figure 9.5, can help balance the paving machine speed with plant production. Essentially, the goal is to avoid prolonged waiting times for the paver, which might necessitate reducing its speed to align with the mix supply rate.

9.5 The Placing Operation

Before starting paving operations, a string line must be placed along the edge of the proposed pavement to provide horizontal alignment control for the paver operator. This ensures a true and uniform line for the pavement edges. A string line is not required when the first course is placed adjacent to a curb section.

9.5.1 The Spreading Operation

The paving process begins once the paver has been thoroughly checked and positioned on the road. The screed should be lowered onto "starting blocks" (shims) that match the thickness of the loose mat to be laid, with thickness control screws adjusted accordingly. A general rule of thumb is to increase the loose thickness by 0.25 inch for every inch of compacted thickness.

As the first load of asphalt mix is spread, it's important to immediately assess the texture of the unrolled surface for uniformity. Adjustments to the screed, tamping bars, vibrators, spreading screws, hopper feed, and other components should be made regularly to ensure the mix spreads evenly to the proper line and grade. A straightedge can be used to check for a smooth surface.

During loading, the truck's wheels must firmly contact the paver's truck push rollers. Many modern pavers feature oscillating push rollers that compensate for any misalignment of the truck, ensuring that the load is centered on the paver. If a paver lacks oscillating push rollers and a truck is skewed, it may cause the spreader to also skew, requiring continual adjustments by the operator. This can lead to an uneven line and poorly compacted joints. Additionally, the rollers must be clean and free to rotate for the smooth forward travel of the paver.

If any segregation of the mix is observed, the spreading operation must be halted immediately and not resumed until the cause has been identified and rectified. To minimize segregation, the paver operator should avoid frequently folding the hopper wings. The frequency of emptying the wings depends on the mix delivery rate, mix temperature, and environmental conditions. On cooler days, the wings may need to be emptied more often than on warmer days. It might be advisable to leave the mix on the wings until the end of the day and then remove and discard any cool, hardened mix that cannot be effectively broken down and incorporated into the augers and under the screed.

If the specifications dictate that the mix be laid at a certain depth, the compacted depth should also be monitored to establish a correlation ratio between the loose and compacted depths. This ensures the final pavement meets the required specifications for thickness and density.

9.5.2 Fundamentals of Screed Operation

The screed unit, operating without automatic control, is connected to the tractor unit via two tow arms that pivot around a hinge point (tow point) located just past the midpoint of the paver. In manual operation, these tow arms are fixed at the tow point. The fundamental principle of screed operation is that it "floats" on the mix deposited in front by the augers. It moves up or down, seeking a level where the forces acting on it are balanced and its flat bottom surface runs approximately parallel to the direction of the pull.

While the paver moves, the pull force (P) at the pivot point consistently exceeds the horizontal resistance (H) exerted on the screed plate. To increase the mat's thickness, the screed is tilted upwards, allowing more material to accumulate under it. This results in a vertical uplift (V) that surpasses the screed's weight (W), causing the screed plate to rise until V decreases and equals W again. At this equilibrium, the vertical motion ceases, and the screed moves horizontally, parallel to the direction of pull. Adjustments to the mat's thickness are made by tilting the screed plate using the depth control screws or by vertically moving the pivot point of the tow arm, assuming all other factors remain consistent.

"Nulling out the screed" refers to adjusting the screed angle at both ends so that the screed plate rests flatly on the starting blocks or a similar surface. When both the front and rear edges of the screed plate are in firm contact with a surface, the adjusting screws should have a small degree of free rotation, indicating that the screed's angle of attack is in a neutral or flat position. This is considered the nulled out position. From here, it is standard practice to increase the angle of attack by turning each depth screw handle about one full turn (depending on the paver's make) from the nulled position to start paving. Continuous checks and adjustments are necessary to achieve the desired mat depth.

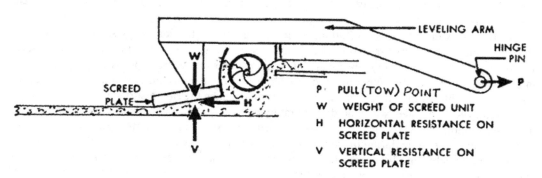

Figure 9.7 Forces Acting on the Screed During Paving Operations

The screed's weight also applies a compacting force as it travels over the asphalt mix. The angle of attack adjustment slightly raises the front edge of the screed, enabling it to climb sufficiently to match its compaction level. The precise angle needed can vary due to differences in the mix, the temperature of the mix, and the paving speed. These factors affect the required angle of attack, which must be finely tuned to ensure a smooth and even asphalt mat.

9.6 Compaction of Asphalt Pavements

Compaction is essential for compressing the asphalt mix into a denser volume by pressing binder-coated aggregate particles closer together, thus reducing air voids and increasing the density of the mixture.

Understanding the need for proper compaction becomes clear when considering the adverse effects of air, water, and traffic on an under-compacted pavement. Voids in an under-compacted mix are typically interconnected, allowing air and water to penetrate the pavement. This intrusion accelerates the oxidation of the asphalt binder, making the pavement brittle and prone to failure due to repeated traffic loads. Water trapped within the pavement can also expand upon freezing, leading to premature failure.

All asphalt mixes must be compacted to a minimum percentage of their maximum specific gravity (G_{mm}) as specified.

Inadequately compacted pavements during construction fail to achieve their designed strength and may deform under traffic loads. Conversely, if the mix is over-compacted and void content is excessively reduced, the pavement may become unstable due to traffic or thermal expansion. The ideal as-constructed void content for dense-graded mixes should be around 8% or less, ensuring the voids are not interconnected. Excessively high air void content can lead to raveling and disintegration of the pavement, while too low a content can cause flushing and instability.

Compaction effectively arranges aggregate particles in a dense configuration where the asphalt binder can hold them, achieving two primary objectives: (i) enhancing the mix's strength and resistance to rutting; and (ii) sealing the mix to prevent the penetration of water and air, which accelerates aging and can cause freeze-thaw damage and stripping.

Compaction Process: Immediately after the asphalt mixture is laid and adjusted for surface and edge irregularities, it should be uniformly and thoroughly compacted. The mix must be compressed to the specified degree for the type of mixture used. The compaction equipment must be approved by the Engineer and should provide enough force and coverage to achieve the required density while the mix is still workable. If uniform density is not achieved throughout the material's depth, adjustments to the type or weight of compaction equipment may be necessary, even if previously approved.

For final wearing surfaces, excluding open-graded asphalt friction courses, use at least two steel wheel tandem rollers for compaction. If a roller malfunctions, paving may continue for that workday provided satisfactory results are maintained. Pneumatic-tired rollers may be used for intermediate rolling. Limit rolling for open-graded friction courses to one pass with a tandem steel wheel roller, with additional passes only as needed to enhance the riding surface.

Vibratory Rollers: Steel wheel tandem vibratory rollers, specifically designed for asphalt pavement, are suitable for layers one inch or thicker during breakdown and intermediate phases. Avoid using the vibratory mode during finish rolling or on sensitive mixes and courses. Vibratory rollers should have variable amplitude and frequency settings tailored for asphalt compaction and must automatically disengage vibration before stopping. The use of vibratory rollers may be restricted if they cause damage to the pavement or surrounding infrastructure.

Rolling Equipment Restrictions: Avoid rollers that cause excessive crushing of aggregate or displacement of the mix. In areas inaccessible to standard rollers, use hand tampers, small rollers, or other approved methods for compaction. Rollers must be in good condition, capable of reversing without backlash, and operated at speeds that prevent mixture displacement. Equip steel-wheel rollers with wetting devices to prevent sticking.

Compaction should start immediately after the mix is laid and shaped, conducted at the highest temperature that the mix can support without displacement. Ensure uniform density across the entire section and complete all compaction phases before the mix cools below a workable temperature. Finish rolling should effectively remove any marks left by the initial compaction stages.

9.6.1 Types of Asphalt Rollers

Three fundamental types of rollers are used for the compaction of asphalt pavements: steel-wheeled rollers, pneumatic-tired rollers, and vibratory rollers.

Steel-Wheeled Rollers come in two primary configurations: three-wheel and tandem.

9.6.1.1 Three-Wheel Rollers

Three-Wheel Rollers are equipped with two drive wheels on a single axle and a steering drum. The drive wheels typically measure about 5 feet in diameter and 18 to 24 inches wide, while the steering drum is smaller in diameter but wider. These rollers, which vary in weight from 5 to 14 tons, are primarily used for the initial rolling of asphalt mixtures.

Figure 9.8 Three-Wheeled Roller

9.6.1.2 Tandem rollers

Tandem rollers can be two-axle or three-axle. Two-axle tandem rollers weigh between 3 to 14 tons or more and often include wheels to which ballast can be added for increased weight. Three-axle tandem rollers range from 10 to over 20 tons, with the center axle designed to allow a significant portion of the roller's weight to be applied as needed, especially useful for compacting high spots.

9.6.1.3 Pneumatic-Tired Rollers

Pneumatic-Tired Rollers are self-propelled pneumatic-tired rollers feature two to eight front wheels and four to eight rear wheels, with oscillating axles that allow the wheels to move up and down. These rollers vary in weight from 3 to 35 tons, with the ability to add ballast to increase weight. Advanced models include an "inflation-on-the-run" system that can adjust tire pressure during operation, crucial for meeting various compacting conditions.

9.6.1.4 Vibratory Rollers

Vibratory rollers are equipped with one or two smooth-surfaced steel drums, ranging from 3 to 5 feet in diameter and 4 to 6 feet in width, specifically designed for compacting asphalt pavements. Their static weight ranges from 1.5 to 17 tons. Some large tandem models also allow for vibrating the third axle. These rollers are generally suitable for all types of asphalt mixtures, assuming the correct amplitude, frequency, and speed settings are used (Figure 9.11).

Figure 9.9 Two Axle Tandem, Steel Wheel Roller

Figure 9.10 Self-Propelled Pneumatic-Tired Roller

Figure 9.11 Self-Propelled Tandem Vibrating Roller

Vibratory rollers are permissible on all asphalt layers 1 inch (25mm) or thicker during break-down and intermediate phases but are not allowed in vibratory mode during finish rolling or on open-graded asphalt friction courses or any layer less than 1 inch thick.

Table 9.1 Recommended Settings on Vibratory Rollers

Type Mix	Frequency (VPM)	Amplitude	Roller Speed
B25.0C, I19.0C 3" thickness	3000–3200	high/low amplitude	2–3 mph (normal walking speed)
S9.5X, S4.75A 1.5" thickness	3000–3400	low amplitude	2–3 mph (normal walking speed)
Consult manufacturer's recommendations for proper settings.			

When operating vibratory rollers, three crucial factors must be considered:

- Frequency (vibrations per minute)
- Amplitude (height of bounce)
- Roller speed

Recommended settings on vibratory rollers for different mixes are shown in Table 9.1.

9.6.2 Rolling and Compaction Procedures

Compaction of pavement materials should commence immediately after the asphalt mix is spread, struck off, and shaped to the required specifications, including width, depth, cross-section, and adjusted edge irregularities. The mix must be compacted to the necessary degree for the specific type of mixture used. The goal is to achieve uniform density across the entire section. Compaction should be performed at the highest temperature so that the mix can support the rollers without shifting horizontally. Ensure all compaction phases, including breakdown and intermediate rolling, are completed before the mixture cools below a workable temperature. Finish rolling is also crucial to eliminate marks left by the compaction process (AI, 2003).

Asphalt mixtures generally compact effectively if spread and rolled at temperatures that ensure proper asphalt viscosity. To maximize compaction, rolling should commence as soon as possible after spreading to avoid premature cooling and should be carefully managed to prevent surface roughening. Delaying compaction could lead to premature cooling, resulting in inadequate density and potential long-term issues, such as cracking or water infiltration (Figure 9.12).

9.6.2.1 Roller Contact Dynamics

The roller wheel should settle into the mix until the area of contact between the wheel and the mix, multiplied by the mix's resistance, equals the weight on the roller wheel. If the mix is sufficiently firm, the roller will not cause any horizontal displacement. However, if horizontal displacement occurs, it may result in crawling of the mix ahead of the roller and ridging on either side of the roller path, leading to a rough and uneven surface.

9.6.2.2 Temperature and Compaction

Mix temperature is a critical factor affecting compaction. Compaction is most effective while the asphalt binder remains fluid enough to act as a lubricant; once it cools and acts more like an

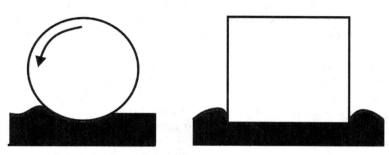

(a) RIDGES AND WAVES CAUSED BY HORIZONTAL DISPLACEMENT OF MIX DURING ROLLING.

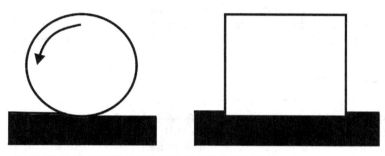

(b) MATERIAL COMPACTED BY ROLLER WITHOUT HORIZONTAL DISPLACEMENT OF MIX.

Figure 9.12 Impressions Made by Roller Wheel on Freshly Spread Asphalt Pavement

adhesive, further compaction becomes challenging. The optimal rolling temperature depends on the internal particle friction of the aggregates, mix gradation, and asphalt viscosity.

During rolling, roller wheels should be moistened only enough to prevent material pick-up; substances like diesel fuel, kerosene, or fuel oil should not be used as they can damage the mix. Rollers should move at a controlled, uniform speed – not exceeding three mph for steel-wheeled rollers and 5 mph for pneumatic-tired rollers. The rolling line should not be suddenly altered, nor the direction reversed abruptly, as these actions can displace the mix.

9.6.2.3 Rolling Sequence:

- Transverse joints
- Longitudinal joints when adjoining a previously placed lane
- Initial or breakdown rolling
- Second or intermediate rolling
- Finish rolling

Particular attention must be paid to the construction of transverse and longitudinal joints. Edges should not be left unrolled for more than 15 minutes.

All final wearing surfaces, except the open-graded asphalt friction course, should be compacted using at least two steel wheel tandem rollers. If a roller malfunctions, paving may continue for the remainder of that day, provided satisfactory compaction is still being achieved. Vibratory rollers, suitable for most pavement layers 1 inch or thicker, should be used only during breakdown and intermediate phases. They must not be used in vibratory mode during finish rolling or on layers less than 1 inch thick.

Vibratory rollers should have variable frequency and amplitude capabilities and be equipped with automatic controls to disengage the vibration mechanism before stopping. Rolling an open-graded asphalt friction course involves one pass with a tandem steel wheel roller weighing up to 10 tons, with additional rolling limited to necessary passes to improve the surface.

Pneumatic-Tired Roller Use: Using pneumatic-tired rollers is optional unless specified within contract provisions. Contractors must review all contract specifications to determine roller requirements effectively. The exact number of passes required to achieve adequate density can vary and is determined through careful observation and testing during the initial stages of paving.

Cooling Rates and Compaction Timing: Studies on cooling rates of mixes under various conditions help estimate the time available for achieving satisfactory compaction. Table 9.2 can be referenced to determine the allowable time for compaction based on mix temperature, lift thickness, and base temperature, which then informs the number and types of rollers needed for the project.

9.6.3 Factors Affecting Compaction

Compaction of asphalt pavements is influenced by various factors, which can be grouped into five main categories: (i) mix properties (aggregate, binder, and temperature); (ii) environmental conditions; (iii) layer (lift) thickness; and (iv) subgrade and bases (AASHTO, 2020).

The properties of asphalt binders and aggregates significantly impact the workability of mixes at different temperatures. Aggregate characteristics such as gradation, surface texture, and angularity affect the mix's workability. Coarser aggregates or those with a rough texture decrease workability and require a more significant compaction effort to achieve target density. Natural sands and fines in the mix also influence compaction; too much sand can make the mix tender and prone to displacement under heavy rollers, while excessive fines can make the mix gummy and difficult to compact.

Table 9.2 Cessation Requirements (Asphalt Mix)

Recommended Minimum Laydown Temperatures for Various Thicknesses						
	1/2"	*3/4"*	*1"*	*1-1/2"*	*2"*	*3" +*
Base Temp.	*Mix Temp.*	*Mix Temp.*	*Mix Temp.*	*Mix Temp.*	*Mix Temp.*	*Mix Temp.*
°F	°F	°F	°F	°F	°F	°F
20 – 32	----	----	----	----	----	285
32 – 40	----	----	----	305	295	280
40 – 50	----	----	310	300	285	275
50 – 60	----	310	300	295	280	270
60 – 70	310	300	290	285	275	265
70 – 80	300	290	285	280	270	265
80 – 90	290	280	275	270	265	260
90 +	280	275	270	265	260	255
Rolling Time (minutes)	*4*	*6*	*8*	*12*	*15*	*15*

Asphalt binder viscosity is crucial for compaction. At room temperature, asphalt binder is nearly solid, but at 265–350°F, it becomes fluid, acting as a lubricant that allows aggregate particles to move past each other during compaction. As the mix cools, the binder stiffens, making further compaction challenging. The asphalt content in the mix also affects workability; more asphalt provides better lubrication at compaction temperatures, but too much can make the mix tender.

9.6.3.1 Mix Temperature

Mix temperature profoundly affects compaction, with ideal compaction occurring when the mix is fluid enough to allow aggregate rearrangement but still capable of supporting roller weight without deformation. The optimal compaction temperature range is generally 185–350°F. Within this range, higher temperatures facilitate better densification during the initial rolling phase.

9.6.3.2 Environmental Effects

Environmental conditions such as air temperature, humidity, wind, and the temperature of the base impact the mix's cooling rate and, consequently, the window for effective compaction. Cool, windy conditions, or high humidity can accelerate cooling, necessitating rapid compaction following placement.

9.6.3.3 Layer Thickness

Thicker asphalt layers retain heat longer, extending the time available for effective compaction. This characteristic can be advantageous when compacting stable mixes that are inherently difficult to compress or in cooler weather conditions that accelerate the cooling of thin mats.

9.6.3.4 Subgrade and Bases

The firmness and stability of the subgrade and base layers are critical for effective compaction. Subgrades that deform under construction traffic may require additional compaction or stabilization efforts (e.g., with lime or Portland cement) to provide a suitable foundation for the asphalt layers. Inadequately stabilized subgrades may necessitate thicker pavement structures to distribute loads and prevent deformation.

9.6.3.5 Compaction Equipment and Procedures

Various rollers are used for the compaction, each suited to different mix types and compaction stages. Roller operators and technicians must understand the mechanics of compaction and adjust their equipment and techniques based on the mix properties and environmental conditions. Techniques include adjusting roller speed, pattern, and the number and type of passes to achieve the desired pavement density and smoothness.

Proper compaction is crucial not only for pavement durability but also for its performance under traffic loads. As such, each factor must be carefully considered and appropriately managed to ensure the quality and longevity of the asphalt pavement.

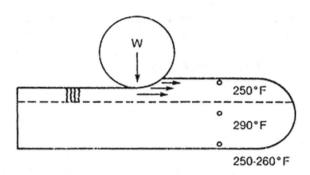

Figure 9.13 Heat Checking (Side View)

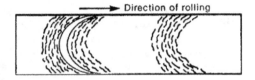

Figure 9.14 Heat Checking (Top View)

9.6.4 Three Phases of Rolling

The compaction process of asphalt pavements consists of three fundamental phases, each critical to achieving a durable and smooth roadway: (i) breakdown phase; (ii) intermediate phase; and (iii) finish phase.

Breakdown Rolling: This initial phase is crucial and typically involves steel-wheeled rollers. Both static-weight and vibratory tandem rollers can be used. Vibratory rollers are permitted on all mixes but only for final wearing surfaces with a thickness of 1 inch or greater. The roller's weight, usually between 8 and 12 tons, is selected based on the mix's temperature, thickness, and stability.

Historically, it was recommended to begin compaction on the lower side of the pavement lane and move upward. However, this technique is generally unnecessary with more stable modern mixes and advanced compaction equipment unless the pavement has extreme superelevation or the mix lift is unusually thick. For adjoining lanes, the longitudinal joint should be compacted by placing most of the roller on the newly laid mat and overlapping the edge of the previous lane by about 6 inches to ensure adequate compaction.

Intermediate Rolling: Following the breakdown phase, intermediate rolling should be executed while the asphalt mix remains plastic and above the minimum effective compaction temperature, ideally between 225–250°F. This phase can involve pneumatic-tired rollers, as well as static and vibratory steel-wheeled rollers. Pneumatic-tired rollers offer several advantages: they deliver uniform compaction, improve surface sealing to reduce permeability, and help orient the aggregate particles for enhanced stability. The tires may be kept hot and treated with a non-foaming detergent or water-soluble oil to prevent asphalt from sticking.

During this phase, pneumatic-tired rollers should make at least three passes, and care must be taken to avoid displacing the mix during turns. Vibratory tandem rollers, with appropriate static weight, frequency, and amplitude, can achieve the required densities with fewer passes than static rollers or combinations of the two.

Finish Rolling: The final phase aims primarily to enhance the surface quality. It involves static-weight steel-wheeled tandem rollers or non-vibrating vibratory tandems performed while the material is still warm enough to smooth out roller marks. Only the necessary number of passes should be made to achieve a smooth finish without over-rolling, which can decrease the mix density. Vibratory rollers cannot operate in vibratory mode during this phase to avoid damaging the finished surface.

Overall, the proper execution of these three rolling phases is vital for optimizing the lifespan and performance of asphalt pavements, ensuring they meet the required specifications and withstand the demands of traffic and environmental conditions.

9.6.5 Roller Patterns

The goal of an effective rolling pattern is to achieve the most uniform compaction across the entire lane width. Since rollers vary in width, there is no one-size-fits-all pattern, and each roller's optimal pattern should be determined to ensure uniform compaction. Here, a roller "pass" is defined as one trip of the roller in one direction over any one spot in the pavement, while a "coverage" refers to the number of passes needed to cover the entire laydown width of pavement.

Rolling Pattern Considerations: The rolling pattern involves not just the number of passes but also the location of the initial pass, the sequence of succeeding passes, and the overlap between them. The breakdown rolling speed should generally not exceed about 3 mph to avoid disturbing the asphalt mix. Sharp turns and sudden starts or stops should also be avoided to maintain the integrity of the freshly laid mix.

For thin lifts, a suggested pattern for static steel-tired rollers might look like the one in Figure 9.15. The operation should start from the pavement edge on the lower side, following closely

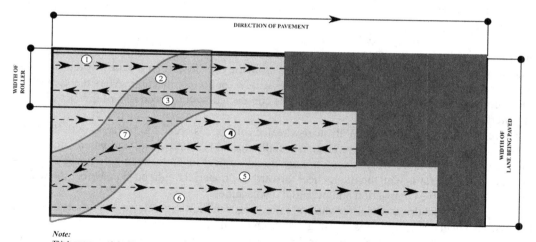

Note:
This is a recommended rolling pattern. Every pass of the roller should proceed straight into the compacted mix and return in the same pass. After the required passes are completed, the roller should move to the outside of the pavement on cooled material and repeat the procedure

Figure 9.15 Correct Rolling Pattern

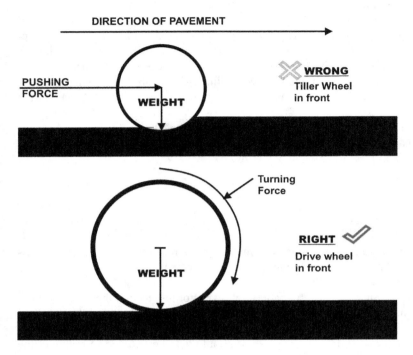

Figure 9.16 Forces Acting When Tiller Wheel or Drive Wheel is Forward

behind the paver. The roller moves forward, then reverses along the same path back to the start, then shifts over and advances again closely behind the paver, repeating this forward and reverse pattern until the entire width of the asphalt has been covered, each time overlapping the previous path by 3 to 4 inches.

For thicker lifts, rolling should begin 12 to 15 inches from the lower unsupported edge, progressing toward the center of the pavement until a degree of stability is achieved. The roller then gradually moves toward the edges. Starting away from the edge helps confine the mix during initial passes, reducing lateral movement and helping achieve better compaction. After the center has been compacted to some extent, the edges can be compacted with reduced risk of the mix displacing laterally.

Drive Wheel Considerations: With steel-wheeled rollers, it's crucial to progress with the drive wheel forward in the direction of paving, especially during breakdown rolling, where most compaction occurs. The drive wheel exerts a more direct vertical load, enhancing compaction effectiveness. If rolling starts with the tiller wheel forward, the material might push up before the wheel, causing less effective compaction and potential displacement.

Exceptions to the drive-wheel-forward rule may occur in high superelevations or steep grades. If the drive wheel chatters or slips, causing displacement of the mix, it may be necessary to reverse the roller so that the tiller wheel compacts the material first, allowing the drive wheel to pass over a partially compacted surface without causing damage or roughness.

Adapting roller patterns to the specific conditions and characteristics of the paving project is key to achieving optimal compaction and pavement quality.

9.7 Questions

1. Describe the correct asphalt mix delivery procedure to a paver and discuss the consequences and risks of deviation from this procedure.
2. Explain the functions and crucial application rates of prime and tack coats in paving, including their impact on milled surfaces.
3. Discuss the role of the paver in initial asphalt mix compaction and the potential drawbacks of operating the paver at incorrect speeds.
4. Outline the purpose and operation of a Material Transfer Vehicle in paving, including how it influences the paving process.
5. Detail the importance of using a string line in paving and how often paver operators should manage hopper wings.
6. Describe how to check the compacted depth of the mix and explain the primary goal of compaction in pavement construction.
7. Discuss the implications of air void reduction in the asphalt mix and the critical nature of achieving the required density.
8. Address the consequences of under-compaction and the role of asphalt binder oxidation in pavement failure.
9. Define the minimum compaction percentage for mixes and the main objectives achieved through proper compaction.
10. Explain the immediate steps to take after spreading and shaping the asphalt mixture, emphasizing the need for engineer-approved compaction equipment.
11. Discuss the operational conditions for continuing paving with malfunctioning steel-wheel rollers and restrictions on vibratory roller use.
12. Address the factors and adjustments needed when uniform density is not achieved and the importance of avoiding excessive aggregate crushing.
13. Explore the types and operation of rollers in compaction, emphasizing their role in preventing asphalt mixture displacement.
14. Detail the differences between three-wheel and tandem rollers, including the special features of pneumatic-tired rollers.
15. Clarify the appropriate use of vibratory rollers, focusing on the critical factors for their operation.
16. Discuss the significance of immediate compaction post-material spreading and the conditions for achieving optimal compaction.
17. Explain the critical compaction temperatures and how they are determined, including the effects of mix stability at high temperatures.
18. Address why certain substances should not be used on roller wheels and the recommended maximum speeds for different rollers.
19. Outline the proper rolling order for newly placed asphalt and the acceptable duration for exposed edges without rolling.
20. Describe the requirements for vibratory rollers regarding frequency, amplitude, and control features.
21. Discuss the optional and mandatory use of pneumatic-tired rollers and the factors influencing asphalt mix cooling rates.
22. Highlight the necessity of immediate rolling under various weather conditions, especially when compacting thin lifts in cooler weather.

23. Examine how increased layer thickness affects mix temperature choices and the importance of a firm, non-yielding subgrade or base.
24. Discuss the different types of compactors used in paving, focusing on their roles in the breakdown, intermediate, and finish rolling phases.
25. Describe the essential aspects of uniform rolling patterns and the recommended approach for thin-lift construction using static steel-tired rollers.

References

AASHTO (2020). *Construction Manual for Highway Construction.* 10th ed. American Association of State Highway and Transportations Officials (AASHTO), Washington, DC.

AI (2003) Manual Series No. 22. *HMA Construction.* Asphalt Institute (AI), Lexington, KY.

NCDOT (2024). *Asphalt QMS Manual.* Materials and Test Unit, North Carolina Department of Transportation (NCDOT), Raleigh, NC.

US Army Corps of Engineers USACE (2000). *Hot-Mix Asphalt Paving Handbook.* US Army Corps of Engineers (USACE), Washington, DC.

Chapter 10

Asphalt Pavement Maintenance and Recycling

10.1 Introduction

Following the completion of roadway construction, as detailed in Chapter 9, attention turns to management and preservation. Pavement preservation employs a proactive strategy for maintaining existing highways, helping highway agencies avoid costly, time-intensive rehabilitation and reconstruction projects, along with the traffic disruptions they cause. Timely preservation efforts lead to enhanced safety and mobility, reduced traffic congestion, and smoother, longer-lasting pavements for the traveling public. A comprehensive pavement preservation program includes three key components: preventive maintenance, minor rehabilitation (nonstructural), and certain routine maintenance activities, illustrated in Figure 10.1.

Implementing an effective pavement preservation program brings numerous advantages to government agencies, including safeguarding investments in highway infrastructure, improving pavement performance, ensuring cost-efficiency, extending the pavement's lifespan, minimizing user delays, and enhancing safety and mobility. A treatment qualifies as pavement preservation if its primary objective is to rejuvenate the current system's functionality and extend its service life, rather than to increase its capacity or strength.

Asphalt recycling plays a pivotal role in pavement rehabilitation, significantly reducing the environmental impact and energy consumption compared to traditional pavement reconstruction methods. The adoption of asphalt recycling and reclamation has increased, recognized for its technical viability and environmental benefits. Asphalt is many countries' most frequently recycled material, highlighting its critical role in sustainable pavement management.

Societal awareness of the environmental impacts of various types of development has significantly increased. Many countries have enacted laws requiring that a specific proportion of materials, particularly those used in roadway construction and rehabilitation, be recycled or include recycled components. Demonstrating the technical feasibility of asphalt recycling, its ability to reduce energy consumption, and the use of non-renewable resources such as crude oil and granular materials highlights the cost benefits associated with this practice. These efforts contribute to advancing one of society's goals of adopting environmentally responsible construction practices.

10.1.1 Pavement Maintenance and Rehabilitation Strategies

Pavement preservation involves a long-term, network-level strategy to enhance pavement performance through cost-effective practices. Endorsed by the FHWA Pavement Preservation Expert Task Group, these practices aim to extend pavement life, improve safety, and meet motorist expectations. The focus is on maintaining pavements in good condition to prevent significant

DOI: 10.1201/9781003197768-10

Figure 10.1 Component of Pavement Preservation

Table 10.1 Pavement Preservation Guidelines

Type of Activity	Increase capacity	Increase Strength	Reduce Aging	Restore Serviceability
New construction	X	X	X	X
Reconstruction	X	X	X	X
Structural overlay		X	X	X
Major (heavy) rehabilitation		X	X	X
Minor (light) rehabilitation			X	X
Preventive maintenance			X	X
Routine maintenance				X
Corrective (reactive) maintenance				X
Catastrophic maintenance				X

Source: https://www.fhwa.dot.gov/pavement/preservation/091205.cfm

deterioration. Timely and practical treatments can restore pavements close to their original condition, delaying costly rehabilitation or reconstruction. Over the life of the pavement, the total cost of regular preservation treatments is substantially lower than full reconstruction and more economical than significant rehabilitation. Additionally, this approach minimizes traffic disruptions compared to the extensive closures required for reconstruction projects.

Preventive maintenance represents a planned strategy of cost-effective treatments for an existing roadway system and its components aimed at preserving the system, slowing future decay, and maintaining or enhancing its functional state without significantly increasing structural capacity. As defined by the AASHTO Standing Committee on Highways in 1997, this concept is typically applied to pavements in good condition with a significant remaining service life. As a crucial element of pavement preservation, it focuses on extending service life through efficient treatments to the surface or near-surface of structurally sound pavements. Preventive treatments include methods such as asphalt crack sealing, chip sealing, slurry or micro-surfacing, thin and ultra-thin hot-mix asphalt overlays, concrete joint sealing, diamond grinding, dowel-bar retrofit, and specific concrete repairs (e.g., for edge spalls or corner breaks) to enhance slab functionality.

Pavement rehabilitation involves structural improvements that prolong the service life of an existing pavement and/or enhance its load-bearing capacity, as outlined by the AASHTO Highway Subcommittee on Maintenance. These improvements, through restoration treatments and structural overlays, aim to extend the lifespan of existing pavements by restoring their structural capacity, addressing age-related and environmental surface cracking, or increasing pavement thickness for better load support under current or expected traffic conditions. This leads to two sub-categories: those that restore structural capacity and those that increase it.

Minor rehabilitation includes non-structural improvements to address top-down surface cracking in flexible pavements caused by environmental factors. Due to their non-structural nature, these techniques are classified under pavement preservation.

Major rehabilitation entails structural enhancements that prolong the existing pavement's service life and/or improve its load-bearing capability, focusing on significant structural improvements, as per the AASHTO Highway Subcommittee on Maintenance.

Routine maintenance, defined by the AASHTO Highway Subcommittee on Maintenance, involves regular work to maintain and preserve the highway system or to address specific issues that restore it to an adequate service level. Performed typically daily, these activities aim to keep the highway system in optimal condition. Tasks include cleaning roadside ditches and structures, maintaining pavement markings, filling cracks, patching potholes, and applying isolated overlays. Critical to this category, crack filling involves sealing cracks with bituminous material to reduce water infiltration and reinforce the pavement. While specific routine maintenance tasks can be considered preservation, they are generally conducted in-house and are not typically eligible for Federal-aid funding.

Corrective maintenance involves reactive measures to address deficiencies that compromise a facility's safety and efficiency, particularly in pavement sections, aiming to restore pavements to adequate service levels and respond to unforeseen issues.

Catastrophic maintenance describes activities generally necessary to return a roadway facility to a minimum level of service while a permanent restoration is being designed and scheduled. This may be required after concrete pavement blow-ups, road washouts, avalanches, or rockslides.

Pavement reconstruction involves completely replacing the existing pavement with a new or improved structure, typically requiring the removal of the old pavement. This is necessary when the existing pavement has failed or become outdated. New or recycled materials may be utilized.

10.2 Pavement Maintenance Methods

10.2.1 Crack Sealing and Filling

Crack sealing involves creating a reservoir in the crack using a saw or mechanical router and then using compressed air (or a heat lance) to remove debris. The crack is then filled with a specified sealant material. This method prevents water infiltration into the pavement structure and minimizes the disruption of thermal movement due to incompressible materials within the crack. The proper preparation of the crack is crucial in crack sealing, enhancing its effectiveness as a preventive maintenance strategy.

Crack filling differs from sealing because it generally does not involve routing or sawing before treatment. The process includes clearing the crack of debris using compressed air and then filling it with a sealant. Unlike crack sealing, crack filling does not completely seal the crack against water infiltration but reduces the amount of water and incompressible material entering the pavement structure and subbase.

Both crack sealing and filling are localized treatment methods that prevent water and debris from entering cracks in asphalt and concrete pavements. Local highway agencies commonly prioritize these methods as essential tools in their pavement maintenance programs. Figure 10.2 illustrates the crack sealing process, highlighting its role as a cost-effective component of pavement preventive maintenance.

Figure 10.2 Crack Sealing

10.2.1.1 Features and Benefits of Crack Sealing and Filling:

- Regular crack filling can maintain or even enhance a pavement's condition rating.
- Sealing cracks in asphalt-surfaced roads helps prevent moisture and incompressibles from infiltrating the pavement structure, a primary cause of pavement deterioration.
- The benefits of crack sealing may not be immediately noticeable but will become apparent several years later, as sealed pavements typically exhibit less deterioration than unsealed ones.
- Crack sealing offers an excellent return on investment for pavement maintenance funds (FHWA, 2019).

Preparation and Equipment for Crack Sealing: Before initiating crack sealing or filling, it is essential to review construction documents, which may include special provisions, manufacturer's sealant installation instructions, agency application requirements, and the sealant material safety data sheet. The project review should verify the appropriateness of the joint size, pavement conditions, joint design, and sealant type for the specific climate and conditions. It should also ensure that methods for cutting and cleaning the joints, as well as removing old sealant materials, are suitable.

For hot-applied sealants, an indirectly heated double boiler-type melter with adequate agitation is used. The melter should be in good working order with all heating, agitation, and pumping systems, valves, and thermostats functioning properly and thermostatically controlled. Temperature gauges must be calibrated and checked for accuracy. The correct size of wand tips for the desired application should be available, and the melter must be adequately sized for the project. The air and/or surface temperature must meet manufacturer and agency requirements, typically 4°C (40°F) and rising, for sawing and sealing. Sealing operations should not proceed if rain is imminent or if there is any moisture on the surface or in the joint. Operators must wear the required personal protective equipment, and the setup of traffic control devices must comply with the Federal Manual on Uniform Traffic Control Devices (MUTCD) or local agency requirements. Joint sealing can also be applied to Portland cement concrete pavements.

10.2.2 Chip Seal, Fog Seal, Slurry Seal, and Microsurfacing

Chip seal is a pavement maintenance technique that involves the application of a thin film of heated asphalt liquid on the road surface, followed by the placement of small aggregates ("chips"). After application, the chips are compacted to enhance adherence to the asphalt, and excess stone is swept away. This method protects the pavement from the effects of sun and water, increases skid resistance, and fills small surface defects. Prior to implementing a chip seal, project documents must be reviewed to assess existing pavement conditions, including the extent of rutting, cracking, and whether crack sealing is necessary, as well as any issues of bleeding or flushing. It is crucial to ensure that the asphalt emulsions used are compatible with the aggregate, that the aggregate chips are uniform in size and free of excess fines, and that the pavement surface is clean and dry before application. The process checklist includes checking the chip spreader, haul trucks, rollers, broom, application temperature, application rate, and traffic control measures (Rajagopal, 2010).

Fog seal is a low-cost method involving a light application of a diluted slow-setting asphalt emulsion to an aged (oxidized) pavement surface to restore flexibility. Fog seals can delay the need for more intensive surface treatments or overlays. Unlike other surface treatments, fog seals do not use an aggregate cover, requiring a very low emulsion application rate to avoid tackiness and skid resistance reduction. The emulsion is typically diluted to facilitate precise control of this low application rate, using either anionic or cationic slow-setting emulsions. Fog seals are most effective when applied early in the pavement's life cycle, with repeat applications at appropriate intervals. They are beneficial for protecting structurally sound pavements by sealing voids with an asphalt-rich film, though they are typically short-lived.

Slurry seal shares similarities with fog seals but includes aggregates in the mix, forming a "slurry." Polymers are often added to the asphalt emulsion to enhance the mixture's properties. Slurry seals are applied approximately ¼ to ⅜ inch thick and serve to seal the pavement surface and refresh its texture.

Microsurfacing is akin to slurry seal but differs mainly in the setting process and the composition of the mixture, which includes water, asphalt emulsion, fine aggregate, and chemical additives. Like slurry seal, polymers are typically added to improve the mixture's properties. The key distinction lies in the curing mechanism; microsurfacing uses chemical additives in the emulsion that allow it to harden without relying solely on water evaporation, enabling it to set more quickly than slurry seal. This makes microsurfacing suitable for use in shaded areas or heavily trafficked roads where a faster return to service is necessary.

For all pavement maintenance methods mentioned, it is essential to review all specifications and project documents carefully to choose the appropriate maintenance method. Equipment must be checked and calibrated, emulsion application rates should be accurately determined and tested, and a comprehensive traffic control plan must be implemented to ensure safety and efficiency during application.

10.2.3 Diamond Grinding

Diamond grinding is a pavement preservation method that involves removing a thin layer from pavement surfaces using diamond saw blades. This technique enhances the smoothness and levelness of the pavement by creating grooves that uniformly break off the concrete between them, resulting in a finely textured surface. This not only provides a safer and quieter travel experience but also improves ride quality. The blades used in diamond grinding are constructed from industrial diamonds and metallurgical powder, tailored to the hardness of the aggregate material. For harder aggregates, a blade with a soft bond is preferred to efficiently expose the diamonds necessary for effective cutting. Conversely, a hard bond blade is more suitable for grinding softer aggregates. Diamond grinding is distinct from milling; while milling involves chipping away concrete through impact, diamond grinding uses a precision cutting process to achieve a controlled and consistent surface texture.

10.3 Asphalt Recycling Technology

In recent years, several advancements have been made in the rehabilitation and recycling of existing roadways, addressing a wide range of road defects such as rutting, cracking, profiling, and polishing. One effective approach is the combination of different asphalt recycling methods for roadway rehabilitation projects. For instance, the upper portion of an existing roadway can be removed via cold planing, creating reclaimed asphalt pavement (RAP) that can be stockpiled and reused in the asphalt plant. After preparing the cold-planed surface, it can be overlaid with hot-mix asphalt (HMA), incorporating the RAP from the milled-off layer.

Additionally, the existing pavement could undergo treatments like hot in-place recycling (HIR), cold in-place recycling (CIR), or full-depth recycling (FDR). These methods not only mitigate or eliminate reflective cracking but also add structural strength to marginal materials that may have been used in the original construction. These advancements significantly enhance the durability and lifespan of rehabilitated roads (ARRA, 2004).

10.3.1 Cold Milling

Recent advancements in the rehabilitation and recycling of existing roadways have significantly enhanced approaches to addressing road defects such as rutting, cracking, profiling, and polishing. A particularly effective method involves combining various asphalt recycling techniques within a single roadway rehabilitation project. For example, the upper layer of an existing roadway may be removed through cold planing (also known as CP or cold milling), which produces reclaimed asphalt pavement (RAP) that can be stockpiled and reused at an asphalt plant. After preparing the cold-planed surface, it can be overlaid with hot-mix asphalt (HMA), incorporating the RAP from the milled layer.

Additional treatments such as HIR, CIR, or FDR may be applied to the existing pavement. These methods not only help mitigate or eliminate reflective cracking but also enhance the structural strength of the pavement, even if marginal materials were used in the original construction. These advancements significantly improve the durability and lifespan of rehabilitated roads.

Cold milling, also known as pavement milling, asphalt milling, or profiling, is a crucial process in pavement rehabilitation. It involves removing part or all of a paved area's surface, such as roads, bridges, or parking lots. The depth of milling varies, ranging from light surface smoothing to full-depth removal, depending on the project's requirements. This method is often favored over repaving because it preserves existing structures and geometries, and it is critical to refer to project-specific special provisions in contracts, as they may dictate specific measurement and payment methods that differ from standard specifications. Figure 10.3 shows cold milling of a local road and airport runway.

Figure 10.3 Top: Cold Milling of a Local Road and Bottom: Cold Milled Airport Runways before Overlay

Cold milling machines provide several advantages, including rapid pavement removal with minimal traffic disruption – often allowing traffic to resume almost immediately after milling. This method is energy-efficient, and the milled material is typically reusable without additional processing.

Milling effectively resolves pavement issues like rutting, washboarding, pushing, shoving, and bleeding, offering time and cost savings. It also improves the surface texture for bonding new asphalt, maintains appropriate guardrail and curb heights, and ensures proper drainage by restoring the correct profiles and cross-sections.

Milling machines must be equipped with an electronic control system to maintain the desired longitudinal profile and cross slope. Various methods, such as mobile grade references, string lines, or slope control systems, are employed based on contract requirements. The result should be a uniform surface suitable for traffic, with minimal damage to the underlying structure. The equipment should allow for efficient loading and reuse of milled material without excessive segregation.

As specified in the plans, the milled surface should conform to the designated longitudinal profile and cross-section. It is crucial to clean the milled surface of loose materials, and the disposal of oversized pavement pieces within the right of way is prohibited. Lastly, the milling operation should minimize dust emissions and ensure safety for all adjacent persons and properties, including the traveling public.

10.3.2 Cold In-Place Recycling (CIR)

Cold in-place recycling (CIR) offers a cost-effective method for refurbishing damaged pavements, especially when budget constraints are a concern. It recycles the top 2–5 inches of asphalt through a continuous train operation, making it ideal for roads with a pavement condition index (PCI) of 70 or below. The process starts by milling the existing asphalt up to a depth of 125 mm. The reclaimed asphalt pavement (RAP) is then crushed to a maximum size of 37.5 mm and mixed with a rejuvenating emulsion. This mixture is reapplied to the road, forming a robust, recycled subgrade layer, which is typically topped with hot-mix asphalt or another surface treatment. CIR can be implemented as either full-depth or partial-depth, with full-depth CIR also referred to as full-depth reclamation (FDR). Figure 10.4 present the details of CIR.

10.3.2.1 Process in Detail

The CIR process includes several key steps:

- Milling: Removing the existing asphalt layer up to a depth of 125 mm
- Crushing: Breaking down the reclaimed asphalt pavement (RAP) to a maximum size of 37.5 mm
- Mixing: Blending a rejuvenating emulsion with the RAP. Virgin aggregate may be added as necessary, according to the mix design
- Laying Down: Applying the treated material back onto the road with a paver or grader, forming a restored subgrade layer, which is typically topped with hot-mix asphalt or other surface treatments

CIR is an environmentally friendly option, cutting greenhouse gas emissions by 90% compared to traditional mill and fill methods. This process also reduces the need for material trucking and hauling, offering significant economic and environmental benefits.

10.3.2.2 Detailed Process Steps for Cold In-Place Recycling

- Reclamation: The existing bituminous concrete pavement is being reclaimed.
- Transformation: The reclaimed pavement is transformed into a calibrated bituminous aggregate.
- Addition of Components: If necessary, corrective aggregate and new binder are added.
- Mixing: All components, including possible additives like cement or emulsified asphalt, are thoroughly mixed.

Figure 10.4a Left: Mill – Deteriorated Pavement is Milled 2–5 inches from the Pavement's Surface. Right: Mix – As the Surface is Milled, the Single-unit Train Mixes Additives such as Cement or Emulsified Asphalt

Source: Courtesy of Rock-Solid Stabilization & Reclamation, Inc.

Figure 10.4b Left: After the New Mixture is Laid, a Compactor is Driven across the Surface and Right: Apply a Wearing Course of Hot Mix or Slurry Pavement

Source: Courtesy of Rock-Solid Stabilization & Reclamation, Inc.

- Laying Down and Compaction: The new mixture is laid down, aerated, and compacted to ensure stability.
- Curing: The mixture undergoes a period of curing to develop adequate strength.
- Final Application: A wearing course is applied to finish the process and ensure durability.

10.3.2.3 Advantages of CIR

- Conservation: Saves energy and reduces the need for new pavement materials
- Environmental Protection: Reduces waste production and pollutant emissions
- Cost-Effectiveness: Provides substantial savings over the pavement's life cycle

- Traffic Management: Minimizes traffic disruptions during the rehabilitation process
- Structural Improvements: Improve pavement surface smoothness, decrease reflective cracking, and accommodate heavy traffic loads

Project Selection for (CIR) Selecting appropriate projects for CIR requires comprehensive field investigations, including

Work Site Investigation: Evaluates site constraints and resource availability. An assessment may reveal that a cold, in-place recycled course could serve as a leveling or base course for pavement overlays. Constraints such as surface elevation (e.g., curb height on urban roadways, underpass clearance) might suggest that a cold, in-place recycled course topped with a thin-wearing course is more suitable than milling and hot-mix paving. Additionally, an assessment of local resources might indicate poor quality or scarcity of local aggregates, making CIR a preferable option due to its utilization of in-place aggregates.

Visual Inspection: Aims to determine the type and severity of pavement distress. Suitable for a range of deteriorating bituminous pavements, CIR can be considered when the following distresses are observed:

- Pavement Cracking: Including age, thermal, fatigue, and reflective cracking
- Permanent Deformation: Such as rutting due to unstable bituminous mixtures, shoving, and rough pavement
- Loss of Integrity in Bituminous Pavement: Including raveling, potholing, stripping, flushing, and loss of bond between layers

Pavement Evaluation: Provides crucial data such as layer thickness, roughness, and subgrade condition. CIR demands a minimum bituminous pavement thickness of 60 mm. Extensive base or subgrade issues, or severe deformation, may disqualify a pavement from being a candidate for CIR or may require additional corrective operations as part of the pavement design.

Additional Considerations: Standard CIR might not correct certain issues like rutting, shoving, and flushing, which could indicate an excess of bitumen. These conditions may necessitate the addition of a corrective aggregate to improve the gradation of the existing mixture and reduce the bitumen/mineral aggregate ratio. Full-depth reclamation might be considered for pavements with these types of distresses. When binder stripping from aggregate is observed, the addition of an antistripping agent may be necessary, though this remains a high-risk application for CIR and requires a detailed assessment to confirm feasibility.

10.3.2.3 Equipment and Construction Procedures for CIR

CIR utilizes a range of specialized equipment configurations, from single-unit systems to multi-unit recycling trains, each designed for efficient processing and application.

Recycling Equipment: Equipment varies based on the integration and separation of key operations

- Reclamation: Removing the existing bituminous concrete pavement
- Transformation: Converting the reclaimed pavement into a calibrated bituminous aggregate
- Addition and Mixing: Incorporating an emulsion and thoroughly mixing all components
- Lay down: Applying the new mixture on the roadway

Options include single-unit systems that perform reclamation, sizing, mixing, and laydown sequentially within one machine, and multi-unit trains that distribute these tasks across several specialized pieces of equipment:

- Single-Unit Systems: These compact the entire process into one machine where the cutting drum reclaims, sizes, and mixes in additives, while a standard screed attached to the rear handles laydown.
- Multi-Unit Recycling Trains: Comprised of separate units for each major task: a standard milling machine for reclamation, a mobile screen/crusher for sizing, a mobile pugmill for mixing, and a standard paver for laydown.
- Grade and Slope Control: Essential to ensuring that the newly recycled surface achieves the desired level of smoothness and proper drainage.
- Segregation Control: Maintains uniformity in the recycled mixture, crucial for achieving consistent quality and performance across the pavement
- Compaction: A critical aspect of CIR, ensuring the durability and longevity of the pavement. Effective compaction requires careful control of rolling patterns and moisture content to achieve optimal density and bonding of the material.

These equipment options and construction procedures are designed to adapt to various site conditions and project requirements, ensuring that CIR can be effectively implemented to rehabilitate deteriorating pavements.

10.3.2.4 Compaction and Curing in CIR

Compaction: Successful CIR operations heavily depend on effective compaction, which is more demanding for recycled mixtures than for standard hot bituminous mixtures. Due to high internal friction, which causes the mixture to fluff up, appropriate compaction equipment and techniques are essential.

- Equipment: Typically, one heavy pneumatic roller along with one double drum vibrating roller is used. The double drum vibrating roller is especially effective due to its ability to deliver higher compaction energy at high amplitude and low frequency.
- Rolling Patterns: These are determined through trials in test sections to ensure optimal compaction.
- Moisture and Density Monitoring: Nuclear gauges are employed to monitor both the moisture content and the density of the mixture during compaction, ensuring the mixture meets the necessary specifications.

Moisture Management: Moisture content is crucial for proper compaction.

- Insufficient Water: If the mixture is too dry, it becomes harsh and resistant to compaction.
- Excess Water: Too much water prevents effective compaction by filling potential air voids and introducing excess fluids.
- Aeration: Often, the mixture contains more water than needed for effective compaction, necessitating aeration to allow excess moisture to escape before compaction begins.
- Timing: Compaction generally starts as the emulsion breaks, potentially after a delay of 20 minutes to an hour post-placement to allow initial water content adjustments. The optimal total water content for compaction typically ranges from 3.5% to 4.5%.

Curing: The recycled mixture requires time to cure and develop internal cohesion before the application of a wearing course.

- Curing Period: A standard curing period of 14 days is generally recommended to ensure the mixture achieves sufficient strength and stability.
- Moisture Content during Curing: While low moisture content is often a criterion for determining adequate curing, it can be misleading if rain increases moisture levels, suggesting insufficient curing when the material may actually be ready. As a practical indicator, the ability to extract a complete core from the mat with relative ease often signifies that the material has developed adequate internal cohesion and is ready to be covered with a wearing course.

This structured approach to compaction and curing ensures the CIR process yields a durable and stable pavement, prepared to handle subsequent layers and traffic.

10.3.2.5 Pavement and Mix Design for CIR

The design process for CIR involves both bituminous mixture design and pavement design, taking into account structural requirements, pavement profiles, traffic conditions, and the selection of an appropriate wearing course.

10.3.2.6 Structural Design

- Structural Coefficients: Currently, there is no universally accepted structural coefficient for CIR. The structural capacity of recycled CIR material depends on factors such as the nature of the in-place bituminous material, the type and amount of added binder, and the duration of curing.
- Assumptions by Agencies: In the United States, some road agencies treat the structural capacity of recycled mixtures as equivalent to that of standard hot bituminous mixtures. Others, including the State of California, assign AASHTO layer coefficients ranging from 0.25 to 0.40, or a gravel base equivalency (GBE) of 1.5 to 2.5, without differentiating between full-depth reclamation (FDR) and standard CIR.

10.3.2.7 Pavement Evaluation

- Considerations: Essential considerations include the need for improved or new drainage systems, the structural adequacy to determine if additional granular or RAP material should be added, and the appropriate surface treatment or hot-mix asphalt overlay based on traffic and existing conditions.

10.3.2.8 Wearing Course Selection

- Selection Criteria: The choice of wearing course is influenced by the structural design assumptions and is critical for sealing the surface and potentially reinforcing the pavement. Local experience often guides the selection, with chip seals, slurry seals, bituminous concrete overlays, and open-graded emulsion mixes commonly used. In regions like France, specific technologies are recommended based on daily traffic volumes.

10.3.2.9 Base Stabilization

- Objective: To enhance the bearing capacity of the roadbed, which is achieved by incorporating emulsified asphalt and possibly other fillers. This increases the resistance of the road to weather effects and helps minimize movement or rutting in the base layer.
- Full Depth Reclamation (FDR): This method involves pulverizing the asphalt-wearing layer and underlying base material to depths of up to 300 mm, with the addition of emulsified asphalts, fillers, and aggregates to maximize the bearing capacity. FDR is particularly effective for roads that have undergone multiple maintenance actions and are now experiencing increased traffic loads.

10.3.2.10 Economic and Environmental Benefits

- Cost-Effectiveness: CIR offers a more economical alternative to traditional pavement rehabilitation methods, providing significant savings and reduced environmental impact by reusing all existing materials and operating at lower temperatures.

10.3.2.11 Challenges and Limitations

- Expertise Requirement: Successful implementation of CIR demands extensive knowledge and precision in execution.
- Compaction Energy: Requires more compaction energy than standard hot bituminous mixtures due to the unique properties of the recycled mixture.
- Profile Evaluation: A detailed evaluation is necessary to achieve the desired pavement smoothness and proper cross-slope.
- Curing Time: Requires a sufficient curing period to allow the mixture to develop internal cohesion before the wearing course is applied.
- Wearing Course Necessity: The selection and application of a suitable wearing course are essential to protect the recycled layer and extend the pavement's lifespan.

This comprehensive approach ensures that CIR projects are optimally designed and implemented to achieve long-lasting, sustainable, and cost-effective pavement solutions.

10.3.3 Full-Depth Reclamation (FDR)

Overview of Full-Depth Reclamation: FDR is an innovative pavement rehabilitation method that involves the comprehensive recycling of the existing pavement structure and some of the underlying materials. This technique crushes, pulverizes, or blends the materials to create a stabilized base course. Additives such as asphalt emulsion, foamed asphalt, Portland cement, or lime can further enhance the base's strength.

Evolution and Significance: Traditionally, deteriorating roadways were simply overlaid with new material layers. As these roadways further degraded, maintenance became inefficient and costly. FDR offers a sustainable alternative by reusing onsite materials and reducing waste from conventional reconstruction. Advanced equipment like high-horsepower road reclaimers and sophisticated additive systems have made FDR a preferred method globally due to its effectiveness, cost savings, and shorter construction periods.

Process and Technique: FDR involves penetrating through the asphalt into the base layers with a rotor or cutting head, eliminating deep crack patterns and preventing future reflective cracking. Reclamation typically reaches depths from 6 inches to over 12 inches, producing a homogeneous and structurally improved base material.

Additives and Material Enhancement: FDR facilitates the integration of various stabilizing agents.

- Bituminous Agents: Enhance flexibility and fatigue resistance
- Portland Cement or Fly Ash: Increases compressive strength
- Liquid Calcium Chloride: Used in freeze/thaw conditions for better moisture control
- Lime: Reduces plasticity and enhances load-bearing capabilities
- Granular Material: Improves gradation or reduces the content of the asphaltic binder

Advantages and Applications: FDR uniquely eliminates existing crack patterns, significantly reducing future maintenance expenses. It is especially beneficial for roads that have transitioned from dirt or gravel to paved surfaces and now face increased traffic and load demands. FDR is applicable to a variety of infrastructure, including city streets, medium-volume roads, interstates, and airports, forming a robust base for new asphalt overlays.

Equipment and Methodology: The FDR process uses an array of equipment, including reclaimers, motor graders, compactors, and water trucks, and encompasses four main approaches:

- Pulverization: Simple and cost-effective, ideal for lower-structure demands like parking lots
- Mechanical Stabilization: Adds imported granular materials for enhanced structural strength
- Bituminous Stabilization: Uses bituminous additives for improved strength and moisture resistance
- Chemical Stabilization: Employs chemical additives, either dry or wet, for enhanced bonding of material particles

The equipment operates even on complex sites such as interstate highways and regional airports. Instead of laying new pavement over distressed asphalt, a new overlay is applied on top of a stabilized, uniformly compacted base. In some scenarios, part of the asphalt layer is removed, and the remaining portion is pulverized and mixed with the base along with stabilizing additives, which is particularly useful for adjusting to existing curb and gutter configurations or maintaining specific grades. With the right design and execution, FDR not only resolves existing pavement issues but also contributes to the structural integrity of the roadway, allowing for potentially thinner asphalt overlays. Figures 10.5 and 10.6 show the stabilization methods used in FDR.

10.3.3.1 Full-Depth Reclamation Disciplines and Techniques

Each FDR discipline utilizes essential equipment, which may require additional machinery when using stabilizers. The construction sequence also varies based on the type of stabilizer used.

Figure 10.5 FDR Lime Stabilization

Figure 10.6 FDR Emulsion Stabilization

10.3.3.2 Process Overview

- Pulverization of In-situ Pavement: The initial step involves pulverizing the existing pavement layers and blending the underlying material to achieve the specified sizing. This is controlled by the reclaimer operator adjusting the machine's speed, rotor speed, gradation control beam, and positions of the mixing chamber doors.
- Moisture Addition: Proper moisture is added, usually through the machine's integrated fluid injection system during pulverization, ensuring even water distribution.
- Breakdown Compaction: Compaction immediately follows pulverization to ensure consistent material density. Equipment used may vary from vibratory pad-foot rollers to heavy pneumatic rollers, depending on the depth and characteristics of the layer.
- Shaping and Grading: A motor grader shapes the material to the correct grade and cross-slope. Additional water may be added to the surface if drying occurs.
- Intermediate and Final Rolling: Following shaping, intermediate rolling is performed with a pneumatic roller or a heavy smooth drum vibratory compactor, and final rolling is carried out using a static steel drum roller.
- Sealing: A fog seal or specified sealer is applied to bond any loose particles and protect the layer until the new surface course is installed.

Pulverization is ideal for projects like parking lots, where structural demands are lower, and also for preparing roadways for new pavement layers.

10.3.3.3 Mechanical Stabilization

This discipline involves incorporating imported granular materials to enhance structural strength.

10.3.3.4 Process Details

- Material Incorporation: During pulverization or a subsequent blending pass, imported granular materials such as crushed aggregates, RAP, or crushed concrete are added.
- Equipment Needs: Additional equipment may include dump trucks and stone spreaders to transport and evenly distribute the granular material.
- Application: Mechanical stabilizers can be spread prior to or during the pulverization pass, either manually with a motor grader or mechanically for consistency. Mechanical stabilization is a cost-effective method to increase the structural integrity of the pavement, ideal for locations needing additional strength without extensive stabilization.

10.3.3.5 Bituminous Stabilization

This discipline adds bituminous stabilizers to enhance the reclaimed material's strength and water resistance. Figures 10.7 and 10.8 show foamed asphalt technology and its use in stabilization of pulverized pavement.

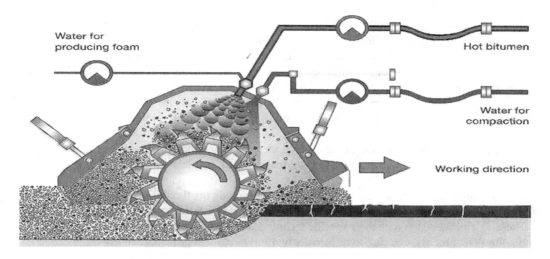

Figure 10.7 Foamed Asphalt Stabilization

Figure 10.8 CIR with Paver-Laid Foamed Asphalt Stabilization

10.3.3.6 Implementation

- Additive Mixing: Bituminous stabilizers are integrated during the pulverization or in a subsequent pass, using the machine's liquid additive injection system.
- Considerations for Multiple Passes: If significant grade or surface adjustments are needed, a shallower initial pulverization followed by intermediate shaping and final additive blending are recommended.
- Asphalt Emulsion: The emulsion process disperses tiny asphalt droplets in water, stabilized by emulsifiers. Proper selection and mixing of the asphalt emulsion are crucial for its effectiveness.
- Curing: Post-mixing, the asphalt emulsion undergoes a breaking process where water evaporates, allowing the asphalt droplets to coalesce and bind the material.

Bituminous stabilization is particularly effective for enhancing the flexibility and fatigue resistance of the base, suitable for use in combination with other granular or chemical additives.

These disciplines, each with their specific techniques and equipment, provide versatile solutions within the FDR process, catering to different structural needs and project goals.

10.3.3.7 Chemical Stabilization in Full-depth Reclamation

10.3.3.7.1 OVERVIEW OF CHEMICAL STABILIZATION

This FDR discipline involves the addition of chemical additives, either in dry or wet form, to stabilize the base material. Common chemical stabilizers include Portland cement, lime, fly ash, and various blends, which act as cementitious or pozzolanic agents. These additives bind the material particles and aggregates within the reclaimed layer, enhancing strength primarily through the chemical cementing process.

10.3.3.7.2 STRENGTH AND FLEXIBILITY CONSIDERATIONS

The strength gain from chemical stabilization depends on the type of reclaimed material and the quantity and type of stabilizers used, as assessed through laboratory testing. However, using too high a percentage of these additives may overly strengthen the material, potentially reducing its flexibility and decreasing its ability to manage repeated loading.

10.3.3.7.3 APPLICATION METHODS

Chemical stabilizers can be applied in several ways:

- Dry Application: Spread in powder form ahead of the reclaiming machine using calibrated spreading units
- Wet Application: Dispersed in slurry form either on the ground before the reclaimer or through a spray bar integrated into the machine's mixing chamber

10.3.3.7.4 ADDITIONAL BENEFITS AND USES

Chemical stabilizers like calcium chloride and magnesium chloride, applied either dry or as a liquid, not only improve strength through cementing but also lower the freezing point of the reclaimed layer. This helps mitigate the impact of freeze/thaw cycles.

10.3.3.7.5 INTEGRATION WITH SUB-GRADE STABILIZATION

While chemical stabilizers are primarily used for reclaiming the pavement layer, they can also address sub-grade deficiencies identified during pre-project evaluations. It is important to differentiate between stabilizing more than 50% of sub-grade material—which is considered sub-grade stabilization rather than FDR—and typical FDR practices.

10.3.3.7.6 PRE-PROJECT EVALUATION AND PAVEMENT CONDITION

A thorough evaluation of the existing pavement condition is critical. This includes:

- Pavement Condition Survey: To determine the types and severities of pavement distress.
- Pavement and Sub-Grade Sampling: Involves collecting core samples or conducting test holes/pits to study the pavement's thickness, material types, and underlying conditions. The California Bearing Ratio (CBR) of granular bases and sub-grade soils is typically measured using a dynamic cone penetrometer (DCP).

10.3.3.7.7 CONSIDERATIONS FOR PROJECT SELECTION

- Pavement Condition: Identify and assess the types and severities of pavement distress through a detailed survey.
- Sub-grade Evaluation: Understand the strength and classification of the supporting sub-grade, as it serves as the foundation for the reclaimed base and surface layers.
- Project Flexibility: FDR is versatile enough to accommodate various pavement structures, from rural roads to busy interstates and airports, making it a viable option across different settings.

10.3.3.7.8 ADVANTAGES OF CHEMICAL STABILIZATION IN FDR

Chemical stabilization within the FDR process is highly effective for enhancing pavement durability and structural integrity. It addresses a wide range of distress signs, such as deep and reflective cracking, heavy pothole patching, severe rutting or shoving, and areas with insufficient base strength. Given its cost-effectiveness and environmental benefits, chemical stabilization in FDR stands out as a sustainable choice for extending the life of road infrastructure while minimizing the use of new materials and reducing overall project costs.

10.3.4 Hot In-Place Recycling (HIR)

Overview of Hot In-Place Recycling: HIR is an on-site, in-place method that revitalizes deteriorated bituminous pavements, significantly reducing the need for new materials. This process can be executed in either a single-pass operation, where the restored pavement is immediately recombined with virgin material, or a two-pass procedure, which involves recompaction of the restored material followed by a delay before applying a new wearing surface.

Benefits of Hot In-Place Recycling: HIR offers a cost-effective maintenance strategy, allowing public works officials to efficiently reuse existing pavement materials. This process underscores the sustainability of asphalt as a construction material that can be economically restored, preserving the original pavement structure rather than covering it with excessive new material or completely removing it.

10.3.4.1 Process Efficiency

- Single-Pass HIR: Involves heating, remixing, and reinstating the existing pavement material along with added new materials in a continuous operation.
- Two-Pass HIR: Starts with heating and reshaping the existing pavement, followed by an interim period for curing before the final new surface layer is applied.

10.3.4.2 Key Improvements Achieved Through HIR

- Crack Repair: Cracks in the pavement are interrupted and filled.
- Aggregate Rejuvenation: Aggregates that have been stripped of bitumen are remixed and recoated.
- Surface Corrections: Ruts and holes are filled, and irregularities such as shoves and bumps are leveled.
- Drainage Re-Establishment: Proper drainage pathways and crowns are reinstated to enhance water management.
- Flexibility Restoration: The aged and brittle pavement is chemically rejuvenated to restore its flexibility.
- Material Modifications: Some variations of HIR allow for adjustments in aggregate gradation and asphalt content.
- Safety Enhancements: The process improves highway safety by enhancing skid resistance.

Table 10.2 Selection of In-Place Recycling Technologies

Pavement Distress	CIR	FDR	HIR
Surface Defects			
Raveling			X
Flushing			X
Slipperiness			X
Deformation			
Corrugations			X
Ruts-shallow			X
Rutting deep	X	X	X
Cracking (Load-Associated)			
Alligator	X	X	
Longitudinal			X
Wheel path		X	
Pavement edge	X	X	
Slippage			X
Cracking (Non-Associated)			
Block (shrinkage)	X	X	
Longitudinal (joint)			X
Transverse (thermal)	X	X	
Reflection	X	X	
Weak Base or Subgrade			
Ride Quality/Roughness			
General unevenness	X		X
Depressions (settlement)	X	X	X
High spots (heaving)	X		X

Cost Efficiency and Environmental Impact: In an era of escalating costs and restricted budgets, HIR provides a significant opportunity to extend limited funds further. It allows for the preservation and enhancement of existing pavements, postpones the need for costly reconstruction, and offers a sustainable method to manage and maintain road infrastructure. By utilizing the existing pavement material and reducing the volume of new materials required, HIR not only cuts costs but also minimizes environmental impact, making it a favored choice among modern road maintenance strategies.

10.4 Questions

1. What are the three key components of a comprehensive pavement preservation program?
2. How does pavement preservation contribute to improved safety and mobility on highways?
3. What is the primary objective of pavement preservation treatments, and how do they differ from methods that increase capacity or strength?
4. Why is asphalt recycling prioritized in pavement rehabilitation, and what are its environmental benefits?
5. Describe the two sub-categories of pavement rehabilitation and give examples of each.
6. What is the difference between minor and major rehabilitation in the context of pavement maintenance?
7. How is routine maintenance different from corrective maintenance in pavement management?
8. What situations typically require catastrophic maintenance, and how does it differ from routine and corrective maintenance?
9. Explain the process and significance of pavement reconstruction. When is it deemed necessary?
10. What are the primary differences between crack sealing and crack filling in pavement maintenance?
11. How does crack sealing contribute to the overall longevity and performance of a pavement?
12. Describe the process and purpose of chip sealing in pavement maintenance.
13. What are the various asphalt recycling methods used in roadway rehabilitation, and how do they contribute to addressing road defects?
14. Explain the process and purpose of cold planing (cold milling) in pavement rehabilitation.
15. How does cold in-place recycling (CIR) differ from other recycling methods, and what are its key steps?
16. Discuss the advantages and challenges of using CIR for pavement rehabilitation.
17. How is the effectiveness of CIR measured, and what factors are considered in selecting suitable pavement for this method?
18. Explain the role of base stabilization in asphalt recycling and its impact on roadbed materials.
19. What considerations are made in the mixture design for cold in-place recycling, and how does it affect the final product?
20. What is full-depth reclamation (FDR), and how does it differ from traditional pavement rehabilitation techniques?
21. Explain the process and technical aspects involved in FDR.
22. What types of additives are commonly used in FDR, and what purposes do they serve?
23. What are the key considerations and steps involved in the pre-project evaluation for FDR?
24. How do chemical and bituminous stabilization techniques enhance the effectiveness of FDR?

References

ARRA (2004). *Basic Asphalt Recycling Manual*. Asphalt Recycling and Reclaiming Association (ARRA). www.arra.org.

FHWA (2019). *Pavement Preservation Checklist Series 1–13*. US Department of Transportation, Federal Highway Administration (FHWA), Washington, DC. https://www.fhwa.dot.gov/pavement/preservation/ppcl00.cfm

Rajagopal, A. (2010). *Effectiveness of Chip Sealing and Micro Surfacing on Pavement Serviceability and Life*. FHWA/OH-2010/8. Ohio Department of Transportation, Columbus, OH.

Chapter 11

Concrete Paving, Preservation, and Recycling

11.1 Concrete Paving

Concrete, with its durability, resilience, and versatility, has long been the material of choice for paving. This section discusses several key fundamentals, including materials used for concrete pavements, construction, and quality and testing procedures.

11.1.1 Materials Used for Concrete Pavements

Concrete is a mixture of mainly cement, water, air, fine and coarse aggregate (Figure 11.1). Admixtures that contribute to concrete strength and impermeability are sometimes added. Curing compounds are applied to the surface to facilitate the curing process. Other components, such as dowel bars and reinforcement, may be added to the mixture during construction.

When mixed with water, Portland cement, as a binder, can set, harden, and bind fine and coarse aggregate to make concrete. The raw materials used in manufacturing Portland cement include lime, iron, silica, alumina, and magnesia. High energy is required, and emissions are released during cement production. Ferrous slags and nonferrous slags contain some mineral sources needed in cement. Tremendous potential exists for some slags to be used as mineral sources in cement manufacture, either for blended cement making or as a raw material in cement clinker and other cementitious applications. It is usually used for reinforced concrete bridges, pavements, and buildings. It is also used for most concrete masonry units and all uses where the concrete is not subject to a particular sulfate hazard or where the heat generated by the hydration of cement is not objectionable. It has excellent resistance to cracking and shrinkage but less resistance to chemical attacks.

Aggregate, often sourced from granite or limestone, constitutes most of the concrete mixture. It acts as the filler that's bound together by the cement paste. Not only does aggregate significantly impact the fresh properties of concrete, such as workability, but it also plays a vital role in its long-term durability. Ensuring the aggregate is well-graded is essential. This means having a broad spectrum of particle sizes. A well-graded aggregate results in fewer gaps between particles, requiring less chemically reactive cement paste to fill these spaces. Moreover, it aids in producing a workable mix using the least amount of water.

Water intended for concrete mixing should be of potable quality. While certain recycled waters may be suitable, any water of uncertain quality should be tested to ensure it does not adversely affect the concrete's strength or setting time.

Chemical admixtures are substances introduced to concrete mixtures to alter specific properties, including air content, water requirements, and setting time. While they can enhance specific

DOI: 10.1201/9781003197768-11

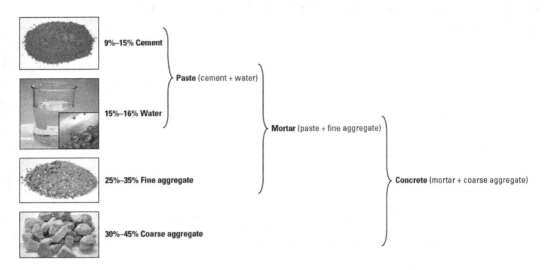

Figure 11.1 Concrete Pavement – Materials

attributes, admixtures should not replace sound concrete proportioning and practices. Given that admixtures can sometimes produce unintended consequences, conducting trial batches using the actual job materials and under natural job conditions is advisable.

After texturing, it's essential to apply curing compounds thoroughly to concrete surfaces to minimize moisture loss. The most frequently used curing compounds are those that form a liquid membrane.

In addition to the primary materials, several components are utilized in concrete mixtures to enhance performance. Dowel bars, inserted across transverse joints, offer vertical support and facilitate load transfer across these joints. On the other hand, tie bars are positioned across longitudinal joints, whether at centerlines or slab meeting points, to prevent slab separation and transfer loads. Slab reinforcement can be incorporated to enhance concrete pavements further. This boosts the concrete's capacity to bear tensile stresses and effectively binds any random transverse cracks that might form within the slab (Taylor, Van Dam, Sutter, & Fick, 2019).

11.1.2 Construction of Concrete Pavements

A typical construction sequence of concrete pavements is shown in Table 11.1.

11.1.2.1 Site Preparation

Most concrete pavement failures stem not from the concrete slab itself but from issues with the underlying materials.

Preparing the Grade: Before laying the concrete pavement, it's essential to adequately prepare and compact the subgrade. The subgrade is the foundational layer of a roadbed, made up of either natural ground, imported materials, or a mix of both. After grading, it is compacted to form the

Table 11.1 Typical Construction Schedule of Concrete Pavements

Concrete Pavement Construction

Start
Site Preparation
Preparing the grade
Establishing the stringline
Placing dowel baskets
Paving Operations
Wetting the grade
Delivering the mix
Placing the concrete
Spreading and consolidating the concrete
Setting header joints
Placing tiebars during construction
Finishing
Texturing
Curing
Insulating
Jointing
Finishing

base for the pavement. A consistent and stable subgrade ensures the pavement's long-term durability. Notably, its uniformity and stability often matter more for pavement performance than its strength. Three primary issues can undermine subgrade stability: expansive soils, frost action, and pumping. These issues can result in localized stress on the pavement and need to be mitigated. To address these challenges, pavement subgrades might require temporary improvement through soil modification or permanent enhancement via soil stabilization, often using additives or binders.

A base layer is typically constructed directly above the subgrade. This base serves as the immediate layer beneath the concrete pavement. Various materials can be used for bases, including granular substances, cement-treated materials, lean concrete, hot-mixed asphalt, or open-graded, highly permeable materials, either stabilized or unstabilized. Similar to subgrades, the primary qualities of a base are its uniformity and firm support. Balance is crucial to ensuring proper drainage and maintaining the base layer's stability.

Establishing the Stringline: A stringline, held up by stakes beside the paving lane, directs the paving equipment horizontally and vertically (Figure 11.2). The paver has an elevation-sensing wand that glides beneath the string and an alignment-sensing wand that rests against the string's inside. Both wands should operate without causing any noticeable deflection to the stringline. Care should be taken during construction to avoid disturbing the stringline, and trucks or equipment should not operate too closely.

Placing Dowel Bars: Dowel bars offer support between neighboring pavement panels (Figure 11.3). Ideally positioned at the slab's mid-depth, they must be meticulously aligned both horizontally and vertically. Securing dowel assemblies in their designated spots is essential to avoid displacement during concrete placement. With specialized tools, dowels can also be inserted in post-concrete applications.

Figure 11.2 Concrete Pavement – Stringline

Figure 11.3 Concrete Pavement – Dowel Bars

11.1.2.2 Paving Operations

Wetting the Grade: A dry subbase can extract water from the bottom of the concrete mixture, potentially leading to stress from varying moisture levels in the slab, which can cause cracking. To mitigate these stresses, it's advisable to moisten the subbase thoroughly just before laying the concrete. The objective of transporting the concrete mix from the plant to the construction site is to deliver a uniform and workable mixture consistently. Ensuring batch-to-batch consistency is crucial.

Delivering the Mix: Surpassing recommended delivery times can compromise the workability of the concrete. Here are the guidelines for various transport methods:

- Dump Trucks: Deliver and place within 30 minutes post-mixing
- Agitor Trucks: 30 minutes if the paddle remains stationary; up to 90 minutes if it's rotating
- Ready-Mix Trucks: Up to 90 minutes

Placing the Concrete: Contractors use either slip-form (Figure 11.4) or fixed-form (Figure 11.5) methods, depending on the specific requirements of the placement. The slip-form method is ideal for smaller projects such as city streets and parking lots. Steel forms used in this method typically have three holes per 10-foot section for pins. The height of the form matches the concrete's thickness. These forms are typically left in place for 8–12 hours, after which they can be removed, cleaned, oiled, and reused. On the other hand, the fixed-form method is used for larger projects, such as highway pavements and airport runways. This method employs low-slump Portland

Figure 11.4 Concrete Pavement – Slipform

Figure 11.5 Concrete Pavement – Fixed Form

cement concrete (PCC), which contains less water, resulting in higher compression and flexural strengths. The fixed-form approach offers high productivity and ensures a smooth riding surface.

Spreading and Consolidating the Concrete: Vibratory screeds and form-riding pavers can efficiently place and consolidate concrete between forms in a single pass. For optimal productivity and a smooth pavement finish, it's crucial to operate pavers at a consistent speed. Moreover, scheduling the right number of concrete delivery trucks is vital.

Setting Header Joints: These are constructed at the end of a pavement section where future pavement construction will continue. They are essential at the conclusion of a pour, a day's work, or when paving is postponed for 30 minutes or longer.

Placing Tiebars during Construction: When paving two or more lanes, tiebars must be set across the centerline or lane lines to inhibit movement along the line. If the paver isn't equipped for automatic tiebar placement, a crew member aboard the paver can manually insert them into the concrete.

Finishing: Once the paver has passed, the surface is typically smoothed to eliminate holes and ensure a tight surface.

Texturing: Following the finishing process, two distinct operations are employed. First, a coarse carpet is dragged to create micro-texture, ensuring adequate surface friction in dry conditions. Secondly, a grinder is utilized to produce macro-texture, which aids in preventing hydroplaning during wet weather.

Curing: Curing involves maintaining optimal moisture and temperature conditions in freshly poured concrete, facilitating hydration and pozzolanic reactions. Typically, immediately after concrete placement, a curing compound is uniformly applied to its entire surface and exposed edges (Figure 11.6), minimizing water evaporation.

Figure 11.6 Concrete Pavement – Curing

Insulating: Should air temperatures plummet on the first night following placement, it's essential to insulate or cover the concrete surface.

Jointing: Joints in a concrete pavement are essential for controlling cracking. Joints are designed to manage stresses in the concrete. A meticulously crafted joint pattern not only prevents random cracking but also ensures that cracks occur at predetermined locations. This, in turn prevents faulting, promoting a longer pavement lifespan. While these joints can be introduced through saw cutting (Figure 11.7) after the concrete is placed, timing is crucial. Cutting too early can lead to raveling, while cutting too late risks random cracking. The optimal cutting moment hinges on the concrete's properties and environmental conditions.

11.2 Concrete Preservation

Our roadway networks, increasingly burdened by age and surging traffic, are often maintained with inadequate funding. State and local highway agencies relentlessly work to keep road conditions meeting expectations, ensuring they can handle today's and tomorrow's demands (Smith, Grogg, Ram, Smith, & Harrington, 2022).

a) Single-blade, walk-behind saw

b) Span-saw

c) Early-entry, walk-behind dry saw

Figure 11.7 Concrete Pavement – Jointing

Without proactive preservation, pavements can quickly degrade; refer to Figures 11.8 and 11.9. This can lead to more expensive repairs, frequent road closures, and higher overall costs. Additionally, road users face inconvenience, and the safety of roadway workers is jeopardized.

To address these challenges, agencies employ proactive preservation strategies that include

- Addressing specific issues such as spalling, cracking, and faulting
- Optimizing slab support and enhancing load transfer capacities
- Ensuring smoother rides
- Preventing water intrusion
- Guarding against foreign materials in cracks or joints
- Prioritizing safety and noise reduction
- Systematically enhancing the overall pavement conditions

However, the efficacy of preservation strategies is contingent upon:

- Timely application on the appropriate pavement
- Adapting to prevailing conditions and constraints in the design
- Adhering to proven construction methodologies

Pavement preservation offers clear benefits: high user satisfaction due to smoother roads, increased safety, fewer traffic disruptions, and a more significant segment of the pavement network in satisfactory condition. This approach also improves pavement durability, cost-efficiency, and enhances sustainability (Smith, Harrington, Pierce, Ram, & Smith, 2014).

As detailed in Tables 11.2 and 11.3, a variety of treatments are available for preserving concrete pavements. These preservation techniques employ varying materials and, in some instances, require no materials at all. They can be applied comprehensively across the entire pavement or targeted specifically where distress or other issues are present. Some methods might necessitate the removal of a portion of the existing pavement, while others involve adding new material.

Table 11.4 presents the distinct capabilities and roles of different pavement preservation treatments, emphasizing their influence on the structural and functional aspects of the existing pavement. These treatments can help prevent or postpone the occurrence of new distresses, decelerate the progression of existing distresses, restore pavement integrity and serviceability, and enhance surface properties for greater user safety and comfort.

Table 11.5 outlines the expected performance lifespans for selected pavement preservation treatments. Realizing these performance durations heavily relies on choosing the right pavement, utilizing durable materials, and adhering to quality construction practices.

11.2.1 Slab Stabilization

Lack of support beneath concrete pavement slabs significantly contributes to pavement deterioration. Slab stabilization refers to injecting material beneath the slab or subbase to fill voids, which helps reduce deflections and related distresses. Since there are numerous causes of support loss, slab stabilization often pairs with other rehabilitation methods such as patching, diamond grinding, and dowel bar retrofit (DBR).

Materials frequently used for slab stabilization include cement-based mixtures – such as limestone dust-cement grouts and pozzolanic cement grouts – and polyurethane. It is crucial to

Figure 11.8 Common Concrete Pavement Distresses

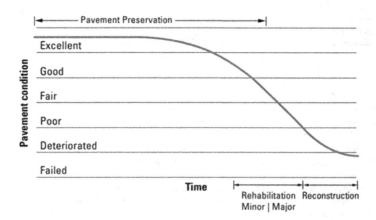

Figure 11.9 General Applicability of Pavement Preservation, Rehabilitation, and Reconstruction Activities

Table 11.2 Primary Concrete Pavement Preservation Treatments

Treatment Type	Treatment Description
Crack sealing	Sawing, power cleaning, and sealing cracks (typically transverse, longitudinal, and corner-break cracks wider than 0.125 in) in concrete pavement using high-quality sealant materials
Diamond grinding	Removal of a thin layer of concrete (typically 0.12 to 0.25 in) from the pavement surface using special equipment fitted with a series of closely spaced diamond saw blades
Diamond grooving	Cutting of narrow, discrete grooves into the pavement surface, either in the longitudinal direction (i.e., in the direction of traffic) or the transverse direction (i.e., perpendicular to the direction of traffic)
Dowel bar retrofit	Placement of dowel bars across joints or cracks in an existing concrete pavement to restore load transfer
Full-depth repair	Cast-in-place or precast concrete repairs that extend through the full thickness of the existing slab, requiring full-depth removal and replacement of full or partial lane-width areas
Joint resealing	Removal of existing deteriorated transverse and/or longitudinal joint sealant, refacing and pressure cleaning of the joint sidewalls, and installation of new material (e.g., liquid sealant and backer rods, preformed compression seal)
Partial-depth repair	Removal of small, shallow (typically up to half of the slab thickness) areas of deteriorated concrete and subsequent replacement with a cementitious or polymeric repair material

understand that slab stabilization doesn't aim to lift the slab. Therefore, monitoring the slab lift is vital during the material injection process to prevent over-grouting, which could damage the slab. For a successful slab stabilization project, using an experienced contractor and ensuring a thorough inspection is imperative.

11.2.2 Slab Jacking

Slab jacking involves the pressurized injection of materials, such as polyurethane or a cement grout mixture, beneath a slab or sequence of slabs. The goal is to restore the pavement to its smooth

Table 11.3 Additional Concrete Pavement Preservation Treatments

Treatment Type	Treatment Description
Concrete overlay	Placement of a thin concrete layer (typically 4–6 in thick) on a milled or prepared surface
Cross-stitching	Placement of deformed tie bars into holes drilled at an angle through cracks (or, in some cases, joints) in an existing concrete pavement
Slab stabilization	Filling of voids beneath concrete slabs by injecting polyurethane, cement grout, asphalt cement, or other suitable materials through drilled holes in the concrete located over the void areas
Slab jacking	Raising of settled concrete slabs to their original elevation by pressure-injecting cement grout or polyurethane materials through drilled holes at carefully patterned locations
Slot-stitching	Grouting of a deformed bar into slots cut across a longitudinal joint or crack
Retrofitted edgedrains (and maintenance)	Cutting of a trench along the pavement edge and placement of a longitudinal edgedrain system (pipe or geocomposite drain, geotextile lining, bedding, and backfill material) in the trench, along with transverse outlets and headwalls

profile gradually. This procedure demands a series of injection holes over the settled slabs, with the slab jacking material being applied alternately to avoid concentrating stress points. Slab jacking is best suited for pavements displaying localized settlement areas while largely remaining crack-free. Settlement is often seen in fill areas, over culverts, and at bridge approaches.

It's worth noting that slab jacking isn't recommended for addressing faulted joints in a project. Methods like DBR and diamond grinding are more efficient for such repairs.

11.2.3 Partial-Depth Repairs

Partial-depth repairs (PDR) involve removing and replacing small, shallow, deteriorated concrete areas less than half the slab's thickness. These repairs enhance the pavement's integrity and ride quality, thus extending its lifespan. Compared to full-depth repairs (FDRs), PDRs are preferable when deterioration is confined mainly to the slab's upper half and existing load transfer devices remain functional. The success of PDRs hinges on the quality of the installation, emphasizing strict compliance with stipulated removal, preparation, placement, and curing procedures. Although the historical performance of PDRs varies, with the right choice of distresses, durable materials, and rigorous construction methods, they can last between 10 to 20 years or even longer. Furthermore, given their smaller scope, PDRs can be more cost-efficient than FDRs.

11.2.4 Full-Depth Repairs

Full-depth repairs (FDR) are a primary preservation method that aims to restore concrete pavements' rideability and structural integrity. These repairs address intermittent distress that can develop over the pavement's lifespan, such as transverse cracking, corner breaks, deteriorated joints, blowups, and punchouts. FDRs are an effective pre-overlay treatment, preparing distressed concrete pavements for structural overlays, whether asphalt or concrete. However, if a concrete pavement exhibits excessive cracking and other structural issues, it might not be suitable for FDRs.

The anticipated performance of FDRs can vary based on the condition of the existing pavement and the specific type of repair application. With proper implementation, FDR projects can have

Table 11.4 Primary Functions of Concrete Pavement Preservation Treatments

Treatment	Seal Pavement /Minimize Pumping	Fill Voids, Restore Support, Address Pavement Deterioration	Remove Moisture Beneath Structure	Prevent Intrusion of Incompressible Materials	Remove/ Reduce Faulting	Improve Texture for Friction	Improve Profile (Lateral Surface Drainage and Ride)	Improve Texture for Noise
Slab stabilization	✓	✓						
Slab jacking		✓			✓		✓	
Partial-depth repair	✓	✓						
Full-depth repair	✓	✓		✓	✓		✓	
Retrofitted edgedrains (and maintenance)			✓					
Dowel bar					✓		✓	
Cross-stitching					✓		✓	
Slot-stitching					✓		✓	
Diamond grinding					✓	✓	✓	✓
Diamond grooving						✓		
Joint resealing	✓							
Crack sealing	✓							
Concrete overlay						✓	✓	✓

Table 11.5 Typical Range of Expected Performance Life for Selected Concrete Pavement Preservation Treatments

Treatment	Typical Range of Expected Performance (Treatment Life)
Slab stabilization	5–10 years
Partial-depth concrete patching	10–20+ years
Full-depth concrete patching	20+ years
Dowel bar retrofit	15–20+ years
Cross-stitching	10–20+ years
Diamond grinding	15–25+ years
Joint resealing	8–16+ years

service lives extending beyond 20 years across diverse climates, traffic conditions, materials, and designs. The durability of FDRs is contingent upon several factors, notably appropriate project selection, effective load transfer design, and proper construction practices.

11.2.5 Retrofitted Edgedrain

Pavement engineers often encounter older concrete pavements exhibiting moisture-related damage. Such damage can arise from initial inadequate drainage, subsurface drainage system wear and tear, or insufficient drainage system maintenance. To rectify these drainage issues, retrofitting the existing pavement with edgedrain is a viable rehabilitation option.

Numerous edgedrain systems, each with its distinct characteristics, are available for drainage retrofit projects. For instance, prefabricated geocomposite edgedrains and aggregate drainage systems are more cost-effective than pipe drains. However, they can pose maintenance challenges, especially if clogging occurs, making them difficult, if not impossible, to clean. Historically, geocomposite edgedrains have had lower hydraulic capacity compared to pipe drains, but advances in materials are gradually leveling the field. Conversely, while pipe edgedrains typically offer superior hydraulic capacities compared to aggregate or other edgedrain systems, they come at a higher installation cost. Irrespective of the chosen drainage system, ensuring proper construction and installation is paramount for sustained efficacy.

11.2.6 Dowel Bar Retrofit, Cross-Stitching, and Slot-Stitching

Dowel bar retrofit (DBR) involves placing dowel bars across existing transverse joints or cracks. This process enhances the slab's capacity to transfer wheel loads between slabs efficiently, minimizes deflections, and prevents faulting. For DBR to be effective, the existing concrete must be in good condition. DBR would not be an appropriate treatment for such a project if there's any deterioration in the slab's lower section.

Cross-stitching is a preservation technique that reinforces nonworking longitudinal joints and cracks that are still relatively intact. This method involves the insertion of tie bars, which are bonded into angled holes drilled across the joint or crack, typically at 35 to 45 degrees relative to the pavement surface. Doing so effectively mitigates vertical and horizontal shifts or expansion of the crack or joint. Consequently, this method maintains a tight joint or crack, ensures efficient load transfer, and decelerates deterioration.

Slot-stitching is derived from the DBR method and is designed to manage longitudinal cracks and joints. This technique involves using deformed tie bars, typically 1 inch in diameter or larger, which are placed in slots cut into the existing concrete pavement. The aim is to maintain a level of interlock and to stop cracks from widening and deteriorating further. While slot-stitching can address wider cracks than cross-stitching, it should be used judiciously on continuously reinforced concrete pavement (CRCP). Due to the shallower depth of the slots in CRCP, the repair material might be more vulnerable to failure in the event of substantial vertical movements.

11.2.7 Diamond Grinding and Grooving

Diamond grinding and grooving are specialized surface restoration techniques to correct specific concrete pavement surface distresses or deficiencies. While they serve distinct purposes, they are often used with other pavement preservation methods such as DBR, PDR, and FDR. Diamond grinding removes a thin layer of the concrete pavement's surface – usually around 0.25 inches – using a self-propelled machine equipped with a series of adjacent diamond saw blades affixed to a rotating shaft. The primary goal is to restore or improve ride quality. On the other hand, diamond grooving is a technique that cuts grooves into the pavement using diamond-edged saw blades, generally spaced 0.75 inches from center to center. This technique forms channels that help drain surface water, effectively decreasing the potential for hydroplaning, a significant factor in wet-weather accidents. It's crucial to note that diamond grooving is best suited for structurally sound pavements that perform adequately.

11.2.8 Joint Resealing and Crack Sealing

Joint resealing and crack sealing are essential maintenance activities for concrete pavements, serving two primary objectives. First, they limit moisture infiltration and deicing chemicals into the pavement structure. Second, they prevent the intrusion of incompressible materials, such as sand and pebbles, into the joints. An added benefit of sealing joints and cracks is the reduction of noise emissions. These noises, often termed "tire slap" or "joint slap," result from tire tread and carcass vibrations when they impact pavement joints. Consequently, joint resealing and crack sealing are routine practices among many state and local roadway agencies to preserve their concrete pavement networks.

Joint resealing is necessary when the existing sealant no longer fulfills its designated functions. On the other hand, crack sealing is a comprehensive process involving thorough crack preparation followed by the application of high-quality materials. This process aims to minimize moisture infiltration and slow down crack deterioration significantly. It is most effective for concrete pavements with minor structural wear. Crack sealing can be applied to random transverse and longitudinal cracks up to 0.5 inches or wider if there's minimal spalling or faulting. Typically, cracks narrower than 0.11 inches are not sealed.

11.2.9 Concrete Overlays

When placed on existing concrete or composite (asphalt on concrete) pavements, concrete overlays offer a sustainable solution to prolonging pavement life. Typically, they can extend a pavement's lifespan by 15 to 30 years, making them a valuable tool for agencies committed to pavement preservation.

There are two primary types of concrete overlays: bonded and unbonded. Bonded overlays work with the existing pavement to form a unified, monolithic structure. They not only eliminate surface distress but also add additional structural value to the pavement system. On the other hand, unbonded overlays add structural strength to the pavement system but are not designed to bond with the existing pavement.

Beyond enhancing durability, concrete overlays bring forth several additional benefits. They conserve the environment by revitalizing existing pavement structures, reducing waste, and decreasing the volume of materials dumped in landfills. In terms of construction, concrete overlays cause fewer user delays compared to complete reconstruction. Furthermore, these overlays improve the riding experience by enhancing the pavement's quality, reducing noise, and improving albedo and friction (NCDOT, 2015).

11.3 Concrete Recycling

Concrete recycling involves reusing broken-down concrete to produce new aggregate for fresh concrete. Reusing old concrete reduces the need for virgin aggregate, conserving natural resources, reducing emissions from transportation, and minimizing the amount of concrete dumped in landfills.

11.3.1 General Concrete Recycling

The typical concrete recycling process includes the following steps:

1. Collection of Waste Concrete: Contractors or municipalities collect waste concrete from demolition sites.
2. Crushing: Large chunks of concrete are broken down into smaller pieces using machinery such as portable crushers or stationary crushers.
3. Removing Contaminants: After crushing, processes are implemented to remove any non-concrete materials. Magnets might be used to remove metal, while air blowers can remove plastic and other lightweight contaminants.
4. Size Grading: Once contaminants are removed, the crushed concrete is sized to meet specifications for its intended use.
5. Stockpiling: The graded, crushed concrete is then stockpiled for transportation or direct use.
6. Reuse: Crushed concrete, often called recycled concrete aggregate (RCA), can be used in various applications. The most common use is as an aggregate replacement in new concrete, but it can also be used as a base material under roads or buildings, as fill material, or even as aggregate in roadbeds.

Major benefits of concrete recycling are:

- Environmental Conservation: Natural resources are conserved when old concrete is recycled. The process also reduces the burden on landfills, as fewer waste materials are abandoned.
- Energy and Emissions: The extraction and transportation of new aggregates are energy-intensive. By using recycled materials, there is a reduction in the energy required and associated emissions.

- Economic Benefits: Contractors can often save money using recycled concrete since the costs of extracting, processing, and transporting new aggregates are eliminated or reduced. Additionally, charges at landfills can be avoided.
- Durability and Quality: RCA can be of similar quality to natural aggregates when processed correctly. Some studies even suggest that specific properties, such as resistance to shrinking and cracking, might be improved using recycled aggregates.

It should be noted that there are limitations associated with concrete recycling, such as

- Potential Contaminants: The quality of recycled concrete depends on the original quality and the processing. Sometimes, contaminants or different types of concrete can mix, leading to lower-quality RCA.
- Transportation: Transportation costs can impact the economic feasibility of concrete recycling, especially if the recycling facility is located far from the project site.
- Limited Research: While significant evidence supports the quality and durability of RCA in certain applications, there is still limited long-term research on its use in critical structures.
- Specification Barriers: Some regions or organizations may have building codes or standards that restrict the use of RCA in specific applications.

As sustainability becomes a primary concern in modern construction, recycling and reusing materials, including concrete, is becoming increasingly vital. However, like any material, recycled concrete must be based on proper knowledge, quality assurance, and understanding of its properties.

11.3.2 Use RCA in Fresh Concrete

Recycled concrete aggregate is a granular material processed by removing reinforcing steel, crushing, and processing Portland cement concrete for reuse in construction as virgin natural aggregate. RCA's properties are different from natural aggregate, mainly because the resultant crushed material is composed of both natural aggregates in demolished concrete and reclaimed mortar, significantly affecting the properties and behavior of materials produced with RCA unless specific steps are taken in the design and construction process. The composition of RCA can be highly variable and may contain contaminants such as clay, joint sealant, asphalt, or other construction and demolition wastes. However, when its characteristics are adequately considered and accounted for, RCA can be used effectively in concrete.

The practical use of RCA as an aggregate in new concrete can be traced back to the 1940s internationally. Normative documents or standard specifications have been used in many countries and organizations, for example, in Germany: DIN 4226-100; UK: BS 8500-2; Brazil: NBR 15.116; Japan: BCSJ-97; Hong Kong: WBTC 12-2002; the International Union of Laboratories and Experts in Construction Materials, Systems and Structures (RILEM): RILEM-1994; and Cement and Concrete Australia. It has been reported that using RCA in concrete can maximize the economic benefits by matching or exceeding the technical requirements for concrete containing natural aggregate. It is widespread worldwide that partial replacement of 30%-50% of natural aggregate in concrete is used in sidewalks, curbs, and gutters; also for structural concrete with mix adjustments and inferior permeability and shrinkage properties.

In the United States, the quantity of RCA production is increasing as the nation's civil infra-structures are aging and being reconstructed. RCA has been used in various paving layers. New standard specifications and guidelines have been enacted in some states to allow the use of RCA in new concrete. The production of RCA typically includes the following steps: evaluation of the source concrete, preparation of the concrete structure (bridge, pavement) for demolition, concrete breaking and removal, removal of the embedded steel, crushing and sizing, evaluation and sorting (separating good from bad), and stockpiling.

11.3.2.1 Processing of Recycled Concrete Aggregate

Although the traditional method and the same basic equipment to process virgin aggregates can be used to crush, size, and stockpile RCA, the selection of the crushing process can affect the amount of mortar that clings to the RCA particles and, therefore, the properties of the RCA. Jaw crushers generally are more effective at producing higher quantities of RCA, but typically result in rela-tively high amounts of reclaimed mortar in RCA particles. Impact crushers can be lower in produc-tivity but more effective in removing mortar from RCA; therefore, the coarse RCA is more similar to virgin aggregate. Figure 11.10 shows the plant RCA processing procedures. The production of RCA can be completed on-site by one piece of integrated equipment. Figure 11.11 presents the Terex Finlay J-1170 compact and tracked jaw crusher for crushing, screening, and magnetic sepa-ration. Concrete debris can be crushed, screened into two adjustable sizes, and undergo magnetic separation simultaneously within a single machine. Transportation cost to the plant is incurred as part of production costs.

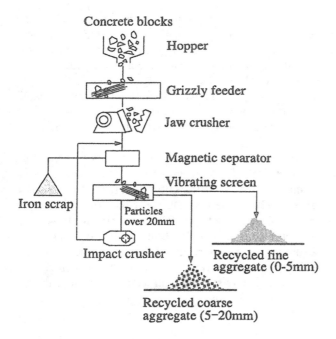

Figure 11.10 Flow Chart for Plant Processing of RCA

Figure 11.11 Terex Finlay J-1170 Compact and Tracked Jaw Crusher

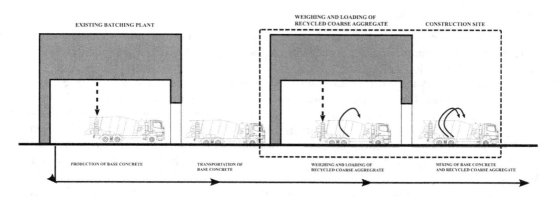

Figure 11.12 Onsite Production Procedure of RCA Concrete

The advantages of onsite processing include using the processed aggregate nearby, allowing the coarse RCA to be added to the ready-mix concrete trucks that have base concrete mixes (partial coarse aggregate), thereby saving transportation costs; and the integrated process uses one crusher, which increases productivity and recovery rate. Coarse aggregate can be recovered up to 77% of the processed volume. Adding RCA as coarse aggregate in concrete is illustrated in Figure 11.12.

11.3.2.2 Properties of Recycled Concrete Aggregate

RCA has several unique characteristics and properties that must be considered during the mix design and construction stages. These properties include lower specific gravity, which decreases with an increasing amount of reclaimed mortar; higher absorption, which increases with an increasing amount of reclaimed mortar; greater angularity; increased abrasion loss, which increases with an increasing amount of reclaimed mortar; presence of unhydrated cement, which may alter its behavior and complicate stockpiling, especially the fines (passing #4 sieve); and the fines produced during the crushing operation are angular, which tends to make RCA concrete mixtures very harsh and challenging to work with.

Not all RCA is appropriate for use in concrete. For example, RCA made from concrete exhibiting materials-related distress (MRD), such as alkali-silica reactivity (ASR) or D-cracking, may not be used in concrete unless certain mitigation methods are employed. Additionally, RCA may have high chloride content due to extended exposure to deicing chemicals, making it unsuitable for reinforced concrete use. The low abrasion resistance may lead to poor performance in applications where intimate aggregate interlock is relied upon for load transfer (for example, in undoweled joints or at transverse cracks of reinforced slabs).

The crushing process to generate RCA exposes unhydrated fines, which can lead to cementation when exposed to water or particularly humid conditions, thus changing the RCA's physical properties. Contrary to most virgin aggregates, RCAs typically fail the sulfate soundness test (ASTM-C88 2018) using sodium sulfate but tend to perform well using magnesium sulfate in a limited study.

RCAs may contain as much as 40% mortar, which affects their deformation properties, such as elasticity, creep, and shrinkage. More water may be needed to enhance workability.

11.3.2.3 Mix Design of Concrete Containing RCA

In principle, the mix design of recycled aggregate concrete is not different from conventional concrete, and the same mix design procedures can be used. In practice, slight modifications are required. These include:

When coarse RCA is used with natural sand, it may be assumed at the design stage that the free water-to-cement (W/C) ratio required for a particular compressive strength will be the same for RCA concrete as for conventional concrete. If trial mixes show that the compressive strength is lower than required, an adjustment of the W/C should be made, up to 10 liters/m^3 (or 5%) higher than for conventional concrete. In some cases, if the free water content of RCA concrete is increased, the cement content may also need to be higher to maintain the same W/C ratio.

The unit weights of concrete made using RCA are within 85% to 95% of the original concrete mixture. Air contents of RCA concrete are up to 0.6% or higher. The optimum ratio of fine-to-coarse aggregate is the same for RCA as for concrete made from virgin materials.

Trial mixtures are mandatory. When concrete is recycled more than once, the reduction in the residual mortar is believed to make the recycled RCA concrete perform better than the conventional RCA concrete.

The use of coarse RCA (up to 30%) is normally recommended. Still, the addition of a superplasticizer is often considered necessary for achieving the required workability of new concrete. More than 50% RCA may cause higher shrinkage of the concrete.

11.3.2.4 Applications of Concrete Containing Recycled Concrete Aggregate

RCA concrete performs well in nonstructural concrete applications such as curb and gutters, valley gutter, sidewalks, concrete barriers, driveways, temporary pavements, and interchange ramps with ADT less than 250, as well as on pavement types such as JRCP and CRCP. Fine RCA is more limited in new concrete mixtures. Some states specify that coarse RCA can only be used for lower-priority applications. Some specifications state that RCA may not be used in concrete for mainline pavements or ramps with an ADT equal to or over 250, concrete base courses, bridges, box or slab culverts, headwalls, retaining walls, prestressed concrete, or other heavily reinforced concrete.

11.3.2.5 Workability of Fresh Concrete

RCA concrete workability is strongly affected by the shape and texture of the coarse recycled aggregate surface; in the hardened state, the elasticity modulus and strength of RCA concrete are comparable to those of conventional concrete, or even better if the same W/C ratio is considered. Recycling concrete aggregates negatively influences shrinkage strain, regardless of the W/C used. On the contrary, creep results appear less sensitive to the W/C ratio.

It is reported that to maintain the functionality of RCA concrete, the water content in the concrete mix may need to be ~10% higher than what is needed to make natural concrete, while other researchers reported different results.

11.3.2.6 Strength Related Properties

Research has confirmed that using RCA to substitute virgin aggregates leads to concrete having lower strengths and higher permeability. Concrete with 100% coarse RCA has 20-25% less compressive strength than conventional concrete at 28 days at the same W/C ratio and cement quantity. Concrete made with 100% RCA requires a high amount of cement to achieve high compressive strength. However, concrete made with 25% RCA can achieve the same mechanical properties as conventional concrete, employing the same quantity of cement and an equal effective W/C ratio. Lower RCA addition can achieve similar strength to natural aggregate concrete. Medium compressive strength concrete made with 50% or 100% of RCA needs a 4-10% lower effective W/C ratio and 5-10% more cement than conventional concrete to achieve the same compressive strength at 28 days.

The modulus of elasticity of RCA concrete is lower than that of conventional concrete. However, the tensile strength of recycled aggregate concrete can be higher than that of traditional concrete. Results from different researchers exhibited similar conclusions. Research also shows that when fly ash is used, the free shrinkage of RCA concrete is reduced, and 48 MPa concrete can be made in a laboratory containing RCA with a low W/C ratio.

The use of fly ash in RCA concrete enhances its physical and mechanical properties, thus mitigating the worsening effect of the recycled aggregates. The results show that specific gravity and compressive strength decrease, while the water absorption rate increases. To increase the strength of RCA concrete, researchers used steel slag and NaOH.

11.3.2.7 Sustainability Benefit of Using RCA in Concrete

Concrete is the second most consumed material on Earth after water. To produce cement, the major constituent material of concrete, produces greenhouse emissions of 2.5 billion tons of CO_2

per year, and 160 million tons of CO_2 is made due to aggregate production. Refer to Figure 11.13 for CO_2 generated during cement manufacturing. Total CO_2 emissions from the concrete sector are about 5.9 billion tons, including cement, aggregate, steel, and transportation.

The many economic, environmental, and societal benefits of concrete recycling and using RCA in concrete include the following:

- Lower reliance on virgin quarried aggregates
- Reduced energy consumption
- Reduced use of landfill space
- Reduced greenhouse gas emissions
- Time savings associated with hauling time reductions
- Recaptured value of prior investments in concrete paving materials

Economic design is critical to all infrastructure projects. Utilizing RCA as aggregate for new construction can minimize costs and reserve virgin aggregate resources, thereby decreasing environmental pollution from concrete waste. Using RCA can save approximately 11% on the construction budget.

The sustainability benefits of recycling concrete pavements can be quantified using life-cycle cost analysis (LCCA), life-cycle assessment (LCA), and rating systems. LCCA is an economic analysis technique that is principally used to quantify the financial component of sustainability. LCA is most suitable for analyzing and quantifying the environmental impacts of a specific project or strategy over a life cycle. Rating systems rely heavily on providing incentives (points and recognition) for addressing a broad set of sustainability best practices. The approach, assumptions, and analysis techniques used by each tool are different, but various aspects of sustainability can be quantified when utilized alone or in concert. Stakeholders' goals should be carefully considered before selecting one or more approaches. Each of these types of tools provides one or more means of incorporating recycling-related activities and material choices into the analysis

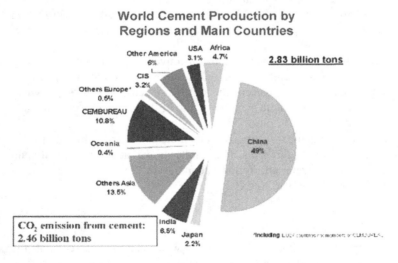

Figure 11.13 Cement Production and CO_2 Emission

and evaluation, provides guidance, and potentially rewards (recognition). As outlined in the case studies presented, several agencies have successfully used these tools to justify and support concrete recycling activities. More extensive use of these tools could incentivize stakeholders to use concrete recycling more frequently in pavement construction, moving towards a more sustainable highway infrastructure.

The research found that concrete containing recycled concrete aggregates has 13–27% lower thermal conductivity than the reported literature values for dry concrete with approximately the same density. Concrete sandwich panel walls containing RCA have a 16.6% lower U-value than similar precast concrete wall panels on the market and have higher thermal mass than metal sandwich panels, resulting in 11% less energy needed for cooling and 22% less energy required for heating in buildings. This research showed that using RCA as recycled aggregate in concrete mixes can improve the sound insulation of precast wall systems.

Financial analyses conclude that using RCA can create a sustainable end-use for concrete waste, reduce the demand for natural aggregates, and lead to natural resource preservation. A probabilistic estimation of the price difference between RCA concrete and ordinary concrete concludes that ready-mixed concrete plants having aggregate feeding mechanisms with front-end loaders would be an appropriate entry for the industrial-scale manufacturing of RCA concrete.

11.3.2.8 Standards and Specifications for Recycled Concrete Aggregate Use in Concrete

Some state DOTs have specifications for RCA use in concrete. For example, the TxDOT 2004 Standard Specifications for Construction and Maintenance of Highways, Streets, and Bridges specify nonhazardous recyclable material (NRM) for TxDOT projects. RCA must meet the requirements of departmental material specification DMS-1 1000, "Evaluating and Using Nonhazardous Recyclable Materials Guidelines." Item 421, Hydraulic Cement Concrete, specifies that RCA can be used as a coarse aggregate if it meets the specification requirements. TxDOT does not allow RCA in structural concrete.

California Department of General Services, Division of the State Architect (DSA), issued an Interpretation of Regulations Document, IR 19-4, which clarifies the use and acceptance of RCA on projects under the purview of the Division of the State Architect: Coarse and fine RCA may be used only in exposed minor concrete applications such as sidewalks, curbs, gutters, parking strip, and pavement in an amount not to exceed 50% of the total dry aggregate mass. RCA should be thoroughly cleaned and washed before use, must not contain any deleterious materials, must meet the requirements of the California Building Code and its referenced standards, i.e., ASTM C33, and must satisfy specific project requirements. California DOT allows up to 100% of RCA in minor and lean concrete. If the source is approved, South Carolina DOT allows 100% of coarse RCA in PCC.

The MDOT Standard Specifications of Construction, in Aggregate Section 902.03 Part B, 902.04, and 902.06, permit the statewide use of RCA in PCC in Michigan. This allows the use of RCA as coarse aggregate in PCC for curb and gutters, valley gutter, sidewalk, concrete barriers, driveways, temporary pavements, and interchange ramps, and shoulders.

Internationally, many countries have established specifications for RCA use in concrete. In Brazil, specification NBR 15.116, "Recycled aggregates from construction and demolition waste (CDW)", allows the use of RCA only in nonstructural concrete, and both coarse and fine fractions are permitted in concrete production. CDW is separated into four classes (A, B, C, and D). RCA belongs to Class A and can be considered an aggregate for use in concrete.

Germany has two standards; the first refers to the requirements of CDW aggregate, while the second proposes rules for implementing these aggregates in concrete. DIN 4226-100 "Aggregates for Mortar and Concrete - Recycled Aggregates" specifies requirements for aggregates particle density higher than 1,500 kg/m³ for use in mortar and concrete. It also specifies the system of production control and conformity assessment. The particle size must follow the requirements specified in DIN 4226-1: "Aggregates for concrete and mortar – normal density and high-density aggregates." Using DIN 4226-100, CDW aggregates are classed into four types: (i) waste concrete; (ii) CDW; (iii) masonry waste; and (iv) mixed material. The first type, waste concrete, is mainly composed of recycled concrete.

Hong Kong's Works Bureau Technical Circular, WBTC No. 12/2002, outlines using recycled aggregates in concrete production and constructing the base and subbase of road pavement. Two alternatives are suggested in this specification for using RCA in concrete. RCA's complete replacement of natural aggregate can be used for concrete in less demanding structures, such as benches, flowerbeds, or cyclopean concrete. Only 20% replacement of natural aggregate by RCA is allowed in structural concrete with a 28-day compressive strength range of 25–35 MPa.

In Japan, the Building Contractors Society of Japan issued a "Proposed standard for the use of recycled aggregates and RCA". RCA is allowed to be used in concrete. This document does not limit the use of masonry material.

The Japanese guideline "TR A006 2000 – Concrete using recycled aggregate" provides information on using RCA in concrete. The main principle is that the mortar component will decrease quality, which becomes evident as higher water absorption increases when the mortar content increases. Classification is based on the amount of mortar. The mortar phase increases water absorption. With a higher mortar amount, the quality of recycled aggregate is lowered. The amount of absorption should be less than 7% for coarse aggregate and less than 10% for fine aggregate. This regulation also includes the methods for quality control.

In Japan, recycled aggregate is recommended only for concrete without frost attacks. The amount of RCA is determined on a case-by-case basis. It can be 100% coarse aggregate or 50% of both coarse and fine aggregates.

A quality assurance system was created for the supply and utilization of coarse RCA to meet existing product standards in Australia to produce concrete from clean, uncontaminated crushed concrete (particle density >2,100 kg/m³, including <2% of brick, stony material, or other forms of contaminants). Physical contaminant levels typically less than 2% were achievable under existing manufacturing practices. The need for assessing chemical contaminant levels was recommended. Such pollutants can potentially alter concrete rheology, setting characteristics, and durability. Class 1 RCA was deemed suitable for producing plain, unreinforced, and reinforced concrete up to and including 40 MPa concrete, with no mandatory limits on RCA substitution levels. Extra care must be taken to ensure satisfactory compliance with acceptance criteria based on a standard deviation of compressive strength test results. The issue of concrete durability was considered of great concern regarding chemical contaminants such as sulfur-based residues that can induce deleterious expansive reactions and the impact of chloride contaminants on the corrosion of embedded steel reinforcement. Medium- and long-term field durability tests were recommended.

The RILEM – TC 121-DRG recommendation is a specification that deals with recycled coarse aggregates with a minimum size of 4 mm for concrete. As the properties of RCA fines differ vastly from those of natural sand, there is no specification for the fine fraction in the RILEM specification. It classes the recycled coarse aggregates and indicates the scope of application for concrete containing these RCA classes in terms of acceptable environmental exposure classes and concrete strength classes. Recycled coarse aggregates are classed as follows: Type I – aggregates that are implicitly understood to originate primarily from masonry rubble; Type II – aggregates that are implicitly understood to originate mainly from concrete rubble; Type III – aggregates that are implicitly understood to consist of a blend of recycled and natural aggregate; the composition shall have at least 80% natural aggregate and up to 10% Type I aggregate.

Within the United Kingdom, the British Standard BS 8500-2:2002 "Concrete – Complementary British Standard to BS EN 206-1 – Part 2: Specification for constituent materials and concrete" specifies requirements for coarse RCA only, excluding the use of fine RCA for concrete production.

In some concretes, any amount of RCA is possible in the UK. For instance, low-grade flooring (small garages) with no reinforcement and some pavement curbs can include high amounts of RCA. In other allowable cases, the amount of RCA is usually limited to 20% of the total aggregate, though 30% often has only minor effects on the critical concrete properties in low-exposure classes.

In the Netherlands, the Dutch center CUR has developed specifications for using RCA. In 1984, a specification was released for aggregates from crushing concrete. In 1986, CUR developed a specification for using recycled aggregate generated from masonry. Subsequently, another specification was developed for the use of crushed mortar as aggregate.

In Portugal, the National Laboratory of Civil Engineering (LNEC) issued E 471: 2006 in 2006, a prepared pre-norm, prE 469 "Guide for the Use of Recycled Coarse Aggregates in Hydraulic Binder Concrete," which classes the coarse RCA covered by NP EN 12620 "Aggregates for concrete" and establishes the minimum requirements that they must meet in order to be used in the manufacture of hydraulic binder concrete. Recycled aggregate requirements and their applications are not shown for fine RCA since they have a high percentage of particles with dimensions less than 0.063 mm and greater water absorption capacity, making it difficult to control the workability and impairing the mechanical strength of concrete containing that fraction.

In Belgium, a recommendation for using recycled aggregate in concrete was compiled by a working group in 1990. A series of information on the document was presented. This recommendation is divided into three parts: the first contains the requirements of the recycled aggregate, the second regulates the scope, and the third is related to the calculations of the coefficients and their characteristics. This standard defines RCA only for coarse fractions. It excludes the fine fraction, as in several other specifications.

In Switzerland, OT 70085 – The Swiss document was published in 2006, Objective Technique OT 70085 "Instruction Technique. Utilisation de Materiaux de Construction Minéraux secondaires dans la Construction d'Abris." It creates a wide range of RCA applications, with different approaches depending on user demands. This application is regulated together with the standard SIA 162/4, 1994, "Béton de Recyclage." The document establishes requirements to be met by RCA as well as their application conditions.

In Russia, the former Soviet Union introduced a specification in 1984, developed by a scientific research institute, for using RCA in plain and reinforced concrete. The scope of the standard specifies that the replacement ratio of natural aggregate by RCA can reach 100% if the concrete is used in foundations or reinforced concrete with strength below 15 MPa. If the replacement ratio is not more than 50%, it can be applied to concrete structures with a strength of over 20 MPa.

Finland has a national specification on the use of aggregate in concrete, 43 2008. This specification allows the use of recycled aggregate. If recycled aggregates are used, it must be proven beforehand that they are suitable for the specific intended use. Relevant preliminary testing is needed. Requirements on RCA can be set based on the standard [EN 12620 + A1] (Aggregates for Concrete).

It is suggested that a specification for using RCA in Finland be made. Without comprehensive guidelines, the use of recycled aggregates in concrete would also be unsound and very limited in the future. National specifications should be based on local climatic conditions and all other local circumstances. The aim should be the value-added sustainable application of recycled aggregates. Also, using RCA in more comprehensive applications should be studied and promoted. Considerable attention is required to control construction and demolition waste processing and subsequent sorting, crushing, separation, and grading of aggregates for use in concrete and possibly other materials, especially cement-based materials.

European Standard: The Technical Committee, CEN/TC 154, "Aggregates," has developed an amendment, currently known as EN 12620:2002/PRA.1: 2006. This standard will be a European regulation, changing the current EN 12620:2002 and its national versions. The standard establishes requirements for the composition of the coarse RCA, beyond water absorption and density. This amendment also includes a clause reserved for alkali-silica reactions, establishing that all RCA should be classed as potentially reactive unless it is specified that they are not reactive (Wang, Hollar, & Poole, 2018).

11.4 Questions

1. What are the main topics covered in this chapter?
2. How does the chapter describe the role of concrete in modern roadway infrastructure?
3. What materials are discussed as being crucial for the composition of concrete pavements?
4. Can you explain the significance of aggregates in the concrete mixture and their impact on concrete properties?
5. What role do admixtures play in concrete pavement construction, and what are some of their potential effects?
6. The chapter details various concrete paving practices. What are some key aspects of site preparation for concrete pavements?
7. How are dowel bars and tie bars used in concrete pavement construction, and what purpose do they serve?
8. What preservation strategies are discussed in the chapter for maintaining aging concrete infrastructures?
9. How does the chapter address the environmental and economic benefits of concrete recycling?
10. Describe the process of recycling concrete, as outlined in the chapter, and its potential applications.
11. What challenges and limitations are associated with concrete recycling?
12. How do preservation treatments like diamond grinding and joint resealing contribute to the longevity of concrete pavements?
13. What is the anticipated performance lifespan for various concrete pavement preservation treatments mentioned in the chapter?
14. In terms of sustainability, why are the recycling and reuse of concrete materials increasingly important in modern construction?
15. How does the chapter contribute to the understanding of advancements and complexities in concrete paving, preservation, and recycling?
16. Explain how the process of creating recycled concrete aggregate (RCA) differs from the production of virgin natural aggregate, particularly in terms of the materials involved and the processing steps.

17. Discuss the main reasons why RCA properties differ from those of natural aggregate. How do these differences affect the use and behavior of RCA in concrete construction?
18. Describe the environmental benefits of using RCA in concrete production. How does this practice contribute to sustainability in the construction industry?
19. Explain the challenges associated with the workability and strength properties of concrete made with RCA compared to that made with natural aggregate. How can these challenges be addressed in the mix design process?
20. Why is it important to conduct trial mixtures when designing concrete mixes containing RCA, and what potential adjustments might be necessary based on the outcomes of these trials?
21. Discuss the varied applications of RCA in concrete construction, including the limitations and specific uses based on the structural requirements and the type of construction involved.

References

ASTM (2018) ASTM C88. *Standard Test Method for Soundness of Aggregates by Use of Sodium Sulfate or Magnesium Soundness Test.* American Society for Testing and Materials, (ASTM), West Conshohocken, PA.

NCDOT (2015). *Partial and Full Depth Repair Manual* North Carolina Department of Transportation (NCDOT). Raleigh, NC

Smith, K., Grogg, M., Ram, P., Smith, K., & Harrington, D. (2022). *Concrete Pavement Preservation Guide.* 3rd ed. National Concrete Pavement Technology Center, Federal Highway Administration. Washington, DC.

Smith, K., Harrington, D., Pierce, L., Ram, P., & Smith, K. (2014). *Concrete Pavement Preservation Guide.* 2nd ed. FHWA-HIF-14–014. US Department of Transportation, Federal Highway Administration, National Concrete Pavement Technology Center. Washoington, DC.

Taylor, P., Van Dam, T., Sutter, L., & Fick, G. (2019). *Integrated Materials and Construction Practices for Concrete Pavement: A State-of-the-Practice Manual.* 2nd ed. Federal Highway Administration, Iowa State University, Iowa Department of Transportation, Ames, IA.

Wang, G., Hollar, D., & Poole, R. (2018). *Using Recycled Concrete Aggregate in Nonstructural Concrete on NCDOT Projects in Eastern NC,* FHWA/NC/2017–06 Report. North Carolina Department of Transportation, Raleigh, NC.

Chapter 12

Transportation Asset Management

12.1 Introduction

The American Association of State Highway and Transportation Officials (AASHTO) defines Transportation Asset Management as a

> strategic and systematic process of operating, maintaining, upgrading, and expanding physical assets effectively throughout their life cycle. It aims to help a government entity focus on the business processes of resource allocation and efficient utilization with the objective of better decision making based upon quality information and well-defined objectives.[1]

Just like a person wants to prolong life and get the best value out of their assets like houses and cars, transportation agencies also want to extend the life of their highways. The Federal Highway Administration (FHWA) estimates that the replacement value of the country's pavement and bridge assets alone is over $5 Trillion.[2] While these are the agency's most expensive assets, that cost does not include the pipes, guardrails, signs, and other ancillary assets.

Transportation agencies must decide how to allocate their limited funds best, whether to repave a road, replace a bridge, upgrade substandard guardrails, replace failing pipes, or any other combination of projects. The bigger the highway system, the more complicated these decisions become. Small states like Utah, with a 3,700-mile highway network, struggle with these concepts. Big states like North Carolina and Texas, with over 80,000-mile networks and over $5 billion budgets, go through multi-layered analyses with many variables. States must make decisions based on traffic volumes, motorist safety, bridge and pavement health indexes, location, and costs, to name just a few of the variables considered.

The concept of transportation asset management (TAM) has grown steadily in the last 15 years, with Federal legislation playing a significant role in advancing it. The Moving Ahead for Progress in the 21st Century Act (MAP-21), signed into law in 2012, was considered a milestone for transportation agencies, requiring performance-based transportation programs. MAP-21 was the first federal legislation to require transportation agencies to apply performance management principles when making investment decisions. It defined national minimum performance standards for pavements and bridges, which agencies receiving federal money must meet, and it was the first legislation to tie outcome-based performance measures to funding (Hartle, et al., 2002).

To help states understand the importance of asset management and comply with MAP-21, the AASHTO developed detailed guides and other materials to assist states with developing their programs. Private industry also seeks to advance the practice by developing asset management software that makes many decisions quickly, providing engineers with a potential list of projects.

DOI: 10.1201/9781003197768-12

Using engineering judgment, highway officials can then take these lists of projects and fine-tune them to deliver the best program for their citizens.

Management systems, including pavement management, bridge management, and maintenance management, assist agencies in making decisions within each of their own "organizational silos" and meanwhile provide information to upper-level management to establish agency priorities and performance targets that are aligned with investment decisions. Essentially, management systems serve as the basis for the information to be used to make decisions in asset management.

12.2 Asset Management Principles

A sound asset management program must be based on principles that underpin the agency's activities and guide project decisions. These core principles ensure that scarce transportation funds are spent effectively and on projects that significantly improve the transportation network.[3] The principles agencies must put their focus on include:

Policy-Driven: Asset management programs should capture and respond to the agency's policy objectives. Policy objectives allow transportation agencies to communicate with legislators, stakeholders, and the general public what is essential, and asset management provides meaningful information about how changes in the transportation system support these objectives. Trust is a critical value that agencies must cultivate with their stakeholders, and communicating how asset management drives policy decisions builds confidence.

Performance-Based: The asset management program should have concrete objectives translated into system performance measures used for day-to-day operations and longer-term strategic management. Performance measures such as the number of cracks per mile in pavements or linear feet of damaged bridge rail are typical measures that can drive project decisions. Using performance data to support the management of assets enables agencies to select and deliver projects that achieve their objectives.

Risk-Based: Risk management is critical to an asset management plan. The risk of a road being closed can significantly affect travel and even strand motorists. This risk varies greatly depending on the type and location of a road. A large, heavily traveled interstate with hundreds of thousand vehicles a day would have significant impacts if it suddenly closed because of a pavement or bridge failure. On the other hand, a small, rural two-lane road with a few hundred cars a day carries a far lower risk if it has to be closed. Agencies perform these risk assessment scenarios to make better decisions on the allocation of resources and project selection.

Strategically Aligned with Agency Priorities: Transportation asset management measures should be aligned with agency priorities to ensure that the investments made extend the asset's life. For example, if an agency's priority is smooth pavements, measuring pavement smoothness on the nationally accepted scale of the international roughness index (IRI) and selecting projects that improve this number would strategically prioritize funding.

Transparent: Once an asset management plan is implemented, the results, compared to the stated performance measures, should be monitored and reported for impact and effectiveness. Actual performance outcomes should influence agency goals, resource allocation, and project decisions. Being transparent with stakeholders about performance outcomes is crucial to ensure long-term support and trust.

Information-Driven/Evidence-Based: Decision support tools such as computerized management systems should be used to access, analyze, and track data. Once an agency begins an asset management program, the amount of data can become overwhelming, but computer systems can make easy work of the analysis. The performance outcomes must be an integral part of the analysis and decision-making.

Continuously Improved: Asset management processes should be reviewed and improved on a regular basis. This provides managers with sufficient information to understand problems and suggest solutions. The agency should be committed to regular, ongoing processes of monitoring and reporting results.

12.3 The Asset Management Process

Transportation agencies with hundreds of millions or even billion-dollar budgets cannot spend those funds effectively if they don't know what assets they own and their condition. Transportation agencies are constantly pressured to do more with less, and their stakeholders evaluate every decision. Agencies should implement a cyclical asset management process to build trust and meet the above principles. When assets are inventoried and evaluated, projects are developed, and plans are executed, then the process repeats. This process can be described as the "E" asset management cycle, as shown in Figure 12.1.

Evaluate Assets: The first step in any asset management process is to evaluate the assets owned by the agency. What assets does the agency own? How many miles of pavement do they maintain? Is the pavement asphalt or concrete? How many bridges do they have? Are they concrete or steel? What sizes are the pipes? States typically own a large number of bridges, guardrails, pipes, and signs in addition to miles and miles of pavements. It is critical to understand this asset portfolio. Different types of bridges require different maintenance strategies. Different types of guardrails require different end treatments. Likewise, drainage pipes come in various materials and sizes, requiring different end treatments and cleaning schedules. Sign sizes vary from small roadside

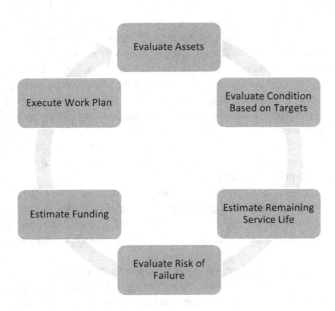

Figure 12.1 The "E" Asset Management Cycle

signs to large overhead sign structures with different grades of sign sheeting. These sign differences can lead to different life spans.

Evaluate Conditions Based on Targets: Once an agency knows what assets it owns, it must determine their condition. Understanding which assets are in good condition and which are in poor condition helps prioritize spending. Agencies must define condition targets and thresholds to determine good, fair, and poor conditions. It is generally accepted that threshold conditions for pipes to maintain proper drainage should have less than 25% of the diameter of the pipe blocked to be in good condition, while pipes that have more than 50% of the diameter of the pipe blocked by sediment and debris are considered in poor condition. With those condition thresholds defined, a transportation agency might set a performance target that they want at least 80% of their storm drainage in good condition. Having evaluated their system, they can now compare the number of pipes clogged to that performance target and develop a work plan to clean their pipes and achieve that target.

Estimate Remaining Service Life: All assets have a useful life and wear out over time. Pavements fail and must be replaced. Bridges weaken and can no longer carry their design loads. Determining the remaining serviceable life of an asset helps determine what treatments should be performed and the estimated cost. For example, pavements that show minor fatigue and cracking can be treated with lower-cost overlays, whereas older pavements that have severe cracking and rutting must be milled and replaced, costing significantly more. Just like a homeowner will decide on household repairs based on the age of things like the HVAC system, transportation agencies will base funding decisions on the age and remaining service life of their assets.

Evaluate the Risk of Failure: Agencies must constantly evaluate risk. The risk of a closed road can significantly affect travel, causing significant disruptions to other roads beyond the one affected by the closure. The consequences of a roadway going out of service due to a catastrophic failure can vary, and these risks must be factored into any asset management plan. Evaluating these consequences must be included in asset management strategies, emphasizing keeping major routes in optimum condition.

Estimate Funding Needs Based on a Work Plan: Once the assets have been inspected, conditions evaluated, and remaining service life and risk of failure have been determined, a work plan can be developed. This work plan should include preservation activities for those assets still in good condition, maintenance activities for those assets in fair condition, and rehabilitation activities for those assets in poor condition with little remaining service life. These treatment strategies and their associated costs are then calculated into a total work plan. In most cases, the estimated funding plan will far exceed the budget, and engineers must make hard choices on how to optimize the allocated resources to achieve the best outcome.

Execute the Work Plan: Once a work plan is developed, the execution of that work plan becomes the final step in the process. Project-level managers are tasked with completing the identified preservation, maintenance, and rehabilitation work on the assets in their area. In theory, any work on the assets should improve their condition, but to confirm this, the asset management cycle starts over again with the evaluation step.

12.4 Common Asset Management Elements

12.4.1 Introduction

No matter the type of asset or asset management system, there are some common elements. Transportation agencies think on three levels: strategic, network, and project. These three levels drive decisions throughout the organization.

At the strategic level, long-term decisions are made to address an organization's highest goals and objectives. Individual decision-makers include elected officials, transportation boards and commissions, city councils, and agency upper management. Strategic decisions align the network and project-level goals with agency and stakeholder priorities; they are less structured than decisions made at network and project levels because information at the strategic level is more speculative.

At the network level, the ultimate needs of the entire roadway network are identified and multi-year improvement programs are developed to support the agency's planning, programming, budgeting, and analysis efforts. Alternate strategies that incorporate a mix of preventive maintenance, rehabilitation, and reconstruction projects are considered, and the results of the network-level analysis are reported to decision-makers to assist them in establishing realistic performance targets and evaluating investment options. Generally, information at the network level is moderate in terms of sophistication, and network-level decisions are the primary focus of pavement management activities (AASHTO, 2012).

At the project level, specific decisions concentrating on a particular section of the roadway network are made to provide the most cost-effective and feasible design, maintenance, rehabilitation, and reconstruction strategies possible that cover a relatively shorter timeframe, for instance, the first two or three years of a multi-year improvement program developed at the network level. Typically, at the project level, more detailed information can be collected to determine in situ conditions (AASHTO, 1993).

The types of decisions made at each level and the level of detail of the data used to make decisions are included in Table 12.1.

Table 12.1 Levels of a Pavement Management System

PMS Level	Examples of Job Titles at This Level	Types of Decisions/Activities	Range of Assets Considered	Level of Detail	Breadth of Decisions
Strategic	• Legislator • Commissioner • Chief Engineer • Council Member	• Performance targets • Funding allocations • Pavement preservation strategy	All assets statewide	Low	Broad
Network	• Asset Manager • Pavement Management Engineer • District Engineer	• Project and treatment recommendations for a multi-year plan • Funding needed to achieve performance targets • Consequences of different investment strategies	A single type of asset or a range of assets in a geographic area	Moderate	Moderate
Project	• Design Engineer • Construction Engineer • Materials Engineer • Operations Engineer	• Maintenance activities for current funding year • Pavement rehabilitation thickness design • Material type selection • Life-cycle costing	Specific assets in a specific area	High	Focused

Source: AASHTO, Pavement Management Guide, 2nd ed., 2012

12.4.2 Inputs

Condition data of the asset is a key component of a data-driven management process because it enables agencies to describe current performance, determine maintenance and rehabilitation needs, predict future conditions, evaluate the impacts of treatments, prioritize work, and help optimize funding strategies. The extent of the data, the level of detail collected, and the data collection tools used can differ from one agency to another. Data collected at the project level are typically detailed and exhaustive, may use different destructive and nondestructive methods, and are used to investigate specific performance problems or to develop an ideal rehabilitation design. On the other hand, network-level data are usually less detailed and are collected on a large portion of the agency's network, which can be used to identify and prioritize treatment needs and determine and allocate funds (Zimmerman, 2017).

Data quantity and the quality of condition data collected directly determine the management system's effectiveness. Generally, the greater the amount of data collected, the higher the cost of data collection, but the more informative the decisions can be. Understandably, asset condition data must be collected regularly to keep the management system up to date and reflect the current actual condition. However, a balance between data quantity, agency needs, and associated resources must be achieved (Flintsch & McGhee, 2009).

Data consistency is one of the most recent issues the agencies face, caused by different survey procedures and the change of data collection equipment or vendors. It can significantly impact the overall network conditions and the value of historical data. The following recommendations are provided to help agencies handle this issue:

- Establish and utilize data collection quality management programs that include the calibration of equipment, field calibration before and during data collection, and QA checks to determine the completeness and reasonableness of the data before entering it into a database
- Recognize that variability is inherent to each type of condition in the collected data. Without understanding the degree of variability associated with that data, data from different survey years or equipment should not be compared
- Store historical condition information with the dates the inspection dates were performed and the surface types linked. This enables the raw data to be used to calculate new indices if the method of computing a condition index changes or new technology is used to collect the information

Preserving historical survey procedures is not recommended when significant changes to data collection methodologies are implemented. When these changes occur, such as the transition from manual to automated surveys, agencies should recognize that correlations with historical data may not be meaningful and that it may be better time to begin rebuilding the historical records from where the change occurred. It is often easier to explain one period of inconsistency in the data being reported due to the change in methodology than to unnecessarily correlate very different condition surveys.

12.4.3 Data Management

Effective data management is essential for a successful asset management system. This data can be managed using a spreadsheet, a relational database, or a geographic information system (GIS).

Key factors that need to be considered when managing data include

- Data sources and individuals responsible for maintaining the data
- Methods and frequency of data collection
- Location referencing systems associated with the data.
- Data structure, format, and size
- Methods of data transmittal, processing, and storage
- Use of data (e.g., in business processes and in relation to other user needs)
- Applications that draw data from existing databases (e.g., bridge management systems and pavement management systems)
- Types of reports that are currently produced or are needed

12.4.4 Analysis Parameters

Assets deteriorate over time, and conditions become worse. Even with maintenance and preservation treatments, conditions will deteriorate, albeit at a slower rate. This deterioration is tracked/modeled by deterioration curves.

Deterioration models can be used to predict future conditions, determine the appropriate timing for a preservation program, identify an ideal treatment strategy in the network, conduct a needs analysis, and establish performance criteria for performance specifications and warranty contracts. To develop reliable deterioration models, the agency should have data on the materials the asset is made of, the construction technique used, climate, traffic loading, and the time in a given condition. Adequate and complete data sources, along with regular updates to the data, will produce more accurate models that predict reasonable values. An appropriate model form that fits the data and reflects the deterioration trend of the asset should be used, and the reliability of the models should be evaluated.

Deterioration models can predict the amount and severity of individual distress, pavement performance rating indices, or composite indices such as the pavement condition index (PCI) for pavements and the bridge rating index (BRI) for bridges. There are four types of deterioration models:

- Deterministic models are developed based on statistical analysis results. They can predict a single dependent variable using one or more independent variables.
- Probabilistic models predict probabilities that an asset deteriorates from one condition to another.
- Bayesian models can be developed using both objective and subjective data.
- Expert-based models are developed either informally or formally. An individual or a group of individuals may informally develop an equation that describes the deterioration process. Alternatively, a panel of experts may formally come to a consensus on changes in performance over time and then perform a regression to develop performance models.

Deterioration models can also be classified as family models and site-specific models. Family models are developed to represent the deterioration trend for a group of roadway sections that share similar performance characteristics. Site-specific models, for example, predict pavement performance of a particular pavement section.

Deterioration models that are developed using a statistical regression analysis processes, the following parameters or procedures can be used to evaluate the goodness-of-fit:

- Standard error of estimate
- Residual plots
- Root mean square error
- T-test
- F-test

12.4.5 Analysis Module

Asset management systems include analysis modules that allow agencies to process and analyze their data and evaluate the consequences of implementing different treatment strategies at varying investment levels. The aim of this analysis is to produce a work plan that maintains the best level of condition for a given price. It should give the agency information on what treatments to apply, where to apply them, and when they should be applied to get the best results. An ideal plan would give the "perfect" answer; however, agencies never have unlimited resources. They must often apply a variety of constraints. Funding is one common constraint, as agencies do not have unlimited funds and must choose the best ways to spend the money they do have. Another constraint may be the types of treatments an agency uses. Some agencies do not allow certain pavement preservation strategies such as chip seals, so although that may be the most effective treatment, it is not one in the agency's toolbox. Analysis models are flexible enough to take these constraints and develop realistic work plans that can then be implemented.

12.4.6 Reporting Module

Asset management systems are an information hub that allows managers at all levels of a transportation agency to communicate asset conditions quickly and effectively. They can generate both standard and ad hoc reports in a variety of formats. Tabular reports, such as Figure 12.2, can be used to give the reader a quick sense of the state of the system, while map-based graphics, such as Figure 12.3, are helpful to show stakeholders what will be happening in their area. All these types of reports allow users to gain insights and probe the data, getting answers to questions from legislators, council members, and others. Dashboards are a type of front-end screen for the software that instantly gives access to key metrics. When used correctly, dashboards can get data directly into the hands of managers who can view trend lines and make adjustments to work plans as needed.

12.4.7 Feedback Loop

Any effective asset management system relies on a feedback loop that updates the management data with new information regularly. This feedback loop includes data from many sources. Maintenance cost data may come from the agency's financial management system where employees enter their labor and equipment hours as well as material costs. Contract preservation and rehabilitation costs with quantities of work performed, completion dates, and treatments used are another source of data. Types of materials used are a key driver as well. These updates are then processed to develop new deterioration models that can predict performance more accurately, ultimately leading to increased reliability of the management system (GASB, 1999).

Pavements on the NHS (2019 HPMS)					
	Lane Miles	Good	Fair	Poor	
State-owned NHS	36,896	45.0%	52.8%	2.3%	
Interstate	14,419	47.9%	50.2%	1.9%	
Non-Interstate NHS	22,477	43.1%	54.4%	2.5%	
Locally-owned NHS					
Non-Interstate NHS	20,803	3.0%	79.0%	17.9%	
Totals					
All NHS	57,699	29.8%	62.2%	7.9%	
Interstate	14,419	47.9%	50.2%	1.9%	
Non-Interstate NHS	43,281	23.8%	66.2%	9.9%	

Figure 12.2 Inventory and Conditions of NHS Pavements in California, by Lane Miles

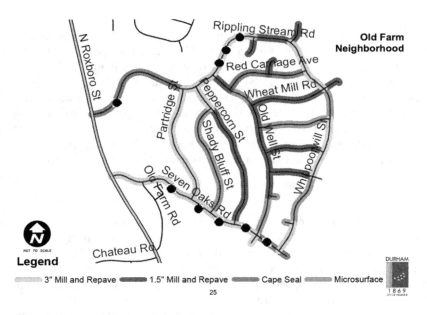

Figure 12.3 City of Durham Resurfacing Map

12.5 Pavement Management

12.5.1 Background

Pavements are the most valuable and largest assets for transportation agencies to manage. According to Highway Statistics 2020, there are more than 8.8 million lane miles of paved roads in the United States, and the total value of these roadways is about $4 trillion.[4] Seeking systematic ways to manage pavements has thus become the top priority within these transportation agencies (FHWA, 2020).

The concept of pavement management is not new. In the mid-1960s, a large quantity of unanticipated pavement failures occurred on the US Interstate and Canadian Highway Systems. A subsequent study found that even though the most appropriate design methods at that time were used, these methods failed to include the impacts of maintenance activities on pavement performance and the life-cycle cost past the initial design and construction period. To find a solution, the National Cooperative Highway Research Program (NCHRP) sponsored a research project, and a parallel study was carried out in Canada. Both efforts concluded that pavement behavior, distress, and performance steps should be included in design. This conclusion remains the backbone of today's pavement management systems (PMS).

In the United States, early pavement management activities included the $30 million American Association of State Highway and Transportation Officials (AASHTO) Road Test in Illinois. In the 1980s, all US states and Canadian provinces had used PMS to some extent. In 1999, the US Governmental Accounting Standards Board (GASB) issued Standard 34, and for the first time, governments were required to report their general infrastructure assets, including roads and bridges. This requirement has since then further motivated transportation agencies to adapt to more effective pavement management principles. By 2017, 49% of the state DOTs and 38% of Canadian provincial ministries of transportation (MOT) were using commercial PMS software to manage their pavement networks.

12.5.2 Pavement Management Systems (PMS)

Pavement management and pavement management system have been defined in several ways. For instance, the definition used by the Organization for Economic Cooperation and Development (OECD, 1987) is:

> Pavement maintenance management is the process of coordinating and controlling a comprehensive set of activities in order to maintain pavements, so as to make the best possible use of resources available, i.e., maximize the benefits for society.

A more commonly used definition was provided by AASHTO in 1993: "A pavement management system is a set of tools or methods that assist decision-makers in finding optimum strategies for providing, evaluating, and maintaining pavements in a serviceable condition over a period of time." An effective PMS enables agencies to "improve the efficiency of decision-making, expand its scope, provide feedback on the consequences of decisions, facilitate the coordination of activities within the agency, and ensure the consistency of decisions made at different management levels within the same organization."

The initial focus of PMSs was to develop the optimum design for a specific roadway section (Dewan & Smith, 2003). It was soon recognized that it is necessary to use a systematic method to determine what, where, and when roadways in a pavement network should be treated. This evolution of PMS leads to discussions of the level of detail needed for effective pavement management practices (Khattak, et al., 2008).

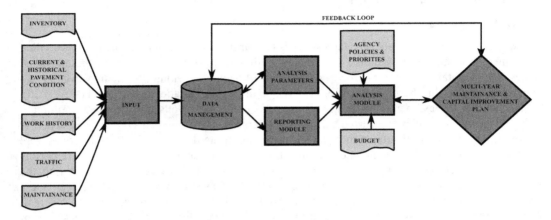

Figure 12.4 PMS Components

Source: AASHTO, Pavement Management Guide, 2nd ed., 2012

12.5.3 Components of a PMS

A PMS is primarily focused on pavement management strategies at the network level. As shown in Figure 12.4, the fundamental components of a network-level PMS are similar to those in any type of asset management system and have been discussed earlier in this chapter. Only items specific to a PMS will be discussed in detail here.

12.5.4 Pavement Inputs

Data is needed as the basis of all decision-making processes within the PMS. The inputs of a typical PMS include general inventory data and pavement condition data. Pavement age is another essential input that can be used to predict future pavement conditions, and it is calculated using the year in which a particular roadway section went through a major construction activity, whether that was the original construction or the last significant treatment. To conduct a more sophisticated analysis, e.g., a relationship between differences in pavement performance and pavement structure characteristics, detailed construction history data is necessary.

12.5.5 Pavement Inventory Data

Pavement inventory data allow agencies to identify, classify, and quantify various aspects of the pavement network. The level of detail per inventory item must be decided by an agency's pavement management goals. Inaccurate or outdated inventory data can adversely affect the performance of a PMS. The minimum amount of inventory information needed to support a PMS includes:

12.5.5.1 Traffic Data

- Segment beginning and end points
- Route designation along with the route type (Interstate, US, County, City)
- Functional classification of the road
- Segment length

- Average pavement width
- Pavement type
- Shoulder type
- Shoulder width
- Number of lanes in each traffic direction

Other inventory information that can be collected to support agencies' decision processes includes

- Layer type
- Layer thickness
- Layer material engineering properties
- Joint spacing
- Transverse joint load transfer
- Subgrade types and material classification
- Drainage information
- Environmental or location information
- Pavement history data
- Cost data
- Ownership information

12.5.5.2 Pavement Condition Data

Pavement condition data is a key component of a data-driven pavement management process because it enables agencies to describe current pavement performance (e.g., the type, amount, and severity of surface distress, structural integrity, ride quality, and skid resistance) and to determine maintenance and rehabilitation needs, predict future pavement conditions, evaluate the impacts of treatments, prioritize work, and help optimize funding strategies. The extent of the data, the level of detail collected, and the data collection tools used can differ from an agency to another. Data collected at the project level are typically detailed and exhaustive, may use different destructive and nondestructive methods, and are used to investigate specific pavement performance problems or to develop an ideal rehabilitation design. On the other hand, network level data are usually less detailed and are collected on a large portion of the agency's pavement network, which can be used to identify and prioritize treatment needs and determine and allocate funds.

Data quantity and quality of pavement condition data collected at the network level directly determine the effectiveness of a PMS. As stated earlier, a balance between data quantity and agency needs and associated resources must be achieved. Additional data collection requirements should also be considered, which include

- The agency's pavement preservation program
- The FHWA Highway Performance Monitoring System (HPMS)
- The FHWA Infrastructure Health Analysis
- Data required to support the agency's PMS activities
- MEPDG calibration and verification activities

High data quality helps establish an agency's confidence in using the system, while inaccurate data leads to poor decisions over time. Common data quality issues include misidentifying a distress or its severity, mislocating a distress, or mismeasuring it. The NCHRP Synthesis 401, Quality Management of Pavement Condition Data Collection, provides detailed guidance to collect high-quality data. This synthesis calls for quality control (QC) and quality assurance (QA) strategies to be implemented within organizations. QC is defined as "actions and considerations necessary to assess and adjust production processes so as to control the level of quality being produced in the end product." Equipment calibration and data collection rater training are examples of QC processes. QA, on the other hand, includes "activities conducted to verify that the collected pavement condition data meet the quality requirements and ensure that the final product is in compliance with the specifications." Performing tests to validate pavement condition data provided by a vendor is an example of a QA process.

Types of pavement conditions normally collected by agencies depend on the unique aspects of the agency's roadway network, including the size, available resources, past practices, and other factors. Typical types include

- Distress: Distress information is essential in selecting appropriate pavement preservation and rehabilitation treatments and in planning for long-term pavement management programs.
- Structural Capacity: Poor structural capacity necessitates major rehabilitation and reconstruction activities.
- Surface Characteristics: A pavement may be free from visible distresses and have good structural capacity, but still have surface characteristics, such as the longitudinal profile or smoothness, surface texture, and noise, that warrant certain surface repairs.
- Pavement distresses are described in two widely used guides: Distress Identification Manual for the Long-Term Pavement Performance Program (LTPP) and Standard Practice for Roads and Parking Lots Pavement Condition Index Surveys (ASTM D6433). Based on these two guides, common distress types that are included in a PMS are shown in Tables 12.2 and 12.3.

Pavement condition data can be collected by many different techniques. NCHRP Synthesis 401 summarizes that:

- Condition data can be collected using manual, automated, and semi-automated (a combination of automated and manual) methods. Today the use of automated data collection approaches is becoming dominant because of the size of the roadway networks, the need for objective measurements, the safety concerns associated with manual approaches, and the complexities associated with generating and analyzing large amounts of data in a timely manner.
- Destructive data collection approaches are more commonly used at the project-level, while nondestructive approaches are more commonly used at the network-level. Destructive approaches, including coring and boring, can reveal layer type and thickness, layer strength and stiffness, material quality, and the location of imperfections. They are, however, slow, generating far fewer data than nondestructive approaches. Data collection may be performed by either an agency's in-house resources, third-party vendors, or both.

12.5.6 Analysis Parameters

Pavement deterioration models play an important role in a PMS. They can be used to predict future pavement conditions, determine the appropriate timing for a PMS program, identify an ideal treatment strategy in the network, conduct a needs analysis, and establish performance criteria for performance specifications and warranty contracts. To develop reliable pavement deterioration models, the agency should have an adequate source of data and maintain it over time so the models continue to predict reasonable values. Although there are many factors that can impact pavement

Table 12.2 Pavement Distresses (FHWA)

Asphalt Concrete Surfaces	Jointed Portland Cement Concrete Surfaces	Continuously Reinforced Concrete Surfaces
Cracking	Cracking	Cracking
Fatigue cracking	Corner breaks	Durability cracking ("D" cracking)
Block cracking	Durability cracking ("D" cracking)	Longitudinal cracking
Edge cracking	Longitudinal cracking	Transverse cracking
Longitudinal cracking	Transverse cracking	
Reflection cracking at joints		Surface Defects
Transverse Cracking	Joint Deficiencies	Map cracking and scaling
	Joint seal damage	Map cracking
Patching and Potholes	Transverse joint seal damage	Scaling
Patch/patch deterioration	Longitudinal joint seal damage	Polished aggregate
Potholes	Spalling of longitudinal joints	Popouts
	Spalling of transverse joints	
Surface Deformation		Miscellaneous Distresses
Rutting	Surface Defects	Blowups
Shoving	Map cracking and scaling	Transverse construction joint deterioration
	Map cracking	Lane-to-shoulder dropoff
Surface Defects	Scaling	Lane-to-shoulder separation
Bleeding	Polished aggregate	Patch/patch deterioration
Polished aggregate	Popouts	Punchouts
Raveling		Spalling of longitudinal joints
	Miscellaneous Distresses	Water bleeding and pumping
Miscellaneous Distresses	Blowups	Longitudinal joint seal damage
Lane-to-shoulder dropoff	Faulting of transverse joints and cracks	
Water bleeding and pumping	Lane-to-shoulder dropoff	
	Lane-to-shoulder separation	
	Patch/patch deterioration	

Source: AASHTO, Pavement Management Guide, 2nd ed., 2012

performance, the most significant variables should be identified and included in pavement deterioration models. An appropriate model form that fits the data and reflects the deterioration trend should be used, and the reliability of the models should be evaluated.

Pavement deterioration models can predict the amount and severity of individual distress, pavement performance rating indices, or composite indices such as the Pavement Condition Index (PCI). There are four types of deterioration models:

- Deterministic models are developed based on statistical analysis results. They can predict a single dependent variable using one or more independent variables.
- Probabilistic models predict probabilities that a pavement deteriorates from one condition to another.
- Bayesian models can be developed using both objective and subjective data.
- Expert-based models are developed either informally or formally. An individual or a group of individuals may informally develop an equation that describes the pavement deterioration process. Alternatively, a panel of experts may formally come to a consensus on changes

Table 12.3 Pavement Distresses (ASTM)

Distress in Asphalt Pavements	Distress in Jointed Concrete Pavements
Alligator cracking (fatigue)	Blowup/buckling
Bleeding	Corner break
Block cracking	Divided slab
Bumps and sags	Durability ("D") cracking
Corrugation	Faulting
Depression	Joint seal damage
Edge cracking	Lane/shoulder dropoff
Joint reflection cracking	Linear cracking (longitudinal, transverse, and diagonal cracks)
Lane/shoulder dropoff	Large patching (more than 5.5 ft^2) and utility cuts
Longitudinal and transverse cracking	Small patching (less than 5.5 ft^2)
Patching and utility cut patching	Polished aggregate
Polished aggregate	Popouts
Potholes	Pumping
Railroad crossing	Punchout
Rutting	Railroad crossing
Shoving	Shrinkage cracks
Slippage cracking	Scaling, map cracking, and crazing
Swell	Corner spalling
Weathering and raveling	Joint spalling
(Ride quality, a separate "distress," is actually an input in determining the severity level of bumps, corrugation, railroad crossings, shoving, and swells)	(Ride quality is an input in determining the severity level of blowup/buckling and railroad crossings)

Source: AASHTO, Pavement Management Guide, 2nd ed., 2012

in pavement performance over time and then perform a regression to develop performance models.

Pavement deterioration models can also be classified as family models and site-specific models. Family models are developed to represent the deterioration trend for a group of roadway sections that share similar performance characteristics. Site-specific models predict the pavement performance of a particular pavement section.

For pavement deterioration models that are developed using a statistical regression analysis processes, the following parameters or procedures can be used to evaluate the goodness-of-fit:

- Standard error of estimate
- Residual plots
- Root mean square error
- T-test
- F-test

12.5.7 Analysis Module

The pavement management analysis module allows agencies to process and analyze pavement data and evaluate the consequences of implementing different pavement management strategies at varying investment levels.

Five different pavement management strategies are defined by FHWA as follows:
Rehabilitation

Type the Paragraph Textconsists of structural enhancements that extend the service life of an existing pavement or improve its load-carrying capability, or both. Rehabilitation activities are considered to be examples of minor rehabilitation when non-structural enhancements are made to existing pavement sections…major rehabilitation has been defined by the AASHTO Highway Subcommittee on Maintenance as structural enhancements that both extend the service life of an existing pavement and/or improve its load-carrying capability.

Preventive Maintenance is defined as a planned strategy of cost-effective treatments to an existing roadway system and its appurtenances that preserves the system, retards future deterioration, and maintains or improves the functional condition of the system (without significantly increasing the structural capacity).

Routine maintenance is "work that is planned and performed on a routine basis to maintain and preserve the condition of the highway system to respond to specific conditions and events that restore the highway system to an adequate level of service."

Corrective Maintenance includes activities that are "performed in response to the development of a deficiency or deficiencies that negatively impact the safe, efficient operation of the facility and the future integrity of the pavement section." These types of activities are generally reactive in nature.

Pavement Preservation is defined as "a program employing a network-level, long-term strategy that enhances pavement performance by using an integrated, cost-effective set of practices that extend pavement life, improve safety, and meet motorist expectations."

The general relationship between these pavement management strategies is presented in Figure 12.5. It should be noted that routine and corrective maintenance activities can be performed at any time during a pavement's life cycle; preventive maintenance activities usually are performed at early stages, and reconstruction at late stages. Apparently, the application of the right treatment at the right time on the right pavement can prolong the service life of the pavement.

A summary of the purposes of pavement management strategies is included in Table 12.4.

In recent years, some state DOTs have implemented pavement management strategies to prolong pavement service life and have reported a high cost-benefit ratio of one to nine (Peshkin, et al., 1999). Most transportation agencies agree that it is more cost-effective to preserve existing pavements than allowing them to deteriorate until a costly treatment, such as rehabilitation or reconstruction, must be performed. This concept is illustrated in Figure 12.6.

Ideally, an agency would be able to fix all pavements that have their performance falling below certain trigger values. In reality, few agencies can accomplish this goal due to budget constraints, and oftentimes, agencies rely on their PMS to help them allocate available funding as appropriately as possible. There are three commonly used methods for project and treatment selection under budget constraints: ranking, prioritization, and optimization. A comparison of these three methods is included in Table 12.5.

12.6 Bridge Management

12.6.1 Background

Bridges are some of society's most iconic and valuable assets. See Figure 12.7. For millennia, bridges have enabled the transportation of people and goods to areas that are otherwise inaccessible,

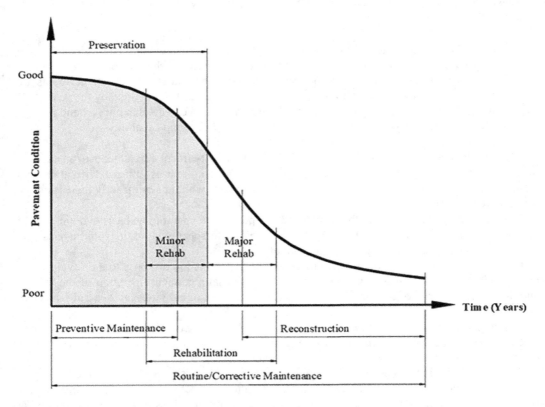

Figure 12.5 Relationship Between Pavement Condition and Pavement Management Strategies

Source: AASHTO, Pavement Management Guide, 2nd ed., 2012

Table 12.4 Pavement Management Strategies by Purpose

Type of Activity	Purpose of Activity				
	Increase Capacity	Increase Strength	Slow Aging	Restore Surface Characteristics	Improve or Restore Functionality
New Construction	X	X	X	X	X
Reconstruction	X	X	X	X	X
Major (Heavy) Rehabilitation		X	X	X	X
Structural Overlay		X	X	X	X
Minor (Light) Rehabilitation			X	X	X
Preventive Maintenance			X	X	X
Routine Maintenance					X
Corrective (Reactive) Maintenance					X
Catastrophic Maintenance					X

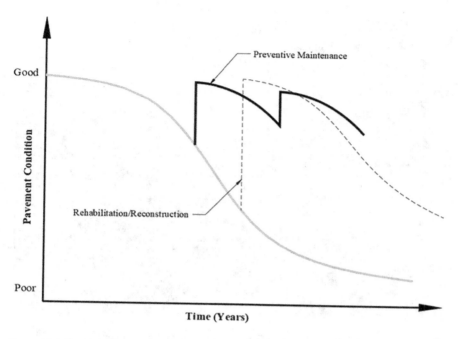

Figure 12.6 Pavement Condition Over Time

Source: AASHTO, Pavement Management Guide, 2nd ed., 2012

Table 12.5 Comparison of Project/Treatment Selection Methods

	Class of Method	*Advantages and Disadvantage*
Ranking	Simple, subjective ranking of projects based on judgment, overall condition index, or decreasing first-year cost (single-or multi-year)	Quick, simple; subject to bias and inconsistency; may be far from optimal
	Ranking based on condition parameters, such as serviceability or distress; can be weighted by traffic (single-or multi-year)	Simple, easy to use; may be far from optimal, particularly if traffic weighting is not used
	Ranking based on condition parameters and traffic, with economic analysis including decreasing present worth-cost or benefit-cost ratio (single-or multi-year)	Reasonably simple, may be closer to optimal
Prioritization	Near-optimization using heuristic approaches, including incremental benefit-cost ratio and marginal cost-effectiveness (maintenance, rehabilitation, and reconstruction timing taken into account); usually conducted as a multi-year analysis	Reasonably simple; suitable for microcomputer environment; close to optimal results
Optimization	Annual optimization by mathematical programming model for a year-by-year basis over the analysis period	Less simple; may be closer to optimal; effects of timing not considered
	Comprehensive optimization by mathematical programming models taking into account the effects of maintenance, rehabilitation, and reconstruction timing	Most complex and computationally demanding; can give an optimal program (maximization of benefits or cost-effectiveness)

Source: AASHTO, Pavement Management Guide, 2nd ed., 2012

Figure 12.7 Iconic Bridges: Brooklyn Bridge (Top Left), Golden Gate Bridge (Bottom Left), Lin Cove Viaduct (Right)

helping to transform communities and bring prosperity and growth to nations. The estimated number of bridges in the United States alone is a staggering 600,000. The average age of the bridges in the United States is 42 years, while the typical lifespan for a bridge is 50 years, meaning that many of the bridges we drive over daily are reaching the end of their usable service life. A recent estimate for the nation's backlog of bridge repair needs is $125 billion. The American Society of Civil Engineers estimates that spending on bridge rehabilitation needs to increase from $14.4 billion annually to $22.7 billion annually, or by 58%, if bridge conditions are to improve. At the current rate of investment, it will take until 2071 to make all of the repairs that are currently necessary, and the additional deterioration over the next 50 years will become overwhelming.

Bridges pose an extraordinary challenge in terms of maintenance, safety, and mobility. Because of their very nature of connecting two sides of a gap, some of the more critical components are difficult to access and, as such, are often overlooked. This connecting of two sides makes them critical for traveler mobility and a significant safety concern because of the lives at stake if a bridge were to collapse.

Our modern bridge inspection program has its roots in the Silver River Bridge collapse in West Virginia in 1967. Figure 12.8 shows newspaper photos from that event. This was not the first bridge collapse in the United States, but to that date, it was the most significant, killing 46 people. Following that collapse, the National Bridge Inspection Standards (NBIS) regulation passed into law on April 27, 1971. This legislation required inspection of bridge components every two years. In the following decades, bridge collapses and their investigations led to further refinements in the NBIS program.

Figure 12.8 Silver River Bridge Collapse

Bridge management focuses on making informed and effective decisions on the operation, maintenance, preservation, replacement, and improvement of bridges within a transportation agency's bridge inventory. Decision making for structures is highly dependent on relevant and quality data and on methodologies and tools for analyzing that data across an inventory of bridges. In this regard, the FHWA promotes the use and understanding of bridge management systems. The modeling and analyses performed by bridge management systems assist bridge owners in making informed and effective decisions that extend the life of their structures.

12.6.2 National Bridge Inspection System (NBIS)

Federal code requires a thorough and complete bridge inspection of all components every two years. This inspection depends upon the bridge inspector's ability to identify and understand the function of these major bridge components, their elements, and their functions. As shown in Figure 12.9, most bridges can be divided into three basic parts: deck, superstructure, and substructure. The NBIS Inspection Program dictates that each of these basic parts and all of the elements that make them up must be inspected every two years.

For decades, these inspections were completed visually by teams of two inspectors. The teams filled out standardized paper forms, took pictures, performed both destructive and non-destructive testing, and estimated the amount of deterioration found. Advances in technology, such as electronic tablets and wearable computers, have also increased the accuracy and speed with which these inspections can be completed. Starting with the passage of draft Federal drone rules in 2019, drone technologies can now be used in some circumstances. This advance has improved the safety of the inspection teams and enabled them to see areas of the bridge that might otherwise be hidden from an onsite inspector.

This data is then input into a bridge management system and used to develop the agency's bridge improvement program and report conditions to the federal government.

12.6.3 Components of a BMS

There are many commercially available bridge management software systems on the market, and because of the NBIS standardization, their inputs and outputs are very similar. Most agencies follow a standardized process in managing their bridges. This process is done on a periodic basis (typically annually), and common workflow steps and features are shown in Figure 12.10, including the following.

Collect Bridge Data: Thincludes is both inventory and condition data and encompasses a variety of bridge attributes, including length, width, material type, and bridge rail types, among just a few.

Define Agency Inputs: Performance measures are pretty uniform, but agencies may have different performance measures or objectives beyond those specified by FHWA. Agencies will also define their maintenance activities and trigger rules based on their climates and desires. These

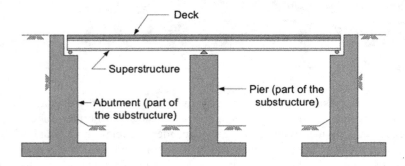

Figure 12.9 Bridge Components

Source: Photo courtesy Bridge Inspector Reference Manual

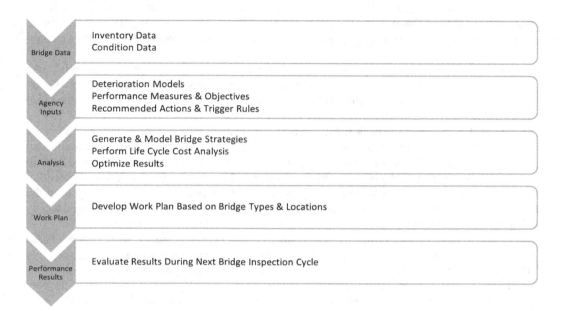

Figure 12.10 Common Workflow Steps

maintenance activities, treatments, and performance measures then have significant input into the structures' deterioration curves.

Perform Analysis: With the bridge data and agency inputs complete, the bridge management system generates scenarios and life-cycle cost analysis calculations to develop an optimized work plan for engineers to review.

Final Work Plan: As engineers review the optimized work plan, they may choose to adjust it based on their knowledge of local conditions. For example, if an engineer sees that two bridges near each other are due to receive preservation treatments in two consecutive years, they may choose to move these treatments to fall in the same year for economies of scale and to reduce traffic impacts.

Review Performance Results: The cycle begins again with the next collection of bridge condition and inventory data.

12.6.4 Inputs

12.6.4.1 Bridge Inventory Data

The bridge management process begins with agencies inventorying and inspecting their bridges based on National Bridge Inventory (NBI) condition standards as well as other agency-defined items. Once the inspection is complete, the bridge and all its individual elements are rated to determine their condition based on NBIS Standard General Condition Ratings (GCR) and element condition ratings.

Bridge management systems are the central repository for bridge inventory and condition assessment data, including bridge inspection and load rating information. These systems, such as Figure 12.11, display screens that review the types of inventory and condition used in BMS, including nationally defined data as well as agency-defined data. A strong BMS can improve agency performance by connecting agency goals and performance measures to asset management decisions.

Inventory data on a structure include standard items such as structure type and material, age, feature spans, height, and functional class. The three major components of a bridge (deck, superstructure, and substructure) are made up of many elements, and each of these elements must also be inventoried.

Bridge decks consist of the deck itself, the bridge railings, the bridge joints, and any drainage or lighting features. Bridge decks can be made of timber, concrete, or steel. Railing systems can be monolithic concrete barriers or open railings made of concrete or aluminum. Bridge decks will have a joint at each span on the bridge, and these joints will have some mechanical or rubber joint designed to give the bridge deck room to expand and contract and channel water away from the substructure components below. Additional drainage along the sides of the bridge, as well as any lighting attached to the bridge, complete the definition of the bridge deck component.

Likewise, bridge superstructures can be made up of timber, concrete, or steel girders. These girders come in many different types, including beams, arches, and prestressed boxes. Depending on the type of girder, there may be other types of cross-bracing components. Bearing assemblies are made up of a variety of components, depending on the type of structure and materials. Key components include bearing plates and bearing assemblies such as rocker bearings, roller bearings, and pin-and-link bearings, to name a few.

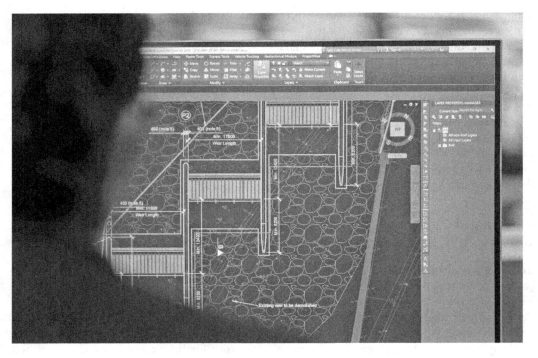

Figure 12.11 Typical Bridge Management System Display

Substructures comprise footings, abutments, and wingwalls at the ends of the bridge, interior piers, or columns with caps that the superstructure sits on. Footings are concrete, and most of the other substructure components are also concrete.

12.6.4.2 Bridge Condition Data

Bridges are designed to carry or resist loads, and it is these design loads that determine the size and configuration of their members. Bridge members are designed to withstand the loads acting on them safely and economically. When bridges begin to deteriorate, these load capacities are diminished and can result in bridge failure.

Because of this, FHWA requires that all of the bridge components listed above be inspected for damage and deterioration. Since condition data can change with each inspection and bridge collapses can happen suddenly, these condition inspections are required every two years. All bridge inspectors must attend inspection courses and be certified through FHWA.

The condition of bridges is rated on a nationally standardized NBI scale of 0–9, with 0 being a failed condition state and 9 being in excellent condition, as shown in Table 12.6. These detailed inspections are conducted on each component of each bridge element and can take a few hours to several days to conduct, depending on the complexity of the structure. These individual element ratings are aggregated into an overall bridge rating.

12.6.5 Analysis Parameters

Element-level bridge condition data from the inspection teams are imported into the BMS and aggregated into recommended strategies for each structure. Basic decision trees shown in Figure 12.12 include three significant categories of work: Preservation/Preventative Maintenance, Rehabilitation, and Replacement.

Preservation and preventative maintenance are those functions that must be completed routinely on bridges. Cyclical activities such as cleaning and washing bridge decks, cleaning expansion joints, and sealing the concrete fall into this category and should be completed routinely. Most states clean bridge decks and expansion joints annually after the winter season to remove salt from snow and ice operations. Other preservation activities include crack sealing of the bridge rails and

Table 12.6 NBI Rating Scale

Rating Number	NBI Description	Performance Measure Classification (23 CFR 490)
9	Excellent Condition	Good
8	Very Good Condition	
7	Good Condition	
6	Satisfactory Condition	Fair
5	Fair Condition	
4	Poor Condition	Poor
3	Serious Condition	
2	Critical Condition	
1	"Imminent" Failure Condition	
0	Failed Condition	

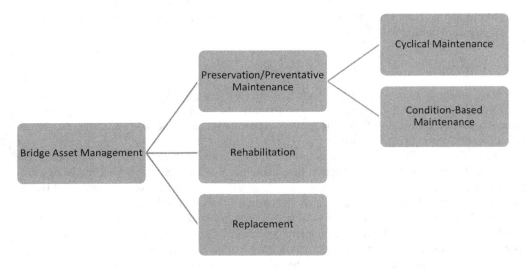

Figure 12.12 Basic Bridge Maintenance Decision Trees

polymer deck overlays. These activities are performed on bridges in good to excellent condition with ratings from 7 to 9.

Also in the preservation and preventative maintenance category is something called condition-based maintenance activities. These activities include replacing spalled concrete, applying epoxy sealers to the superstructure and substructure bridge elements, and replacing deck joints. These work activities keep bridges with ratings from good to satisfactory (ratings from 5 to 6), ranging from deteriorating any further and extending their life.

Rehabilitation activities target bridges in poor condition (rating of 4). These bridges need significant work but are not unsalvageable. Types of activities included here would be hydro-demolition of the deck to the steel reinforcing and replacing it, major railing repairs or replacements, steel and concrete beam repairs, and substructure column repairs.

Replacement of the entire structure is used when it is no longer cost-effective to maintain the structure, or the condition has deteriorated into a serious or critical state (ratings from 2 to 3). Very rarely will a transportation agency find a bridge with a rating of zero to one.

Each of the categories described above is illustrated in Table 12.7.

12.6.6 Analysis Module

The analysis parameters and everyday treatment actions help define the bridge deterioration models. Figure 12.13 below illustrates a typical bridge deterioration curve overlaid with the treatment categories. As with all deterioration curves, each treatment will improve the bridge condition and slow the deterioration over time.

Bridges contain many different elements of many different materials, and as such, the analysis of an individual bridge can involve multiple iterations to achieve the best treatment options. Multiply this nuance across the thousands of bridges a transportation agency is responsible for, and this bridge analysis can become very complicated very quickly.

Table 12.7 Bridge Conditions and Common Actions Associated with Each

Bridge Rating	Description	Common Actions
9	Excellent Condition	Preservation & Cyclic Maintenance Activities
8	Very Good Condition – No problems rated	
7	Good Condition – Minor problems reported	
6	Satisfactory Condition – Structural elements show some minor deterioration	Preservation & Condition-Based Maintenance Activities
5	Fair Condition – All structural elements are sound but may have some minor section loss, cracking, spalling, or scour	
4	Poor Condition – Advanced section loss, spall or scour	Rehabilitation or Replacement
3	Serious Condition – Loss of section, deterioration, spalling, or scour have seriously affected primary structural components	
2	Critical Condition – Advanced deterioration of primary structural elements, fatigue cracking in steel, or shear cracking in concrete may be present	
1	"Imminent" Failure Condition – Major deterioration or section loss in critical structural components or obvious vertical or horizontal movement	
0	Failed Condition – Out of service, bridge is beyond corrective action	

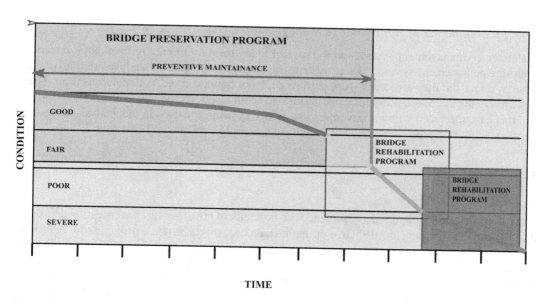

Figure 12.13 Typical Bridge Curve Overlaid with Treatment Categories

The analysis module runs thousands of scenarios on the various elements separately. Using a single bridge as an example, the analysis engine will factor in the individual condition rating for each column or footing for a substructure and the various treatment options for the identified defects to develop an optimal treatment plan. It will then complete a similar analysis for each beam, bearing pad, cross brace, and so forth for the bridge's superstructure, developing its optimal plan before proceeding to complete the same analysis for the deck. Once these calculations are complete for each element, the analysis module develops an entire bridge-optimized plan based on the agency's goals. This process is repeated for all the bridges in the agency's portfolio. This process could never be completed by hand due to the large number of calculations needed, but with bridge management software, it can be completed in a few minutes.

12.7 Maintenance Management

12.7.1 Background

Highway agencies have traditionally focused their asset management efforts on pavement and bridge assets since much Federal guidance exists. Over the last 30 years, transportation agencies have begun to consider the importance of those ancillary assets (e.g., signs and signals, guardrails, culverts, lighting, pavement markings, retaining walls, etc.) that they maintain. These assets contribute a large part to the safety of a roadway. Many agencies have implemented maintenance quality assurance (MQA) programs to help maintenance personnel understand the overall condition of the system and identify and prioritize maintenance needs.

12.7.2 Components of Maintenance Management System

Maintenance management systems on the market today operate differently than either pavement or bridge management systems. These systems inventory assets and run analyses similarly to the other systems, but they also track work performed, labor hours worked, and materials used. These systems are the main repository for work performed on all assets, including pavements and bridges, and then are interfaced with the pavement and bridge systems and used in the analysis. They also interface with the agency's financial management system to generate real cost data.

12.7.3 Inputs

12.7.3.1 Inventory Data

Since there is little standard federal guidance and no standard measures like pavements and bridges, most agencies have developed their own maintenance quality assurance programs to determine which assets are most important to them. Some states have detailed programs inventorying many of the assets they own (pipes, signs, guardrails, roadway lighting, etc.), while others have more streamlined programs tracking only a few assets and using random sampling. Both approaches are valid, and agencies must weigh the cost of collecting large quantities of data against the benefit of having it. States like North Carolina, with 80,000-mile systems and millions of pipes and linear feet of guardrail, decided that the value of the data wasn't worth the expense of collecting a full inventory and thus use a random sampling approach. On the other end of the spectrum are states like Utah, with small highway networks, which decided that the collection of a full inventory was worth the expense.

Using the asset management principles discussed earlier in this chapter, each state decides which assets strategically align with its agency's priorities, communicates clearly the condition of the highway system, and can justify funding needs. South Carolina elected to choose a group of ten assets that they felt best communicated their roadway maintenance needs and developed a condition rating program around those.

12.7.3.2 Condition Data

Minnesota is one state that has a robust MQA program collecting inventory and condition data on a large number of its assets. They use this condition data in a variety of applications including legislative reporting, public information, and driving maintenance operations. Table 12.8 shows some common condition data collected by a majority of states.

When MQA programs began in the 1990s, they were predominantly done as walking surveys with teams of two people. Today many of these condition surveys are conducted by LiDAR trucks traveling at highway speed. If the asset is visible from the roadway, there is a technology that can determine its condition. The one critical feature that must still be surveyed in a mostly manual fashion is drainage structures. Robotic cameras can assist with getting detailed information, but this is still a manual process requiring workers to be exposed to environmental hazards. Such hazards include poisonous vegetation, wildlife, and traffic.

12.7.4 Analysis Parameters

Inventory and condition data are input into the maintenance management system and used to develop a suggested work plan. If the inventory and condition data are based on a sampling method, a specific suggested list of maintenance projects will not be feasible. In this case, the analysis will yield a quantity of work, and it will be up to the user to identify the work locations.

Other inputs an agency may choose to implement would be the amount of money they wish to spend on each type of activity. One agency may choose to spend a large amount of its budget on drainage and pipe repair activities, while the next agency may see a greater need for funding to be shifted to signs and guardrails.

12.7.5 Analysis Module

With the input parameters set, agencies will set their target level of service for each asset class. For example, in order to align with their stated safety goals, an agency may set a high desired level of service (LOS) for the guardrail. The MMS will then analyze current guardrail quantities and conditions and determine the amount of guardrail repair needed to meet that target LOS.

Table 12.8 Common Data Elements Collected by Most States

Asset Class	Asset Type	Condition Measured
Drainage	Pipes	Blocked, flow restricted
	Ditches	Eroded
	Drop Inlets	Blocked, settled, broken grates
Vegetation	Grass Mowing	Height
	Landscape Plantings	Appearance
	Brush	Blocked sight distance
Other Roadside Appurtenances	Sound walls	Functional, clear of vegetation
	Signs	Alignment, Reflectivity
	Guardrails	Damage, Meet current code

This approach is taken for each individual asset in the system until an estimated quantity of work is developed. This estimated work plan must then be vetted by local field engineers who better understand local conditions. They may choose to make adjustments to this plan repeatedly until a final work plan is established and re-entered into MMS.

This final plan will become the basis for the work crews, giving them a projected quantity of work they must perform in order to achieve the stated LOS targets. As work proceeds throughout the year, the performed work quantities are compared to the planned quantities. At the end of the period, new condition data is input, and along with the actual quantities of work performed, a new analysis is run, and a new Estimated Work Plan is developed.

12.7.6 Future of Asset Management Systems

12.7.6.1 Signs, Signals, and Intelligent Transportation Devices

Roadside signs have long been defined as an asset in a maintenance management system, but with the evolution of radio frequency identification devices (RFID) that can provide real-time field data, signs have begun to be classified with other traffic devices in their own traffic management systems.

The traffic management systems currently being developed are designed similarly to other management systems in that they contain inventory and condition data of their assets and use the agency to define goals, objectives, and deterioration data to create an optimized work plan. The major difference in these systems is the ability to have real-time data on the assets. Signs with RFID tags can inform the management system about damage that has occurred or the age of the sign. Signals, message boards, and toll gantries can communicate with the management system, relaying data on outages or timing problems.

12.7.7 Safety Analysis Systems

Safety analyst systems function slightly differently from other management systems. Instead of using asset condition data, these systems incorporate crash data, congestion data, and traffic counts to develop an optimal list of projects that would improve traveler safety and traffic flow. Agency goals and inputs are critical to getting quality output from these systems. Typical project lists could include things like intersection improvements, signal timing changes, widening projects, or roadway realignments.

12.7.8 Cross Asset Optimization

Transportation agencies must make hard decisions about how to spend the funds they have been given. There is never enough money to meet all the pavement, bridge, and roadside needs, and as a result, they must make a myriad of choices. If they choose to spend all the available funds on pavements, the bridges deteriorate, and pipes get clogged. If they choose to spend all the available funds on bridges, then the pavements deteriorate faster. The best solution is to spend some of the funds on all the asset classes, thereby standing a better chance of maintaining the system at an acceptable level. This optimization process can be achieved through various types of analysis depending on the agency's desires and is still futuristic to most transportation agencies (Proctor & Zimmerman, 2016).

12.7.8.1 Cross Asset Tradeoff

The first strategy to achieve the most effective mix of projects is called cross-asset tradeoff. This process moves resources from one asset class to another to maximize the perceived benefit. This perceived benefit is not a rigorous, mathematically based decision but rather an informal perception of one need versus another. For example, a transportation agency might decide to move funding from pavements to bridges based on public opinion or legislative inquiries regarding a perception that bridges are in worse condition.

12.7.8.2 Cross Asset Allocation

Cross-asset allocation is a decision-making process by which resources (funding) to multiple programs or asset classes are distributed based on simultaneously quantified prioritization. It includes analyzing each program's benefits, not just one program versus another.

This type of analysis is the most common and most easily understood. However, it is challenging to apply programmatically because it is hard to quantify the benefits of dissimilar programs. For example, comparing the benefits of a bridge preservation program to those of a pavement preservation program is very complicated. While they are both preservation programs, they each have different performance metrics and goals, making comparing their benefits. Costs are also hard to measure in some cases and rely on broad assumptions. It is also very difficult to quantify intangible benefits such as user costs for delays.

12.7.9 Multi-Criteria Decision Analysis

Multi-criteria decision analysis assigns values to dissimilar attributes and uses total values to prioritize an agency's options. This type of analysis assigns benefits to the various treatments for dissimilar assets, such as pavements and bridges. For example, a low-cost pavement preservation treatment might have an assigned benefit of one, whereas a bridge preservation treatment might have a higher benefit assigned to it. Still, other classes of treatments would have their own benefits assigned, allowing for the comparison of pavements to bridges to maintenance items.

12.7.9.1 California DOT Case Study

The California Department of Transportation (Caltrans) understood the need to develop a method for performing cross-asset optimization. They sought to move away from the traditional method of assigning a percentage of funding to each asset class and instead develop a performance-based cross-asset optimization methodology to support infrastructure programming and budgeting (Hicks, 2021).

The objective of the Caltrans project was to develop and demonstrate a cross-asset optimization methodology to help Caltrans optimize project selection and budget allocations, maximizing the value of the investment, and optimally achieving performance objectives by directing investments where most needed. Caltrans worked with a private engineering firm to develop a novel cross-asset optimization methodology based on a robust multi-objective optimization algorithm. The methodology integrated asset-level, system-level, and program-level analyses in a single framework to support efficient workflows between interdependent decision processes.

Implementation of the methodology was supported by an internal software tool. The software implements a comprehensive data model that embodies key data elements and relationships needed

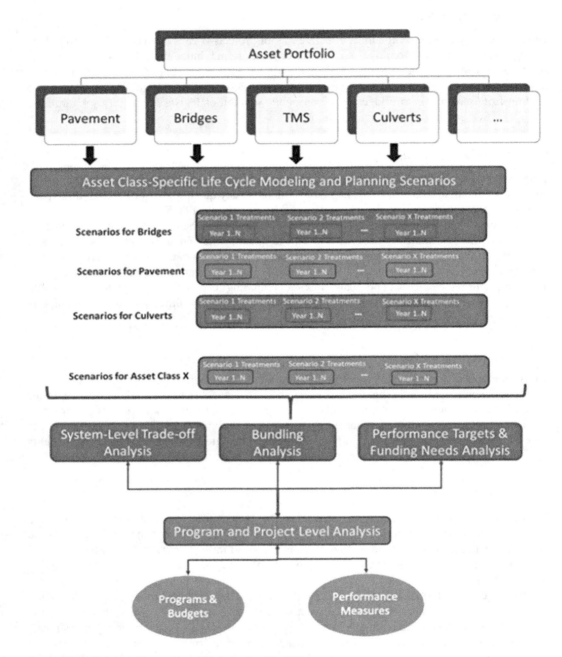

Figure 12.14 Caltrans Cross Asset Optimization Tool Diagram

to model and analyze asset data, needs, life-cycle performance, and programming and budgeting decisions.

Analyzing assets at the portfolio level provided the agency with the opportunity to find investment and work efficiencies through bundling of treatments and coordination of program development, management, and project delivery processes across different asset classes. Bringing a broader perspective to the portfolio planning process over longer planning horizons helps maximize the value of infrastructure investments.

Caltrans found that using this cross-asset optimization methodology and associated software tool can potentially support programming and budgeting decisions at Caltrans in a number of ways, such as:

Efficient development and management of programs and budgets across the entire transportation asset portfolio.

Identification of optimal budget allocation and balanced investment strategies to meet performance objectives and ensure the long-term sustainability of assets.

Quantify trade-offs between funding levels and performance measures for different asset classes and groups across the entire asset portfolio.

Implementation of common performance management, lifecycle models, trade-off analyses, and programming decision models across different districts, thus promoting consistency and transparency in project evaluations, and establishing a quantitative, data-driven, and repeatable process.

12.8 Questions

1. What is the definition of transportation asset management according to the American Association of State Highway and Transportation Officials (AASHTO)?
2. How does the Federal Highway Administration (FHWA) estimate the replacement value of the country's pavement and bridge assets?
3. What major federal legislation impacted transportation asset management in 2012, and what did it require from transportation agencies?
4. Why is policy-driven management important in transportation asset management?
5. Describe the significance of using performance-based principles in asset management.
6. What role does risk management play in transportation asset management?
7. Explain the concept of "strategically aligned with agency priorities" in the context of asset management.
8. How do transportation agencies use information to make decisions in asset management?
9. What is the importance of continuous improvement in the asset management process?
10. Describe the "E" asset management cycle as illustrated in Figure 12.1.
11. What are the common elements across different types of asset management systems?
12. How do the quality and quantity of condition data influence the effectiveness of an asset management system?
13. Discuss the role of deterioration models in predicting future conditions of assets.
14. Explain the different levels of a pavement management system as outlined in Table 12.1.
15. How does the National Bridge Inventory (NBI) condition standard rate bridge conditions?
16. What strategies are involved in bridge management for maintaining and improving bridge conditions?
17. Describe how maintenance management systems differ from pavement and bridge management systems.

18. What is the purpose of the feedback loop in asset management systems?
19. Explain how cross-asset optimization is performed using Caltrans as an example.
20. What are the primary challenges and benefits of implementing a maintenance quality assurance (MQA) program in transportation agencies?

Notes

1 AASHTO, 2022. Transportation Asset Management Guide.
2 AASHTO, 2022. Transportation Asset Management Guide, Subsection 1-1-1
3 AASHTO, 2022, TAM Guide, https://www.tamguide.com/wp-content/uploads/2020/02/TAM_GuideIII _ch01-20200227.pdf
4 Bureau of Economic Analysis, https://apps.bea.gov/iTable/iTable.cfm?ReqID=10&step=2

References

AASHTO (1993). *Guide for Design of Pavement Structures*. American Association of State Highway and Transportation Officials. American Association of State Highway and Transportation Officials (AASHTO), Washington, DC. At https://habib00ugm.wordpress.com/wp-content/uploads/2010/05/aashto1993.pdf

AASHTO (2022). *Transportation Asset Management Guide*. American Association of State Highway and Transportation Officials (AASHTO), Washington, DC. At https://www.transit.dot.gov/sites/fta.dot.gov/ files/2022-03/AASHTO-Transportation-Asset-Management-Guide.pdf

AASHTO (2012). *Pavement Management Guide*, 2nd Ed. American Association of State Highway and Transportation Officials (AASHTO), Washington, DC. file:///C:/Users/wangg/Downloads/PMG-2_Table OfContents.pdf

Hicks, L. (2021). *Cross-Asset Multi-Objective Optimization Approach for Caltrans.* Caltrans Division of Research. Sacramento, CA. At https://dot.ca.gov/-/media/dot-media/programs/research-innovation -system-information/documents/research-results/task3631-rrs-7-21-a11y.pdf

Dewan, S. & Smith, R. (2003). Creating Asset Management Reports from a Local Agency Pavement Management System. *Journal of the Transportation Research Board,* Transportation Research Record (TRB) 979979979979(1):845-5982. DOI:10.3141/1853-02

Hartle, R. A., Ryan, T. W., & Mann, E. (2002). *Bridge Inspectors Reference Manual*, Vol. 1 and 2. US Department of Transportation (USDOT), Washington,DC. At https://rosap.ntl.bts.gov/view/dot/39863

FHWA (2020). *Highway Statistics 2020*. Office of Policy Information, Federal Highway Administration (FHWA), Washington, DC. At https://www.fhwa.dot.gov/policyinformation/statistics/2020/

GASB (1999). Basic Financial Statements-and Management's Discussion and Analysis for State and Local Governments. US Governmental Accounting Standards Board (GASB). Summary of Statement No. 34. Norwalk, CT. At https://www.gasb.org/page/PageContent?pageId=/standards-and-guidance/pronouncements /summary-statement-no-34.html

Flintsch, G. & McGhee, K. (2009). *Quality Management of Pavement Condition Data Collection,* National Cooperative Highway Research Program (NCHRP)Synthesis 401. At https://nap.nationalacademies.org/ read/14325/chapter/1

OECD (1987). *Pavement Management Systems*. Organization of Economic Cooperation and Development (OECD). File Accession Number 00472504. Paris, France.

Peshkin, D.G., Smith, K.D., Zimmerman, K.A., & Geoffroy, D.N. (1999). *Pavement Preventive Maintenance Reference Manual*. Publication FHWA-HI-00-004. National Highway Institute, Federal Highway Administration (FHWA), Washington, DC.

Proctor, G. & Zimmerman, K. (2016). *Asset Management: Defining Cross Asset Decision Making*. American Association of State Highway and Transportation Officials (AASHTO), Washington, DC. At https://www .tam-portal.com/wp-content/uploads/sites/12/2016/01/Cross-Asset-Allocation.pdf

Zimmerman, K.A. (2017). *Pavement Management Systems: Putting Data to Work*. National Cooperative Highway Research Program (NCHRP) Synthesis 501. At https://nap.nationalacademies.org/catalog /24682/pavement-management-systems-putting-data-to-work

Chapter 13

Transportation Infrastructure Construction Administration

13.1 Introduction

13.1.1 Importance of Construction Administration

Highway agencies spend hundreds of millions of dollars annually to improve transportation facilities. Whether it is building a new location roadway from subgrade to final pavement or simply resurfacing an existing roadway, taxpayers expect to receive the best possible value for the money expended.

Hours are spent planning and designing every project detail before it is bid and awarded to a contractor. The materials used must adhere to strict specifications if they are going to perform correctly. The location, lengths, and numbers of features such as pipes and catch basins have been painstakingly quantified so that accurate cost estimates can be calculated and the project built accordingly.

13.1.2 Role of the Owner

The owner is pivotal in a highway construction project as they are the primary entity responsible for initiating, financing, and overseeing its execution. The owner can be a government agency, a private company, or any organization that holds the authority and funding for highway construction.

To ensure that taxpayers are getting a quality product and that contractors are being paid for their work, the owner agency is responsible for administering the contract and providing some inspection services. Agencies can choose to perform this work with their employees or outsource it to private engineering firms that provide these services. Most state highway agencies complete this function with a combination of their staff, private engineering firms, and material testing firms.

13.1.3 Role of the General Contractor

The role of the general contractor in a highway construction project is crucial, as they serve as the primary entity responsible for overseeing and managing the construction process. The general contractor is responsible for completing and coordinating various aspects of the project, including labor, equipment, materials, and subcontractors. They work closely with the highway agency (owner) and other stakeholders to ensure the project is completed on time, within budget, and in accordance with the specified plans and specifications (NCDOT, 2023).

The general contractor is responsible for organizing the construction site, scheduling activities, procuring necessary permits and approvals, and implementing appropriate safety measures. They manage subcontractors, monitor quality control, and address any issues or challenges during

DOI: 10.1201/9781003197768-13

construction. The general contractor acts as a central point of contact, facilitating effective communication among all parties involved and ensuring that the project progresses smoothly from start to finish (FHWA, 2009).

13.2 General Considerations

Contract administration is a fundamental aspect of highway construction projects, and before moving into the more technical aspects, we must first address some all-encompassing, overarching topics that are essential for ensuring project success and compliance with contractual obligations.

13.2.1 Project Safety

One of the first considerations owner agencies and contractors must think about is project safety. That is the safety of the contractor and the traveling public. Highway projects most often happen mere feet from moving traffic traveling at speeds of 55 MPH or greater, where only plastic barrels separate workers from sudden catastrophe.

In December 1970, the Occupational Safety and Health Act of 1970 (Public Law 91-596) was enacted by Congress, mandating occupational safety and health standards for all construction activities. It also provided that existing federal safety standards would become a part of the standards. Each state's department of labor is responsible for administering occupational safety and health standards applicable to most public businesses and private entities, meaning that contractors

Figure 13.1 Work Zone Image
Source: Photo Courtesy of NCDOT and Steve Kite

performing under a construction contract with the transportation agency are required to comply with all provisions of the federal OSHA regulations as well as any applicable state regulations.

For their part, contractors understand the seriousness of safety on their projects and often have robust safety programs employing company safety managers who train and provide guidance within the company on all matters related to health and safety. Contractors want their workers to get home safely at the end of each day and understand the financial implications of worker safety.

Just like worker safety, the safety of the traveling public is a shared responsibility between the contractor and the transportation agency owner of the project. While the contractor is directly involved in executing the construction activities, the transportation agency is responsible for the highway's safety and functionality. Both parties must prioritize safety measures to protect motorists, pedestrians, and other road users throughout the construction process.

The contractor must adhere to stringent safety protocols, implement appropriate traffic control measures, and maintain clear and safe pathways for travelers. On the other hand, the transportation agency owner plays a critical role in overseeing and enforcing safety regulations, conducting regular inspections, and ensuring that the contractor follows established safety guidelines.

13.2.2 Engineer's Authority

All contracts designate the owner's representative, and highway construction projects are no different. Most state transportation agencies designate this person as the "engineer" or duly "authorized representative." As such, the engineer is responsible for ensuring that the project is constructed following the terms of the contract.

The contract includes a variety of documents, including the proposal form, the printed contract form, and all attachments such as the contract bonds, the plans, the standard specifications, all supplemental specifications, and all executed supplemental agreements. The engineer should carefully study all contract components and their relationship to actual field conditions.

The engineer has the authority to alter the contract terms through a supplemental agreement or change order, as well as modify quantities and plan details. Significant changes, however, should not be made without consulting the designer or others as may be appropriate. Each agency details decision-making levels, and the engineer should be aware of what constitutes each decision level. These levels provide guidance for the engineer when significant modifications are needed in quantities that significantly change the project's cost or affect the design. Some common examples would include changes in undercut excavation, borrow, excavation, or subgrade stabilization. In most instances, project plans include designs and details of construction placed in the contract for specific reasons, and they should not be altered without sufficient justification and review to determine that the integrity of the design has not been affected. This type of decision-making requires experience and technical knowledge. Decisions should be based upon due deliberation and at all times within the framework of the contract (FDOT, 2021).

13.2.3 Conformance with Plans and Specifications

Like anything you buy, you expect to get what you pay for. Highway construction projects are the same. Transportation agencies spend millions of dollars locating, designing, obtaining the necessary permits, and contracting a project. As such, they are obligated to the public to build a roadway according to plans and specifications, leading to a long-lasting, low-maintenance facility. Ultimately, adherence to plans and specifications in highway construction helps deliver high-quality and reliable transportation infrastructure that meets the community's needs and enhances overall road safety (FDOT, 2023).

Compliance with plans and specifications in highway construction is of paramount importance due to several crucial reasons. Adherence to plans and specifications ensures that the constructed highway meets the intended design and functionality, guaranteeing a safe and efficient roadway. By following the approved plans, the project can maintain alignment with engineering principles and industry standards, minimizing the risk of design flaws or structural inadequacies. Compliance with plans ensures consistency and uniformity across different sections of the highway, promoting smooth traffic flow and minimizing potential hazards.

It also enables effective coordination among various project stakeholders, including engineers, contractors, and inspectors, fostering clear communication and a shared understanding of project requirements. Compliance with plans and specifications is essential for regulatory and contractual adherence, allowing the project to meet all legal and environmental obligations, avoid penalties, and maintain public trust.

13.2.4 Environmental Issues and Their Significance

Environmental issues are a growing concern, and the impact of highway construction activities cannot be ignored. Significant environmental issues involving the transportation industry include compliance with the Federal Endangered Species Act and other wildlife regulations, protecting forests, general environmental protection, and stormwater management and erosion control.

13.2.4.1 Compliance with Federal Endangered Species Act and Other Wildlife Regulations

Complying with the Federal Endangered Species Act (ESA) and other wildlife regulations is crucial for construction projects. The ESA aims to protect endangered and threatened species and their habitats. Before a project is even designed, transportation owner agencies have to conduct thorough environmental studies documenting the types of wildlife and plants within the project limits. They complete environmental impact assessments (EIA) and work closely with environmental regulatory agencies to obtain the necessary approvals and permits to proceed. These permits will include wetland impacts, threatened and endangered species impacts, or parks or cultural resources impacts, to name a few (FHWA, 2024).

These permits often take years to obtain, and because of this, project design plans are not fully developed until these environmental permit conditions are prepared. Contractors are responsible for complying with all project requirements. By adhering to the ESA and the conditions of all permits and wildlife regulations, construction activities can avoid or minimize harm to affected species, their habitats, and the overall ecological balance.

13.2.4.2 Environmental Protection

Environmental protection encompasses many concerns, including air and water quality, waste management, and energy conservation. Because of the linear nature of highway construction projects, they have the potential to generate pollutants over a vast area and cause significant damage to adjacent properties. To minimize environmental impacts, highway designers and planners focus on preserving and protecting natural habitats and ecosystems. They aim to avoid or minimize disruption to sensitive areas such as wetlands, forests, and endangered species habitats. Designers consider the potential impact on water resources, seeking to reduce water pollution through proper drainage systems and erosion control measures.

Forests are a significant element of environmental protection. They play a vital role in environmental conservation, providing numerous benefits including wildlife habitat and watershed protection. Construction projects must prioritize the protection of forests by implementing measures such as tree preservation, reforestation, and sustainable timber harvesting practices. Highway contracts typically require the contractor to install tree protection fencing around protected areas and include pay items for reforestation. By preserving existing forests and replanting trees, the construction industry helps to maintain biodiversity and safeguard water resources.

13.2.4.3 Stormwater Management and Erosion Control

Highway construction activities can significantly impact stormwater runoff and erosion, leading to sedimentation in water bodies, degradation of aquatic ecosystems, and damage to private property. Implementing effective stormwater management and erosion control measures is crucial and mandated by contract requirements. Construction contracts require sediment basins, silt fences, and erosion control blankets to minimize soil erosion and capture sediment-laden runoff. These items are shown in the construction plans and paid for as part of the contract. Erosion control devices are inspected weekly and after each rainfall event.

By prioritizing environmental compliance, owners and contractors can contribute to sustainable transportation projects, the preservation of biodiversity, and responsible environmental stewardship. Implementing effective project stormwater management and erosion control measures can minimize their environmental footprint, preserve natural resources, and ensure a healthier and more sustainable future for future generations (AASHTO, 2024).

13.3 Construction Documents

13.3.1 Letting Documents

The construction contract includes various documents defining the scope of work for the project and how the contractor will be paid.

When the project is advertised, a set of pre-bid documents will be made publicly available. As shown in Figure 13.2, these documents include the contract proposal, which includes the project's

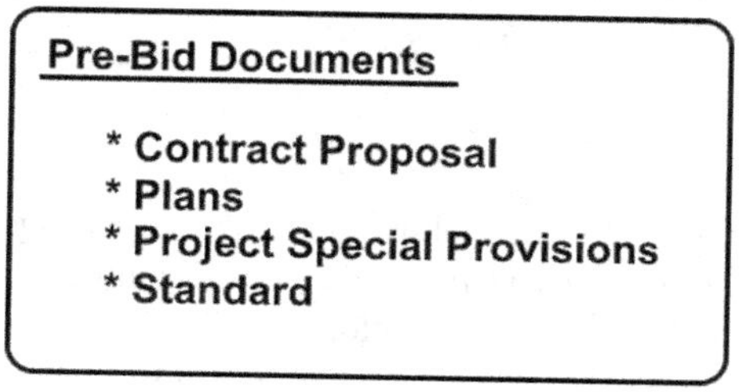

Pre-Bid Documents

* Contract Proposal
* Plans
* Project Special Provisions
* Standard

Figure 13.2 Pre-Bid Documents

special provisions, the plans, and any special environmental permits. At this point, prospective contractors have access to the project details.

The plans detail what work is to be completed, including a detailed summary of quantities, roadway plans, structure plans, traffic control plans, and any other plans or environmental documents. The contract proposal details bid due dates, project completion dates, and other special project provisions. A typical project special provision would be dates the contractor could not be on the road because of a local event or holiday.

The standard specifications and standard drawings are common to all construction projects and are not included in the specific proposal package, but instead are understood to be included. They are standardized documents developed and maintained by transportation agencies or engineering organizations to ensure uniformity and consistency in construction practices across different projects (Uddin, Hudson, & Hass, 2013).

13.3.1.1 Standard Specifications

Standard specifications in highway construction outline the technical requirements, quality standards, and construction methods that must be followed during construction. These specifications cover material standards such as aggregate gradations, testing procedures, construction practices, and safety measures. They provide detailed instructions to contractors on performing the work and ensuring that the completed highway meets the required standards.

13.3.1.2 Standard Drawings

Standard drawings represent various highway elements and components, such as road geometries, pavement cross-sections, signage, traffic signals, and other infrastructure features. They provide visual guidance and detailed dimensions for the construction and placement of these elements. Standard drawings serve as references for design engineers, contractors, and construction crews to ensure the highway is constructed accurately and consistently across different locations.

On the letting date, the contractor submits a bid package with a signed copy of the proposal, a completed bid sheet showing the company's unit price bids, and any required bond documents. By submitting this package, the contractor agrees to the terms of the proposal, plans, and specifications and stands by the prices included in the completed bid sheet. For low-bid contracts, most State DOTs open bids, publicly announcing the total amount bid on all submittals and naming the apparent low bidder. On projects that are not low bid but have a combination of qualifications and price, the contractor submits a technical proposal and a price proposal. These types of selections are more involved, take longer and are often done in private with only the final results being announced.

13.3.2 Award Documents

After the public bid opening, the owner agency will review all bids, looking for irregular pricing, missing documents such as bonds, or invalid signatures/company seals. The selected contractor will be notified once all the checks have been completed and a winning bidder has been determined. A public announcement of the award will usually be made on the owner agency's webpage. At this point, the contract is awarded, and the award documents shown in Figure 13.3 become the binding documents on which the contract will be administered.

Award Documents

* **Signed Contract**
* **Contract**
* **Prices and Quantities**

Figure 13.3 Award Documents

Post Award Documents

* **Executed Change Orders**
* **Pay Records**
* **Inspection Reports**
* **Project Diaries**

Figure 13.4 Post Award Documents

The contractor will have some time to submit any outstanding documents, such as insurance and workers' compensation documentation. Each owner agency has some dispute resolution process, and contractors not selected have a period to file a protest with the owner agency. Agency processes vary widely, and the documentation a protesting contractor can access is limited.

13.3.3 Post-Award Documents

Once the project is awarded and work begins, Figure 13.4 details additional contractually binding documents that are processed. Change orders, or supplemental agreements, are formal written agreements that modify the original contract when there is a need for changes in scope, schedule, or compensation. These documents outline the changes to be made, the impact on project cost and time, and the agreement of both parties to proceed with the changes.

Other types of documentation include pay records, inspection reports, and project diaries that track the progress of work completed and serve as the basis for payment requests by the contractor. They typically include detailed information on completed milestones, quantities, unit prices, and any deductions or retainage as per the contract terms.

13.3.4 Plans and Shop Drawings

13.3.4.1 Highway Construction Plans

Highway construction plans are comprehensive documents that provide a detailed roadmap for highway project construction. These plans serve as a guide for engineers, contractors, and stakeholders involved in the construction process. The specific content and level of detail in highway construction plans will vary depending on the project's scale and complexity. However, some common elements found in these plans include:

- Design Drawings: These drawings illustrate the proposed highway layout, including lane configurations, intersections, interchanges, ramps, bridges, and other infrastructure elements. They provide detailed information on the roadway's alignment, cross-sections, and dimensions, as well as the location and design of various structures and features.
- Schedule of quantities: This section outlines the amounts proposed for various construction activities, including clearing and grading, pipes and drainage structures, pavement, concrete, and bridge items.
- Environmental Considerations: As stated earlier in this chapter, highway construction plans address environmental factors such as erosion and sediment control measures, stormwater management, and protection of natural resources. They may include provisions for environmental permits, mitigation strategies for sensitive habitats, and compliance with applicable regulations.
- Traffic Control: Detailed traffic control plans are included to ensure the safety of workers, the traveling public, and pedestrians. These may consist of temporary traffic shifts, signage, speed limits, and procedures for handling hazardous materials.

Highway construction plans are dynamic documents that undergo revisions and updates throughout the project lifecycle. They serve as a critical reference for project stakeholders, facilitating effective communication, coordination, and adherence to project goals and specifications.

13.3.4.2 Shop Drawings

In highway construction, shop drawings are essential in translating the design intent into detailed technical drawings that guide the fabrication, assembly, and installation of various components and elements. Shop drawings are typically created by subcontractors, manufacturers, or suppliers and provide specific details and dimensions for constructing items such as signage, bridges, guardrails, lighting fixtures, and other highway infrastructure components.

Shop drawings serve as a means of communication between the design team and the subcontractors or suppliers. They visually represent how the designed elements will be fabricated, assembled, and integrated into the overall highway project. They contain precise measurements, dimensions, and specifications that allow subcontractors or manufacturers to fabricate components accurately. They provide detailed information about materials, finishes, fasteners, connections, and installation methods.

Shop drawings serve as a record of the manufacturing and fabrication process. They include details of materials used, manufacturing techniques, and quality control measures implemented during the construction of the components. These drawings can be referenced during inspections, maintenance, or future modifications.

Design professionals must review shop drawings, and this review may take multiple iterations, requiring revisions or modifications to ensure accuracy and alignment with project requirements. This process can take months to complete, so contractors are often encouraged to start the process started as soon as possible.

13.4 Construction Phase

13.4.1 Project Organization and Staffing

Efficient project organization and staffing are crucial in successfully executing highway construction projects. The construction management team involves both the owner agency and the contractor, and they play a pivotal role in overseeing and coordinating various aspects of the project, ensuring its timely completion and adherence to quality standards and budget limitations. The best teams working collaboratively to solve problems achieve the best outcomes (AASHTO, 2020).

13.4.1.1 Project Organization

Effective project organization begins with establishing a well-defined structure that assigns clear roles and responsibilities to team members. In highway construction projects, the construction management team typically comprises professionals from both the contractor and the owner with diverse expertise and responsibilities. Even a small highway construction project has dozens of team members. Some key team members are described below, but this is far from an all-inclusive list.

13.4.1.2 Owner's Project Engineer

The owner's project engineer, referred to in the contract as the engineer, is responsible for ensuring the project is constructed per the project plans and specifications and within budget requirements. They oversee the owner's inspection personnel, process contractors' payments, and communicate with stakeholders. They are the primary point of contact between the contractor, stakeholders, and the public. They ensure that the project objectives are met, monitor progress, and address any issues that arise during construction.

Generally, the engineer will have multiple projects to manage, so they delegate the authority to inspect the work and materials incorporated into it to project personnel known as project inspectors.

13.4.1.3 Owner's Project Inspectors

Project inspectors work closely with the contractor, providing documentary evidence that the work and materials conform to the contract terms. They are thoroughly familiar with the project plans and work daily with the contractor to ensure that the quantities of work completed are documented in the required manner for payment processing. They are expected to bring immediate attention to any work or material that does not conform to the contract provisions. They are familiar with all contract requirements and certified for the various phases of work they are called upon to inspect.

13.4.1.4 Contractor's Project Manager

The contractor's project manager focuses on the on-site execution of the project. They supervise the construction activities, coordinate with subcontractors and suppliers, and manage labor resources. The construction manager ensures that the project is executed in accordance with plans, specifications, and safety regulations. They are responsible for monitoring construction progress, quality control, and timely completion of work, and act as the key point of contact with the owner's engineer.

13.4.1.5 Contractor's Estimator

Although usually not physically on the project, the contractor's estimator is crucial for accurately assessing project costs, including material, labor, and equipment expenses. During the project's bidding phase, they evaluate project plans and specifications, conduct quantity take-offs, solicit quotes from suppliers and subcontractors, and prepare cost estimates. During the project's construction phase, they track project expenses and compare the actual costs to the planned costs, evaluating overruns and underruns and finding ways to stay on budget.

13.4.1.6 Contractor's Quality Control Manager

The contractor's quality control manager is responsible for ensuring that construction activities adhere to established quality standards and specifications. They develop and implement quality control plans, conduct inspections, and monitor compliance with regulatory requirements. The quality control manager identifies and rectifies deficiencies, ensuring the constructed highway meets the owner agency's contract requirements.

13.4.1.7 Responsibilities of the Construction Management Team

This construction management team is central in overseeing and coordinating the project. Each member has different roles and, as a team, they collectively shoulder several responsibilities throughout the highway construction project.

13.4.1.8 Project Planning and Execution

While the contractor is ultimately responsible for the project plan and schedule, the team collaborates on timelines and proposed work activities to allow for better staffing and public information, such as a weekend road closure that will require additional staff and a public information campaign. This communication ensures the project progresses according to the plan and makes adjustments as necessary.

13.4.1.9 Stakeholder Communication

The construction management team maintains effective communication with stakeholders, regulatory authorities, and the community, providing regular updates on project progress, addressing concerns, and ensuring compliance with regulations and permits.

13.4.1.10 Safety and Risk Management

The contractor is also responsible for project safety, but the inclusion of the project owner's staff and subcontractors ensures that everyone on the job site knows the hazards and can make appropriate personal safety decisions. The construction management team should develop and enforce safety protocols, conduct regular safety inspections, and mitigate risks associated with construction activities. This communication can help ensure compliance with safety regulations and promote a safety culture among the workforce.

13.4.2 Construction Management and Supervision

Construction management and supervision are crucial in ensuring the successful execution of highway construction projects. Effective monitoring of construction progress and quality, coordination with contractors and subcontractors, and timely resolution of construction issues and conflicts are essential for the smooth and efficient completion of highway projects.

13.4.3 Monitoring Construction Progress and Quality

Monitoring construction progress and quality is vital to ensure the project stays on track and meets the desired standards. The owner and the contractor are tasked with closely monitoring the implementation of the project plan, schedules, and milestones. The owner's project inspection team conducts regular inspections and assessments to verify the contractor's compliance with engineering specifications, safety regulations, and quality standards. Monitoring progress involves evaluating work completed, identifying deviations or delays, and taking corrective measures to maintain project timelines. By diligently monitoring construction progress and quality, issues can be identified early, and proactive solutions can be implemented to keep the project on schedule and within budget.

13.4.4 Coordinating with Contractors and Subcontractors

Highway construction projects involve various contractors and subcontractors responsible for different aspects. Effective coordination among these entities is crucial for smooth workflow and successful project execution. The prime contractor plays a key role in coordinating their work crews and subcontractors. They facilitate communication, ensuring each party understands their responsibilities and project requirements. This coordination helps establish clear lines of communication, and organizing regular meetings to discuss project progress and challenges ensures effective collaboration. The contractor's project manager is a central point of contact, addressing queries, resolving conflicts, and facilitating timely decision-making. By fostering strong coordination among contractors and subcontractors, the whole construction team can optimize resource utilization and minimize delays.

13.4.5 Addressing Construction Issues and Resolving Conflicts

Construction projects often encounter unforeseen issues and conflicts that can potentially hinder progress. Effective construction management and supervision involve promptly addressing these challenges and resolving conflicts to minimize disruption to the project. The owner, contractors, and subcontractors can find practical solutions by adopting a proactive approach and conducting regular site inspections and assessments to identify emerging issues. Conflict resolution requires effective communication and negotiation skills to reconcile differing viewpoints and interests. Construction managers and supervisors act as mediators, working towards consensus and ensuring that disputes are resolved amicably. Addressing construction issues and resolving conflicts early helps maintain project momentum and minimize delays.

Construction managers and supervisors can proactively identify and mitigate potential risks and deviations by diligently monitoring progress and ensuring compliance with quality standards. Effective coordination among contractors and subcontractors optimizes project workflow and minimizes disruptions. Addressing construction issues and resolving conflicts promptly promotes project efficiency.

13.4.6 Material Procurement and Management

Materials procurement and management are crucial in successfully executing highway construction projects. Ensuring timely delivery of materials, verifying compliance with specifications and standards, and managing material inventory and storage are essential components of efficient construction management.

13.4.6.1 Ensuring Timely Delivery of Materials

Timely delivery of materials is vital to maintaining project schedules and avoiding unnecessary delays. On highway construction projects, space is confined, making it critical that materials are delivered when they are needed. If they are delivered too early, there may be limited storage space, and they could get damaged while in a holding area. If they are delivered late, the project schedule will suffer, causing delays to operations. Construction managers and procurement teams must work closely to identify the required materials and establish procurement schedules. They collaborate with suppliers and vendors to ensure timely ordering, production, and delivery of materials to the construction site. By closely monitoring delivery timelines and proactively addressing any potential delays or supply chain issues, construction managers can mitigate the risk of project disruptions.

13.4.6.2 Verifying Compliance with Specifications and Standards

Highway construction projects require materials that meet specific specifications and standards set by the owner agency to ensure structural integrity, durability, and safety. Construction managers and quality control teams are responsible for verifying compliance with these specifications and standards. They collaborate with suppliers and conduct thorough inspections and testing of materials upon delivery to the project and, in some cases, at the supplier's facilities. Verification may include assessing material characteristics, such as strength, density, and composition, and ensuring adherence to established standards.

13.4.6.3 Managing Material Inventory and Storage

Efficient material inventory and storage management is essential to avoid waste, theft, or damage to materials. The contractor's project manager should implement effective inventory management systems to track the quantity, location, and condition of materials throughout the project. Secure storage is critical to preventing theft, as is maintaining accurate records and ensuring that materials are appropriately accounted for and readily available when needed. This controls project costs and provides the contractor's estimator with accurate data to bid on future jobs.

Construction managers also oversee the establishment of appropriate storage facilities and procedures to protect materials from environmental factors and prevent damage. By efficiently managing material inventory and storage, construction managers optimize resource utilization, reduce material waste, and maintain a well-organized construction site.

Materials procurement and management play a vital role in the success of highway construction projects. Timely delivery of materials and compliance verification help maintain project schedules and avoid costly delays. Efficient material inventory and storage management minimizes waste and maintains an organized construction site. Prioritizing materials procurement and management contributes to the overall efficiency and effectiveness of the project.

13.4.7 Stakeholder Coordination and Communication

Transportation touches every American, and transportation construction projects impact the lives of adjacent property owners, surrounding communities, local businesses, elected officials, regulatory agencies, owner's representatives, and contractor staff. Coordination and communication play a crucial role in successfully executing highway construction projects. Effective engagement with contractors, regulatory agencies, stakeholders, and the public is essential for ensuring smooth project progress, addressing concerns, and maintaining positive community relationships.

13.4.8 Regular Meetings with Contractors, Regulatory Agencies, and Stakeholders

Regular meetings with contractors, regulatory agencies, and stakeholders form the foundation of effective stakeholder coordination. For construction managers and project teams, these meetings foster clear and consistent communication channels. Regular meetings provide a platform for sharing project updates, discussing challenges, and addressing issues or concerns. They help establish a shared understanding of project objectives, timelines, and expectations among all stakeholders.

13.4.9 Addressing Public Concerns and Community Outreach

Highway construction projects often directly impact local communities and the general public. Effective stakeholder coordination involves addressing public concerns and conducting community outreach initiatives. If there are road closures, businesses in the area may be severely impacted. Project teams should actively engage with the community to understand the impacts, address their concerns, and provide information about the project's progress and potential disruptions. Through proactive communication, construction teams can mitigate negative effects on the community, ensure transparency, and build trust. Community outreach programs, such as public meetings, informational sessions, and construction updates, help educate the public about the project's benefits, timelines, and potential inconveniences.

13.4.10 Providing Project Updates to the Owner's Team and Stakeholders

Large, high-profile projects require regular updates to the owner's leadership team and stakeholders, such as legislative liaisons, environmental agencies, and others. These updates are essential for maintaining transparency and ensuring project alignment with transportation agency goals. Construction managers and project teams should provide comprehensive and timely updates on project progress as the contract requires, including milestones achieved, challenges faced, and anticipated timelines. These updates allow the owner's team to track the project's status, address any emerging issues, and make informed decisions regarding resource allocation and project modifications. Timely communication of project updates to stakeholders ensures that all involved parties remain informed and engaged, enabling effective collaboration and the timely resolution of potential obstacles.

Stakeholder coordination and communication are vital components of successful highway construction projects. Regular meetings with contractors, regulatory agencies, and stakeholders address public concerns and build trust. By prioritizing stakeholder coordination and communication, construction teams foster a positive construction environment that benefits everyone (NCDOT, 2024).

13.5 Post-Construction Phase

13.5.1 Project Closeout and Final Documentation

When the highway project nears completion, it transitions into the closeout phase. This closeout phase marks the final stage of work where inspections, compliance verification, and final documentation take place. This critical phase ensures the completed highway meets the required standards, adheres to contract specifications, and maintains accurate as-built records.

13.5.1.1 Conducting Final Inspections and Punch List Completion

During the closeout phase of a highway construction project, the owner's engineer, project inspector, and contractor's project manager conduct a comprehensive final inspection of the project to identify any remaining deficiencies or discrepancies. The punch list created from this inspection details the items that require completion or correction before the project can be officially accepted and final payment can be made. The inspection process includes assessments of the constructed roadway, bridges, safety features, signage, markings, and other crucial elements, ensuring the project was built according to the plans. Contractors typically address the punch list quickly to finish the job, clear the site, and move on to the next project.

13.5.1.2 Verifying Compliance with Plans and Contract Requirements

The owner agency and contractor's staff meticulously review the plans and contract documents throughout the project's life to ensure that all project deliverables align with the agreed-upon specifications, standards, and timelines. At the end of the project, they complete this verification process by examining the completed work and assembling test reports and material certifications. Any deviations from the contract requirements are addressed and rectified before the project closeout. Verifying compliance with the plans and specifications at this point ensures that the completed highway meets the contractual obligations and fulfills the project owner's and stakeholders' expectations.

13.5.1.3 Reviewing and Approving As-Built Drawings

As-built drawings accurately represent the constructed highway, capturing any modifications made during the construction process. During the closeout phase, the owner's engineer and project inspection team complete these drawings, making changes as necessary to create a comprehensive record of the completed project. Typical changes made to as-built drawings include drainage system realignments, driveway relocations, pavement structure modifications, and other items necessary for future maintenance personnel to know. As-built drawings are valuable references for future maintenance, repairs, and expansion work. Their accuracy is critical in ensuring that all elements of the constructed highway are documented correctly.

Conducting final inspections and punch list completion guarantees that the highway is safe and meets all the required contract terms. Creating as-built drawings provides accurate records for future reference and maintenance. By diligently executing the closeout phase and final documentation, construction teams deliver a high-quality highway and establish a solid foundation for future maintenance and enhancement efforts.

13.5.2 Ongoing Maintenance and Operation Considerations

As a highway construction project reaches completion, careful planning and consideration for ongoing maintenance and operation become crucial. Transitioning from the construction phase to the operation and maintenance phase requires effective strategies to ensure the highway's continued functionality, safety, and longevity.

13.5.2.1 Transitioning to the Operation and Maintenance Phase

After a highway construction project, a smooth transition to the operation and maintenance phase is essential to ensure that the highway remains functional and safe for the traveling public. This transition involves handing over the completed infrastructure to the maintenance and operations department for day-to-day management. Effective communication and documentation play significant roles in this process and begin well before the project is completed.

Before construction ever begins, the owner's engineer communicates with maintenance engineers, sharing the design plans and discussing traffic control plans since traffic during the construction project's life will affect the surrounding roadways' day-to-day operations. Near the completion of the project, critical information, such as as-built drawings, material certifications, and maintenance records, is handed over and discussed, allowing the maintenance engineer an opportunity to ask questions. Communication throughout the project's life ensures the maintenance team is well-informed about the construction details, potential risks, and any specific maintenance requirements.

Maintenance personnel are typically invited to tour the project as it nears completion to become familiar with what they will be maintaining. They are also included in the official final inspection to identify any issues they see.

13.5.2.2 Establishing Maintenance Plans and Schedules

Before the project is handed over for maintenance, the owner's maintenance staff will review the project, establishing a comprehensive maintenance plan and schedule to guide the ongoing care of the highway. These maintenance plans outline the specific tasks, schedules, and responsibilities for regular inspections, repairs, and infrastructure upkeep. Some such tasks would include

- Timing of drainage inspections and pipe clearing
- Target date for pavement overlay
- Mowing schedules

Planning for these activities should consider factors such as traffic volume, climate, and expected wear and tear. Regular inspections and preventive maintenance are critical to identifying and address potential issues before they escalate into costly repairs or safety hazards.

Additionally, scheduling routine maintenance ensures that resources and manpower are efficiently allocated. A well-structured maintenance plan promotes the highway's longevity and minimizes traffic flow disruptions.

13.5.2.3 Ensuring Compliance with Warranty Requirements

Highway construction projects often come with warranty requirements from contractors and suppliers. During the closeout phase, construction managers must ensure that all warranty obligations are met and documented appropriately. This includes verifying that construction work and materials meet the specified warranty periods and performance standards. The responsible parties should promptly address and rectify any unresolved issues or defects. Ensuring compliance with warranty requirements not only safeguards the investment made in the highway but also reassures the owner that the contractors or suppliers will resolve any unforeseen issues per the agreed terms. Typical warranties include roadside plantings, pavement markings, and sign sheeting.

Ongoing highway maintenance and operation considerations are vital aspects of the closeout phase of a highway construction project. Carefully considering maintenance needs and warranty requirements during the project closeout contributes to the long-term success of the highway, benefiting the community with safe and reliable transportation infrastructure for years to come.

13.6 Questions

1. Why is construction administration crucial in ensuring taxpayers receive the best possible value for the money spent on transportation facilities?
2. Discuss the significance of the owner's role in a highway construction project. How does the owner's responsibility impact the project's success?
3. What are the key responsibilities of a general contractor in a highway construction project? How do these responsibilities contribute to the project's overall success?
4. Reflect on the importance of safety measures in highway construction projects. How do the Occupational Safety and Health Act of 1970 and subsequent regulations impact construction practices?
5. Discuss the impact of highway construction on the environment and the importance of compliance with the Federal Endangered Species Act and other environmental regulations.
6. How does the engineer's authority affect the construction process, and what are the implications of significant changes made to the project?
7. Why is adherence to plans and specifications crucial in highway construction projects?
8. Discuss the challenges and strategies involved in ensuring timely materials delivery and compliance with specifications in highway construction projects.
9. How do effective coordination and communication with stakeholders contribute to the success of highway construction projects?
10. Reflect on the importance of the project closeout phase, including final inspections, as-built drawings, and ongoing maintenance and operation considerations.
11. Compare and contrast the approaches to ensuring safety for construction workers and the traveling public during highway construction projects.
12. How do environmental considerations, such as stormwater management and erosion control, influence the planning and execution of highway construction projects?
13. Discuss the role of the contractor's quality control manager in a highway construction project. Why is quality control pivotal for the project's success?
14. Identify and discuss the challenges in coordinating with contractors and subcontractors and addressing construction issues. How can these challenges be mitigated?
15. Considering the detailed aspects of highway construction administration, planning, and execution discussed, what innovations or improvements would you suggest for the future of highway construction?

References

AASHTO (2024). Center for Environmental Excellence, American Association of State Highway and Transportation Officials (AASHTO), Washington, DC. At https://environment.transportation.org/

AASHTO (2020). *Guide Specifications for Highway Construction*, 10th ed. American Association of State Highway and Transportation Officials (AASHTO), Washington, DC.

FDOT (2021). *FDOT Design Manual*. Florida Department of Transportation (FDOT). Tallahassee, FL. At https://www.fdot.gov/roadway/fdm/2021-FDM.shtm

FDOT (2023). *FDOT Public Involvement Handbook*. Florida Department of Transportation (FDOT). Tallahassee, FL. At https://fdotwww.blob.core.windows.net/sitefinity/docs/default-source/environment/environment/pubs/public_involvement/pi-handbook_july-2015.pdf?sfvrsn=cf3ea04a_0

FHWA (2024). *Environmental Review Toolkit*, Federal Highway Administration (FHWA), Washington, DC. At https://www.environment.fhwa.dot.gov/

FHWA (2009). *Federal Lands Highway Construction Manual*. Federal Highway Administration (FHWA), Washington, DC. At https://highways.dot.gov/sites/fhwa.dot.gov/files/docs/federal-lands/construction/14836/cm.pdf

NCDOT (2023). *Construction Manual*. North Carolina Department of Transportation (NCDOT). Raleigh, NC. At https://connect.ncdot.gov/projects/construction/Pages/ConstMan.aspx?Method=CM-00-000

NCDOT (2024). *Environmental Policy Document*. North Carolina Department of Transportation (NCDOT), Raleigh, NC. At https://connect.ncdot.gov/resources/Environmental/EPU/Pages/default.aspx

Uddin, W., Hudson, W. R., & Haas, R.(2013). *Public Infrastructure Asset Management*, 2nd ed. McGraw Hill Education, Toronto, Canada. At https://www.accessengineeringlibrary.com/binary/mheaeworks/773541f653948e47/b58b5958e9d904c26fa08c009ef53ed9bfd1cac81b33933977e08c4556a7ba18/book-summary.pdf

Chapter 14

Constructing Connectivity

Building the Bridges, Railways, Airports, and Pipelines of Tomorrow

14.1 Introduction

Transportation infrastructure construction involves a wide range of projects, each with its own unique challenges and requirements. The key sectors include:

- Bridge construction: specialized engineering to span physical obstacles such as rivers, valleys, or roads. Major elements include foundation work, substructure and superstructure construction, and materials like steel and concrete. Earthwork is essential for creating approaches to the bridge.
- Railway construction: Building railways involves laying tracks, constructing stations, and sometimes building tunneling and bridges. Earthwork plays a crucial role in creating a level path for tracks. Specific aspects include laying rails, constructing platforms, and often electrifying the railway line.
- Airport construction: Airports require runways, taxiways, aprons, terminals, and control towers. Like other transportation infrastructure or horizontal construction projects, earthwork is key for leveling the ground for runways. Essential components include constructing long, flat surfaces for runways and taxiways, as well as terminals designed to accomodate both passengers and cargo.
- Pipeline construction: Pipelins are mainly used for transporting liquids and gases. This involves laying pipes underground or underwater. Trenching, a form of earthwork, creates a path for the pipeline. The process also includes welding pipes together and protecting them against environmental factors.

Transportation infrastructure, or often referred to as or horizontal construction, includes roads, runways, and railway tracks that spread over large areas rather than towering vertically like buildings. While all share some common elements, each type of transportation infrastructure has its specific challenges depending on its purpose and the environment in which it's built. Earthwork is a common element in all these types, which involves excavating, moving, and grading soil or rock, which is foundational for creating level surfaces, tunnels, and trenches.

14.2 Bridges

Bridges are a critical part of the transportation system and infrastructure sectors of civil infrastructure. In the United States, the condition of bridge infrastructure remains a significant concern. The American Society of Civil Engineers assigned bridges a C+ on the Report Card for

DOI: 10.1201/9781003197768-14

America's Infrastructure; however, roughly one in four bridges still needs repair. The most recent estimate of the nation's bridge rehabilitation needs totals around $125 billion. To improve the condition, there is a need to increase spending on bridge rehabilitation from $14.4 billion to $22.7 billion annually, or by 58%. In addition to the high cost of rehabilitation needs, the average age of bridges in America continues to rise, and many of these bridges are approaching the end of their 50-year lifespan. With America's population making roughly 188 million trips across a structurally deficient bridge daily, disaster could strike at any time (ASCE, 2021). Inadequate investment in infrastructure can and will produce tragic consequences. It is also crucial for engineers, researchers, and contractors to work together to reduce the risk of failure among bridges. In the construction of new bridges or the rehabilitation of old bridges, new technological and sustainable considerations could be implemented to drastically reduce the ecological footprint of the construction process.

14.2.1 Bridge Design Types

Many bridge designs have been developed over the years. Each offers certain advantages and has certain limitations. The choice of design for a specific project depends on site conditions, economic factors, and other critical considerations. Below are descriptions of several well-known bridge designs:

Suspension Bridges: The principal structural element in suspension bridges consists of large cables supported on towers located relatively close to the ends of the structure (Figure 14.1). The roadway is suspended from the cables, which deliver the structural loads to the towers and end anchorages. Suspension bridges are typically used where long spans, without intermediate substructure supports, are required, as in the case of the Golden Gate Bridge.

Cable-Stayed Bridges: This is a comparatively new type of bridge design in which steel cables are anchored to the tops of high towers and extend down to attachments to the deck in both directions from the tower supports. Cable-stayed bridges are used where it is difficult to provide intermediate supports, where the number of intermediate substructure supports needs to be minimized, and where long clear spans are required.

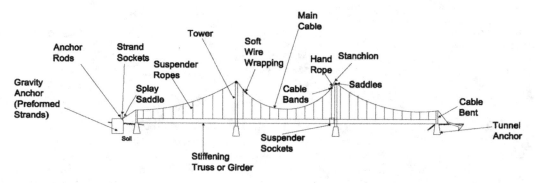

Figure 14.1 Suspension Bridge

Arch Bridges: In arch bridge designs, the roadway may be supported above the top of the arch member. It also may be suspended from arch members located on each side of the roadway. Single arch designs are used when it is not feasible or practical to provide intermediate supports between the ends of the structure. Multiple arch designs have a broad range of applications (Figure 14.2).

Truss Bridges: In truss bridges, the roadway is supported by truss members. It may be located above, between, or at the bottom of the trusses. There are many different types of truss designs. Truss bridges are used where it is necessary to span long distances between foundation supports. Truss bridges are usually are built with steel or timber members (Figure 14.3).

Most modern highway bridges incorporate one of the following design types:

Through/Deck Girder: In girder bridges, the roadway loads are transmitted to significant structural elements called girders, which then transmit the structural loads to the supporting elements (Figure 14.4).

Distributed Beam/Girder: Distributed beam/girder bridges utilize a series of beams or small girders spaced uniformly across the width of the structure to support the deck and transmit the structural loads to the substructure supports. The deck construction typically consists of cast-in-place reinforced concrete, which is usually tied to the beams or girders with shear connectors to form a composite beam slab superstructure.

Box Girder: In box girder bridge designs, the principal structural element consists of a multi-cell box girder, the top flange of which also serves as the deck. Figures 14.5–14.9 illustrate the box girder, flat slab, T-beam, prestressed concrete T-girder, and prestressed concrete I-girder, respectively.

Figure 14.2 Arch Bridge

Figure 14.3 Truss Bridge

Monolithic Cast-In-Place Concrete: There are a variety of cast-in-place concrete bridge designs used in highway bridge construction. They include:

- Flat Slabs: In bridges of this type, a reinforced concrete slab, rectangular in cross-section and of uniform thickness both longitudinally and transversely, forms the superstructure.
- Arched Slabs: Similar to flat slabs, except that the thickness of the slab decreases from a maximum at the supports to a minimum at mid-span and has an arched bottom.
- Composite Beam/Slab: In superstructures of this type, the deck slab is cast monolithically with the longitudinal beams to form elements that are essentially box girders without the bottom flange.
- Rigid Frames: In rigid frame designs, the superstructure is rigidly tied into, instead of bearing on, the supporting elements, and the two components act monolithically in transmitting the structure's dead and live loads to the foundation.

Figure 14.4 Girder Bridge

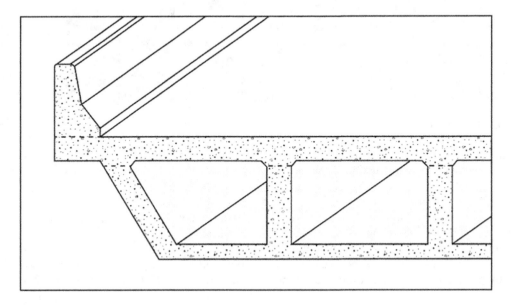

Figure 14.5 Box Girder

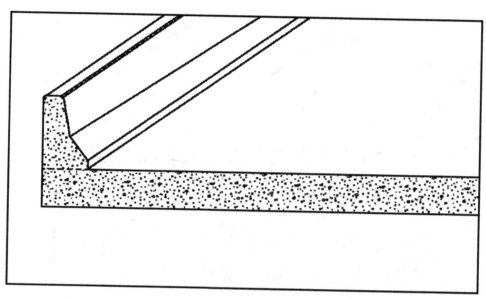

Figure 14.6 Flat Slab

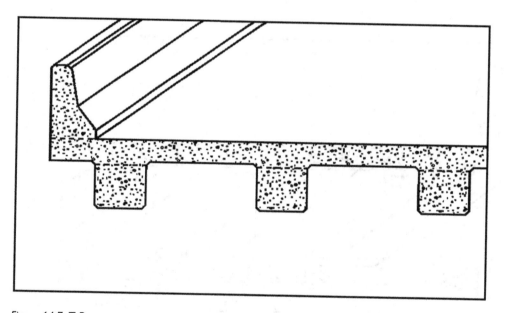

Figure 14.7 T-Beam

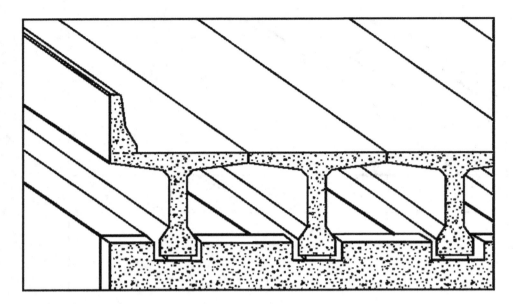

Figure 14.8 Prestressed Concrete T-Girder

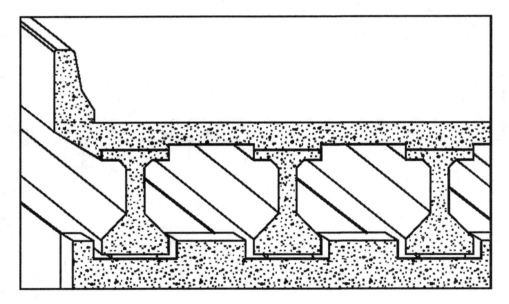

Figure 14.9 Prestressed Concrete I-Girder

14.2.2 Span Types

"Span" can have different meanings depending on its use in design, field construction, or general use. When used in relation to the superstructure, it refers to a segment of the superstructure that crosses from one substructure support to the next. When used in this context, span length is generally considered as the distance between centerlines of bearing from one substructure support unit to the next. When applied to the field construction of substructure elements, it has a slightly different meaning, i.e., span is the unobstructed space or distance between the faces of substructure supports. When used in this context, it is commonly called the "clear span length."

A bridge may comprise one or many spans of the same or differing types. The principal span types are:

- Simple Span: A superstructure span that crosses from one substructure unit to another and stops is a simple span. Some bridges may consist of only one single span; others may consist of multiple spans constructed as a series of simple spans.
- Continuous Span: Continuous spans are segments of a bridge superstructure whose structural members cross over one or more substructure units without a break. It is not uncommon for multi-span bridges to contain both simple and continuous spans.
- Cantilever Span: A cantilever spans is supported only at one end, with the other end projecting beyond the support without any additional external support. In a continuous structure, cantilever spans are often formed by extending segments or sections beyond the points of support, allowing the ends to project freely.
- Suspended Span: Suspended spans have ends that span between bearing points that are not supported on substructure elements, such as the ends of cantilever spans. Suspended spans rely on connections as bearing points.

14.2.3 Material Types

Bridges are often categorized based on the primary materials used for constructing their structural members. In modern highway bridge construction, three principal materials are commonly used, either individually or in combination:

- Steel
- Reinforced concrete
- Timber

Steel is used primarily to form the principal structural members in steel bridges. Typically, reinforced concrete is used in the construction of the bridge deck, although steel decks are occasionally used in movable bridges or where special conditions warrant.

While timber was once a prevalent material for bridge construction, especially on rural roads, it is rarely used today for superstructures on major highways. In modern highway bridge construction, the primary use of timber is for forming and falsework, temporary frameworks that support the concrete during construction.

14.2.4 The Role of Advanced Technology in Bridge Construction

Advanced technology has dramatically impacted the engineering and design of construction processes worldwide. Without advanced technology, bridge construction and the rehabilitation of any project are increasingly challenging. Building mega-complex bridges would be impossible without advanced technologies. On the other hand, if resiliency, risk, and recycling are not considered now, future projects within the infrastructure sector will fail to improve. The current construction methods and procedures will eventually become unsustainable, less resilient, and more vulnerable. As technology continues to improve, the design and construction processes must also improve in terms of quality and sustainability. Now is the time for technological and sustainable considerations to be implemented in the construction and the rehabilitation of bridges. Achievements in bridge technology during the last century directly led to the development of cable-stayed bridge technology and segmental bridge construction. Along with new engineering processes came the implementation of high-performance materials and information technologies. Other technological advances, such as those pertaining to equipment, have drastically reduced the time, money, and resources required to complete tasks.

Cable-stayed bridge technology is a design innovation that offers a better alternative to suspension bridges. Cable-stayed bridges may look similar to suspension bridges, but structurally, they are quite different. A cable-stayed bridge supports the weight of its deck with a series of cables running directly to one or multiple towers (Figure 14.11). In this design, the towers are generally positioned above the piers in the middle of the bridge span and act as significant support for the bridge deck. Suspension bridges also have cables connected to towers; however, these cables ride freely from tower to tower, transmitting the load of the bridge deck onto anchorages at each end. The anchorages at each end of a suspension bridge provide support for the bridge deck. This places a high amount of tension on the cables and an unnecessary amount of stress on the anchorages and bridge deck. Figure 14.10 compares the suspension bridge and cable-stayed bridge.

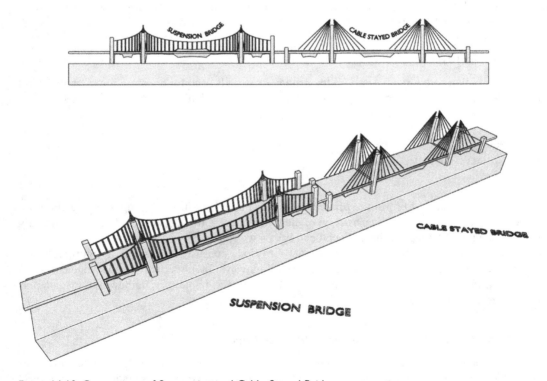

Figure 14.10 Comparison of Suspension and Cable-Stayed Bridge

Figure 14.11 Cable-Stayed Bridge, Kao-Ping River Bridge, Taiwan

A cable-stayed bridge offers greater structural stability than a suspension bridge because it balances the weight of the road deck evenly on each side of its towers. In environments with high wind speeds, cable-stayed bridges are more stable because of the additional support provided to the cables from the towers. Economically, cable-stayed bridges are also the better alternative for bridge construction. A cable-stayed bridge requires less steel cable than a suspension bridge and incorporates more precast concrete sections. With the available equipment and technology in today's construction industry, precast concrete sections can be assembled in record time. Cable-stayed bridges offer many advantages of a suspension bridge but at a fraction of the cost. The life expectancy of this innovation is roughly 100 years. Thanks to improvements in structural modeling, cable-stayed bridges offer a much more cost-efficient structure that uses fewer resources and materials in the construction process. As technologies are developed to help explore the world of bridge design, new and improved engineering and design processes will emerge.

Segmental bridge construction is a relatively newer innovation in bridge design that has already become widely accepted and used within bridge construction. In segmental bridge construction, the bridge is built by adjoining individual precast concrete segments, one piece at a time. The precast concrete segments are generally the exact width of the roadway that will be completed. The individual precast concrete segments are lifted into position and stressed together with strands or bars, producing an aesthetic appeal that is very pleasing (Figure 14.12). A watertight joint is created

Figure 14.12 Segmental Bridge Construction at Piers

Note: LUSAS *Advancing Segmental Bridge Technology* (2017).
http://www.lusas.com/case/bridge/road431.html

between the adjoining segments using epoxy mortar to seal the gap. When using segmental bridge construction, multiple methods and construction techniques are available to choose from. This gives contractors the advantage of selecting the most cost-efficient method for the construction of bridges.

Segmental bridges comprise many precast concrete units, making them very economical for long-span projects. The assembly of a segmental bridge typically occurs in a separate casting yard, and the construction work occurs concurrently, resulting in a faster project completion time. The engineering and design process of segmental bridge construction will be continually improved through innovation and the implementation of high-performance materials.

High-Performance Concrete and Steel: The development of new technology has directly impacted the creation of new high-performance materials. With materials' performance at an all-time high, bridges are becoming more durable and their service life is increasing. Innovation has led to the creation of beams made with high-performance concrete, which provide increased rigidity over traditional concrete. Other significant high-performance material advancements include engineered cementitious composites, fiber-reinforced polymers, and fiber-reinforced concrete.

High-performance concrete is defined as a type of cement-based composite with a specified compressive strength of 17,000 to 22,000 pounds per square inch and other specified durability, tensile ductility, and toughness requirements. High-performance concrete will increase the overall structural stability of a bridge and its predicted lifespan. The service life of a bridge constructed with traditional concrete is roughly 50 years; however, that same bridge would have a service life of approximately 75 to 100 years if entirely constructed with high-performance concrete. The

Figure 14.13 Remote-Control Concrete Pump

unique properties of high-performance concrete allow it to have ductile behavior, meaning it can be deformed without losing strength. High-performance concrete has arguably been the most innovative product in concrete technology over the last 30 years, accomplishing huge strides toward improving the sustainability and resistance of bridges.

High-performance steel provides increased strength and can actually be installed at a much faster rate than regular steel. High-performance steel reduces both construction time and costs. Because it has half the carbon and a fraction of the sulfur content of conventional steel, it can be welded with little to no preheating. Other innovations that reduce maintenance costs. can be as simple as epoxy-coated rebar. Before rebar is sent out to job sites, it is coated with epoxy that protects the steel from rusting and decaying, resulting in a direct increase in the service life of bridges.

Remote control concrete pumping: The pump is designed to be used in bridge applications with a huge span. When it comes time to pour the deck of a bridge, concrete trucks may not have the option of pouring directly onto the deck of the bridge due to the reach constraints of the truck. In these instances, a remote-control pump is brought on-site and connected to the concrete truck by interlocking pipes. As the remote-control pump moves along the bridge during the progress of the pour, individual pipes are removed one at a time until the remote-control pump is no longer needed. At this point, the concrete truck will then finish out the remainder of the deck pour by reaching out with its attached arm. The following figures show how a bridge deck was poured, with the aid of a remote-control pump and crew, on a bridge in Hope, Indiana (Figure 14.13).

3D Use in Bridge Construction: The Institute of Advanced Architecture of Catalonia (IAAC) undertook the great task of accomplishing the first-ever bridge construction using a 3D concrete printer. Many projects have already introduced 3D concrete printing into their processes as a means of reducing labor costs and project duration. The most innovative 3D printers have the ability to produce a fully enclosed structure within a very short amount of time. 3D printers offer extremely cost-effective alternatives to other housing or building options. The 3D-printed bridge created by the IAAC was constructed out of micro-reinforced concrete and designed to be used as a pedestrian crossing. The bridge was printed off-site and had final measurements of 39 feet long and 5.7 feet wide. 3D printing bridges and other structures may be a new concept, but its potential is limitless. Bridge transportation projects can be completed in a much faster, more cost-efficient manner when 3D printing is printed. The following figure is a photo of the installation of the first-ever 3D printed bridge in Alcobendas, Madrid (Figure 14.14).

Virtual Prototyping has transformed bridge design and construction from traditional methods to a sophisticated, technology-based approach. This shift addresses the growing complexity of

Figure 14.14 3D Printed Bridge Installation

Note: 3Ders.org. World's First 3D Printed Pedestrian Bridge (2016).

https://www.3ders.org/articles/20161214-spain-unveils-worlds-first-3d-printed-pedestrian-bridge-made-of-concrete.html

bridges, which now incorporate numerous prefabricated elements, making design and construction more intricate. Information technologies, crucial in this evolution, facilitate storing, retrieving, and transmitting data, significantly reducing rework by identifying design conflicts early and allowing prompt alterations.

At the core of virtual prototyping is the use of tools like computer-aided design (CAD), computer-automated design (CAutoD), and computer-aided engineering (CAE). These technologies enable the validation of bridge designs before physical prototyping, streamlining both construction and the rehabilitation projects. By simulating various construction scenarios, virtual prototyping helps in resource allocation and collision avoidance, allowing construction firms to select the most efficient plan.

The efficiency of planning and scheduling is notably enhanced when integrating virtual prototyping into the process. These technologies produce precise simulations, reflecting the expected timelines of different activities, leading to more effective construction schedules. Additionally, they offer greater visual oversight of the construction and the rehabilitation phases.

Virtual prototypes are gaining traction in the modern construction landscape, where tight schedules and high performance are paramount. They empower engineers to explore various design options rapidly without the expense and time associated with building physical models. This approach not only improves bridge design quality and performance but also aims to revolutionize future planning and construction in the transportation infrastructure sector.

Building Information Modeling (BIM): New technology has introduced building information modeling (BIM) into construction and infrastructure construction. BIM can be used to improve bridge design, construction, and rehabilitation and aid project construction management. BIM technologies could enhance bridge projects' design and construction process.

BIM assists in the design and construction process by aiding in the fabrication and geometry control of precast units. With BIM, the design process is made specific and visually appealing (Figure 14.15). A particular bridge design can be analyzed for collisions among all its components,

Figure 14.15 BIM Model

Source: Ashikian, Ara. *Infrastructure Reimagined* (2016)

http://www.infrastructure-reimagined.com/bim-and-bridges/

and modifications can be easily made until the bridge design becomes structurally feasible. BIM technologies allow a design team to discover and fix design defects and collisions before beginning the construction work on a project. This, in turn, decreases the overall cost and duration of the project.

With BIM, the design process becomes more collaborative, as project participants can communicate directly with BIM team members on the feasibility of different design options. These design options can be quickly evaluated with BIM technologies and collision analysis, and reports can be produced. One of the advantages of BIM is its ability to enhance a design, automatically making changes to one component after changes to a related component are created. With more promotion, BIM can be successfully applied in all bridge construction and the rehabilitation projects.

Using new technologies, such as BIM, can be difficult for project participants unfamiliar with the technology; however, information technologies are the future of design and construction in mega-complex bridge projects, and companies within the infrastructure sector must adapt. It does take effort and time for a contractor to learn and apply BIM technologies, but the benefits for a contractor can be significant. BIM technologies must be promoted in education and training programs so more people become familiar with how to use the technology. The adoption of information technologies within the infrastructure sector will accelerate the growth and improvement of the design and construction process.

14.3 Railways

Railways, with a history stretching back to the early 19th century, have been a cornerstone in the evolution of global transportation. The initial breakthrough came with the introduction of steam locomotives, a pivotal moment marked by George Stephenson's "Rocket" innovation in 1829, which truly ignited the rapid expansion of rail networks worldwide. This expansion became an era of monumental railway projects, such as the Transcontinental Railroad in the United States and the Orient Express in Europe. These developments connected distant cities and countries and revolutionized trade and travel.

Railways have continued to evolve. Modern high-speed rail systems, exemplified by Japan's Shinkansen, France's TGV, and China's CHR, have made travel faster and more efficient. Furthermore, the shift towards electric and hybrid trains reflects a growing commitment to environmental sustainability, reducing the carbon footprint of rail transport.

As the 21st century unfolds, railways remain critical in the global transport network. They are not just a means of transport but also a driver of economic growth and sustainable development, adapting to meet the evolving needs of the modern world.

14.3.1 The Freight and Passenger Rail Network

The US rail network has approximately 140,000 miles and is central to the almost $80 billion freight rail industry. Managed by seven primary Class I railroads, alongside 22 regional and 584 local/short-line railroads, it sustains over 167,000 jobs nationwide. Railways offer distinct advantages over other transport modes. These benefits include reduced road congestion, lower highway fatalities, decreased fuel usage, lower greenhouse gas emissions, and savings in logistics and public infrastructure maintenance.

The US freight railroads are privately owned, and they allocate a significant portion of their revenues – about 19 percent or nearly $25 billion annually – towards maintaining and improving their network, a commitment unmatched by other major transport sectors.

The Federal Railroad Administration (FRA), established in 1967, has played a pivotal role in shaping the nation's passenger rail system. Key legislative acts such as the High-Speed Ground Transportation Act of 1965 and the Rail Passenger Service Act of 1970 laid the foundation for FRA's leadership in intercity passenger rail. More recent legislation, including the Passenger Rail Investment and Improvement Act of 2008 and the Fixing America's Surface Transportation Act of 2015, further integrated passenger rail into the nation's multi-modal transportation framework.

Within the FRA, the Office of Railroad Policy and Development (RPD) is tasked with fostering the development and investment in both passenger and freight rail infrastructure. The RPD also implements policies related to intercity passenger and high-speed rail services. Additionally, the RPD manages the execution of grant agreements with Amtrak, overseeing the federal funds appropriated by Congress for Amtrak's operations, infrastructure, and equipment.

Railway Planning in the US involves developing comprehensive regional rail plans. These plans establish a unified, long-term vision for regional passenger rail service, detailing the necessary infrastructure based on current conditions, future travel demand, and the role of rail within a multi-modal transportation framework. They also connect regional rail systems with local transit, highways, and non-motorized transport options, aiming to provide a cost-effective, seamless journey for travelers.

Southwest Multi-State Rail Planning Study: The Southwest Multi-State Rail Planning Study (SW Study) was the FRA's first regional rail planning study. The SW Study was a great case study that informed FRA's subsequent activities and national rail planning efforts. Representatives from crucial transportation organizations across Arizona, California, Colorado, Nevada, New Mexico, and Utah worked through the challenges of developing multi-state rail plans and outlined a shared preliminary technical vision for high-performance rail (HPR) in the Southwest as part of this study. The study demonstrated an analytical framework for developing early-stage HPR network planning concepts and examining the institutional context for establishing and implementing a long-range rail vision.

NEC FUTURE is FRA's comprehensive plan for improving the Northeast Corridor (NEC) from Washington, DC, to Boston, MA. FRA collaborated with the NEC states, railroads, stakeholders, and the public to define a long-term vision for the corridor's future. In July 2017, FRA issued the Record of Decision, which marked the completion of the Tier 1 environmental review process and contained a description of the Selected Alternative, a vision for the growth of the NEC to guide future investments in the Corridor.

Following the publication of the NEC FUTURE Record of Decision, the NEC Commission launched CONNECT NEC 2035 to provide a roadmap to implement the initial 15-year phase of the NEC FUTURE vision. While CONNECT NEC 2035 will include capacity improvements, trip time reductions, and other enhancements to the NEC, CONNECT NEC 2035 will focus on ensuring safe and reliable commuter, intercity, and freight rail services through state-of-good-repair projects. The target completion date for the CONNECT NEC 2035 plan is fall 2021.

The Southeast Regional Rail Planning Study kicked off in summer 2016 to explore the potential for a high-performance, multi-state, intercity passenger rail network in Virginia, North Carolina, Tennessee, South Carolina, Georgia, and Florida, plus the District of Columbia. During the study, the states were awarded funding to establish a rail governance body, resulting in the establishment of the Southeast Corridor Commission (SEC) in 2020. The SEC adopted the Southeast Regional Rail Planning Study as the framework to guide their work developing the southeast rail network. The Southeast Regional Rail Plan was completed in early 2021 and adopted by the SEC. Implementation of the Southeast Regional Rail Plan would provide access to high-performance rail services for more than 70% of residents in the Southeastern United States, a significant increase from the 55% of Southeast residents with access to long-distance rail service today.

Midwest Regional Rail Planning Study: Starting in the spring of 2017, the Federal Railroad Administration (FRA) embarked on a project to explore the potential for a high-performance, multistate, intercity passenger rail network in the Midwest region. The study built on current rail planning efforts within the 12 states of Illinois, Missouri, Iowa, Michigan, Wisconsin, Ohio, Nebraska, Kansas, South Dakota, North Dakota, Indiana, and Minnesota, and is intended to support both state rail planning efforts and the coordinated planning and implementation of new and improved passenger rail services within the region. In October, 2021, the final report for the study was released jointly by FRA and the Midwest Interstate Passenger Rail Commission (MIPRC), an interstate rail compact formed to promote, coordinate, and support regional improvements to passenger rail service. Additional planning efforts from regional stakeholders will further expand this 40-year framework for the Midwest passenger rail network to include prioritization of corridors and investment projects, an enhanced governance structure, and a focused funding strategy.

14.3.2 Railway Line Planning and Construction

Location: The placement of a railway line is determined by both logistical and geographical factors. Railway routes are often planned to serve populated areas, industrial centers, and harbors. At the same time, there may be estates and private properties to avoid. In less developed regions, where population and commerce are sparse, the main focus shifts to finding the most direct and practical route through large uninhabited areas, with future settlers establishing towns and villages along the line. Natural terrain significantly influences the chosen railway route, including mountains, valleys, and rivers. Major structures like bridges, viaducts, and tunnels add to the original construction costs and create ongoing maintenance requirements. Before construction begins, detailed plans and profiles of the line, including curves, gradients, and major works, must be prepared. While minor deviations can be made during construction, careful initial planning minimizes the need for significant changes due to unforeseen ground conditions or other obstacles.

Gradients: Railways seldom run perfectly level for long distances. Instead, tracks are typically laid on inclined planes or gradients, varying in steepness to accommodate the topography. The gradient is generally expressed as the ratio of the vertical rise or fall per a given horizontal distance, such as 1 in 200, meaning a rise or fall of 1 foot over 200 feet. It can also be represented as a percentage, with a 1% gradient indicating a 1-foot rise over 100 feet. Gradients directly impact the efficiency and cost of railway operation. Steeper gradients demand more power and advanced braking systems, so engineers aim to keep them as moderate as possible while balancing construction costs. For example, a 5% gradient is considered steep for railways and may require special handling.

Curves: Railway curves are defined by their radius of curvature, with larger radii offering gentler curves that can be navigated at higher speeds. Tighter curves with smaller radii require reduced speeds for safe passage. Curves are also described by their degree of curvature, a measure of the central angle formed by the arc over a specified distance. Several factors influence the design and management of curves, including: Radius of Curvature: Larger radii allow higher speeds, while tighter curves require slower speeds. Superelevation (Cant): The outer rail is elevated above the inner rail to counteract centrifugal force, improving safety and comfort. Transition Curves (Easements): These gradually introduce curves, reducing the abrupt change in centrifugal force. Cant Deficiency: This measures the difference between the ideal and actual cant provided. Speed Limits: Curves dictate speed limits, with tighter curves requiring slower speeds.

Track Maintenance: Curves experience more wear and tear due to the forces exerted on the rails, necessitating frequent maintenance.

Earthworks: Earthworks involve excavation (cuttings) and embankments of earth, gravel, and rock. Engineers strive to balance earthworks by using material from cuttings to build embankments. However, surplus material is sometimes disposed of in spoil banks, while additional fill may be borrowed from nearby areas. The choice of materials and techniques depends on the type of soil and rock encountered, with slopes adjusted accordingly to ensure stability.

Culverts and Drains: Proper drainage is crucial to the stability of embankments. Culverts and drains are constructed before earthworks begin to manage water flow and prevent erosion or subsidence. Culverts must be designed to handle both normal water flow and potential flooding, with careful consideration of bed elevation and foundation strength.

Bridges: Bridges, particularly those over rivers and waterways, are significant components of railway construction. Designs must comply with navigational requirements, which may dictate the number and spacing of piers and the clearance height for vessels. Depending on traffic, a railway may use a high-level viaduct that allows trains to pass without interfering with waterway traffic or a low-level viaduct with an opening bridge.

Foundations: The stability of any railway structure depends on the strength of its foundation. Proper site selection and foundation treatment are critical to preventing subsidence or structural failure. In solid ground, foundations are generally straightforward, but care must be taken to avoid building on fill or unstable ground.

Rails and Sleepers: Rails were initially made from wood, but modern railways use steel for strength and durability. Sleepers, which support the rails, are typically made from wood, though stone and longitudinal sleepers have been used in the past. Wooden sleepers are treated with creosote to prevent decay, with proper seasoning being essential for effective preservation. Understanding these principles is essential for engineers and operators to ensure the safe, efficient, and sustainable construction and operation of railway lines.

14.3.3 High-Speed Rail Line

Since its inception in 1967, the Federal Railroad Administration (FRA) has played a crucial role in promoting high-speed rail (HSR) and other intercity passenger services in the United States. This effort began as early as 1964, when the US started exploring HSR around the same time Japan introduced its first Shinkansen line. The timeline below outlines key federal policies and investments that have shaped the progress of this efficient transportation system.

In 1965, the High-Speed Ground Transportation Act, with an initial authorization of $90 million, launched federal efforts to develop advanced HSGT technologies. By 1969, FRA had deployed technologies like the Metroliner and Turbotrain in the Northeast Corridor (NEC) and introduced long-term planning for the region. Throughout the 1970s, FRA's work laid the foundation for modern HSGT, including the establishment of Amtrak in 1971 and the launch of the Northeast Corridor Improvement Project (NECIP) in 1976, aimed at improving NEC infrastructure for reliable HSGT service.

From 1980 to 1991, FRA explored expanding HSR across the US and studied emerging corridors. By the early 1990s, five corridors were officially designated, including the Midwest, Florida, California, Southeast, and Pacific Northwest corridors. Additional corridors were designated in the late 1990s, fueled by acts like the Transportation Equity Act for the 21st Century (TEA-21), which authorized the extension of previously designated corridors and the creation of new ones.

By 2000, two new corridors, including the Northern New England and South Central corridors, were added, with various extensions approved in key regions. In 2008, the Passenger Rail Investment and Improvement Act (PRIIA) provided a framework for HSR development, which was further bolstered by the American Recovery and Reinvestment Act (ARRA) in 2009. This included $8 billion in funding for intercity rail projects, marking a new era for HSR in the US.

The first true high-speed rail line in the southwest is set to become a reality, alongside significant upgrades to existing rail systems. In California, a 520-mile high-speed train system will connect major metropolitan areas, with trains operating at speeds up to 220 mph, reducing travel time from San Francisco to Los Angeles to under three hours. The FRA and the state of California have jointly invested in this transformative infrastructure, including major expansions and signal upgrades at the San Jose–Diridon Station, ensuring the system is equipped for the future of high-speed rail.

14.4 Airport

Airfield pavement provides a smooth, all-weather surface capable of supporting the heavy loads of aircraft on a natural ground base. These pavements are typically designed in layers, each built with enough thickness to handle the applied loads without leading to distress or failure. The Federal Aviation Administration (FAA) offers guidance on airfield pavement design through Advisory Circular AC 150/5320-6E. When conducting airport pavement analysis, it's essential to reference the appropriate advisory circulars and software user guides. Airfield pavement is a layered structure of processed materials. Pavements made of a bituminous material mixed with aggregate on top of high-quality granular layers are known as flexible pavements. When made from a slab of Portland cement concrete (PCC), they are referred to as rigid pavements. Both types are common in airports, with preferences depending on factors like aircraft type, usage frequency, climate, and construction or maintenance costs. The surface course in flexible pavements is a mix of asphalt and aggregate ranging from 2 to 12 inches thick, while rigid pavements consist of an 8 to 24-inch PCC slab. This layer ensures smooth, safe traffic operations, supports loads, and distributes weight to the underlying layers. The base course, which may be treated or untreated, distributes applied loads to the subbase, which in turn supports the overall structure. Subbase requirements depend on the pavement load and soil quality. For rigid pavements, the surface course often rests directly on the subbase. Engineers designing airport pavements focus on the thickness of each layer. Two critical factors in determining the thickness are the soil base and the volume and weight of the traffic. Consequently, the initial steps in pavement analysis involve soil investigation and estimating annual traffic volume.

14.4.1 Soil Investigation and Evaluation

Proper soil identification and evaluation are crucial for designing pavement structures. The subgrade supports the pavement and the loads placed on it. The stronger the subgrade, the thinner the pavement required. Soil investigation typically involves surveys to map soil layers, sampling, and testing to determine properties such as soil type, gradation, and strength. Soil surveys in the US often reference US Geological Survey (USGS) maps, aerial photography, and soil borings. In airfield pavement design, the US Army Corps of Engineers' Unified Soil Classification (USC) System is widely used. Soils are classified into groups such as gravels, sands, silts, and clays, and further categorized based on specific characteristics, including particle size and liquid limits. Each soil type affects pavement design differently, and engineers determine the required pavement thickness using the California Bearing Ratio method.

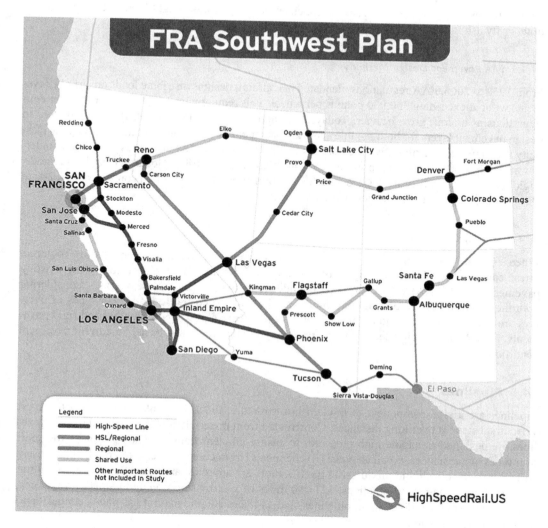

Figure 14.16 The Plan for the First High-Speed Rail Line of the Southwest Region
Source: Courtesy of FRA (2019)

14.4.2 Frost Effects and Subgrade Stabilization

Frost can significantly impact soil strength. Ice lenses that form within the subgrade during winter can lead to uneven heaving, causing pavement distress. As the ice melts during thaw periods, subgrade strength can be compromised due to poor drainage. To address frost-prone soils, the FAA classifies soils into four frost groups, with Group 1 being least susceptible and Group 4 most vulnerable. The design of pavements is adjusted based on frost susceptibility and frost depth, and this is covered in more detail later in the chapter. Subgrade stabilization techniques, including mechanical or chemical stabilization, may be used to improve performance. Mechanical stabilization can involve the use of cobble or shot rock, while chemical stabilization might include adding cement, lime, or fly ash to the soil.

14.4.3 FAA Pavement Design Methods

From 1958 to 2006, FAA regulations mandated that aircraft designs distribute loads on airfield pavements without exceeding 350,000 pounds per wheel. This requirement influenced landing gear configurations as aircraft grew heavier. Today, FAA-approved software like FAARFIELD is used to design pavement layers for both flexible and rigid pavements, considering factors like subgrade elasticity and expected aircraft traffic. Flexible pavements, composed of a bituminous surface and underlying base and subbase layers, prevent water infiltration and distribute loads across the subgrade. Rigid pavements, made of PCC slabs, provide uniform support and are often connected by joints to allow for expansion and contraction, which prevents cracking.

14.4.4 Overlay Pavements

When existing pavements deteriorate or need reinforcement, overlay pavements are required. These can involve adding new layers of concrete or bituminous material on top of the existing pavement. FAA's FAARFIELD program includes tools for designing these overlays, considering both the condition of the original pavement and the anticipated traffic load. Four types of overlays are typically used, including hot-mix asphalt or concrete overlays of flexible and rigid pavements. FAARFIELD helps determine the necessary overlay thickness based on current pavement conditions.

14.5 Pipelines

Pipelines are a vital component of modern infrastructure, functioning like arteries that transport various substances over long distances. Constructed from materials such as steel and plastic, pipelines are designed to handle high pressures to ensure efficient transfer. They primarily transport four types of substances: (i) oil and gas; (ii) chemicals (from corrosive acids to volatile gases); and (iii) slurries or solid-liquid mixtures. In the US, there are approximately 2 million miles of natural gas distribution mains and pipelines, 321,000 miles of gas transmission and gathering pipelines, 175,000 miles of hazardous liquid pipelines, and 114 active liquid natural gas plants connected to natural gas transmission and distribution systems.

Most large-scale pipeline construction occurs in rural or remote areas. Key pipeline construction activities include: (i) route surveying; (ii) mobilizing equipment and personnel; (iii) preparing the right-of-way (ROW); (iv) transporting and storing pipes and materials; (v) stripping topsoil; (vi) grading; (vii) transporting and laying pipe; (viii) welding and checking welds; (ix) installing protective coatings; (x) trenching; (xi) lowering pipes into trenches; (xii) installing block valves and terminus equipment; (xiii) backfilling; and (xiv) restoring the ROW.

14.5.1 Construction Surveying

A construction survey collects detailed data to assist in the design of pipeline sections and to provide access for subsurface equipment, like soil testing tools. The width of the survey cut is minimized to limit impact. Pipeline construction workspace requirements depend on factors such as pipe diameter, equipment size, terrain, and the method of construction. For example, a 75-foot-wide ROW is necessary for installing a 24-inch pipeline in a new ROW, while an 18-meter-wide space is needed for pipelines 4 to 12 inches in diameter. Access to the ROW is usually restricted to existing roads, which may require improvements or special permissions. These roads may include abandoned town roads, railroad ROWs, and farm roads. Repairs for any damage caused during access are typically required. In some cases, road improvements, like grading or laying gravel, are needed to accommodate heavy equipment.

14.5.2 Site Preparation

The entire ROW, including temporary workspace and access routes, must be cleared of trees and brush. Stumps are removed from areas to be graded or excavated, and any underground utilities are exposed and protected. Logs may be salvaged when possible. Clearing is done by bulldozers, while a crew performs necessary grading to accommodate vehicles and equipment. Topsoil is stripped and stored separately from spoil areas, ensuring it remains uncontaminated for later replacement. Grading is then conducted to provide a level working surface for pipe-laying operations.

14.5.3 Loading, Hauling, and Stringing

Pipes are typically stockpiled off-site before being transported to the ROW. Once on site, the pipe is unloaded, inspected, and strung along the ROW using side booms. Special precautions are taken to prevent damage to the pipe or its coating, and quality control inspectors ensure that the correct pipes are placed in accordance with the project's specifications.

14.5.4 Welding

Gas metal arc welding (GMAW) and shielded metal arc welding (SMAW) are commonly used in pipeline construction. GMAW involves a consumable wire that melts to create a weld, while SMAW uses consumable electrodes. Welders must be qualified through a series of tests, including radiographs and destructive testing. Mechanized welding is also used, offering high weld quality with less training required for operators, though it has greater equipment and maintenance demands.

14.5.5 Trenching and Installing

Trenches are dug using wheel ditchers, supplemented by backhoes for more complex areas. Minimum trench dimensions are established to ensure that the pipe is adequately covered and that backfill material surrounds the pipe. After trenching, the pipe is lowered into the trench using side boom tractors, and any required buoyancy control measures are installed. Backfilling follows immediately, with the spoil material compacted around the pipe. The ditch line is left slightly raised to prevent erosion. Cleanup of the ROW and temporary access routes includes removing debris, installing erosion control measures, and restoring the topsoil. Erosion control measures are implemented to protect the pipeline from damage caused by soil erosion. These include revegetation, installation of diversion berms, and use of rock riprap in areas prone to high water flow. The pipeline must be protected against erosion for its entire operational life.

14.6 Questions

1. How do constructing bridges, railways, airports, and pipelines contribute to societal and economic growth?
2. What specific engineering practices are highlighted for bridge construction over large water bodies?
3. Discuss the environmental considerations taken into account during airport construction.
4. How do pipelines transport different substances, and what are the main challenges associated with their construction?
5. What role does technology play in modern infrastructure development, according to the chapter?
6. How are sustainable practices incorporated into the construction of transportation infrastructure?
7. What are some of the unique challenges faced in constructing airports compared to other transportation infrastructures?
8. Describe the design considerations for different types of bridges and their suitability for various locations.
9. How do advancements in materials science impact the construction and longevity of infrastructure projects?
10. In what ways do safety and economic impacts influence the design and construction of pipelines?
11. What are some critical challenges mentioned in maintaining and upgrading existing bridges?
12. Discuss the impact of the structural deficiencies of bridges on public safety and transportation efficiency.
13. How does the chapter illustrate the integration of digital technologies like BIM in infrastructure projects?
14. What specific strategies are recommended for reducing the ecological footprint in bridge construction?
15. How do the design and construction of transportation infrastructure adapt to changes in environmental regulations?
16. What unique challenges do different types of transportation infrastructure construction (like bridges, railways, airports, and pipelines) present, and how do these challenges influence project planning and execution?
17. How do materials like steel, concrete, and timber impact the sustainability and durability of transportation infrastructure? Discuss the trade-offs involved in material selection.
18. How can new technologies and sustainable practices be implemented in bridge construction and the rehabilitation to reduce the ecological footprint?
19. Discuss the impact of high-speed rail systems on transportation efficiency and environmental sustainability. What are the challenges in implementing high-speed rail in different regions?
20. Analyze the significance of rail planning studies like the Southwest Multi-State Rail Planning Study. How do these studies guide the future of railway infrastructure?
21. Explore the importance of pavement design in airport construction. How do different pavement types (like flexible and rigid pavements) cater to an airport's needs?
22. Why is soil investigation crucial in airport construction, and how does it impact airport infrastructure's overall design and safety?

23. Evaluate the overall impact of advanced technology on transportation infrastructure development. How does technology contribute to efficiency, safety, and sustainability?
24. Compare the challenges faced in different transportation infrastructure projects, such as bridges, railways, and airports. How do these challenges dictate the approach and methods used in construction?
25. Predict future trends in transportation infrastructure construction. How will sustainability, technology, and material innovation shape future projects?

References

AAR (2020). *Freight Railroad Capacity and Investment.* Association of American Railroads (AAR), Washington, DC.

AIA (1983). *Aircraft Loading on Airport Pavements. ACN-PCN, Aircraft Classification Numbers for Commercial Turbojet Aircraft.* Aerospace Industries Association (AIA), Washington, DC.

ASBI (2018). *Construction Methods.* American Segmental Bridge Institute (ASBI). At https://www.asbi-assoc.org/index.cfm/segmental-construction/methods

ASCE (2018). *2017 Infrastructure Report Card.* American Society of Civil Engineers (ASCE), Washington, DC. At https://www.infrastructurereportcard.org/cat-item/bridges/

EUB (2005). *Directive 66 Requirements and Procedures for Pipelines.* The Energy and Utilities Board (EUB). Calgary, Alberta, Canada.

FAA (2006). *Airport Pavement Design and Evaluation.* Advisory Circular AC 150/5320-6D. Federal Aviation Administration (FAA), Washington, DC.

FAA (2008). *Airport Pavement Design and Evaluation.* Advisory Circular AC 150/5320-6E. Federal Aviation Administration (FAA), Washington, DC.

FRA (2023). *The Freight Rail Next Work.* Federal Railroad Administration (FRA), US Department of Transportation. At https://railroads.dot.gov/rail-network-development/freight-rail-overview

Liu, W., Guo, H., Li, H., & Li, Y. (2014). Using BIM to improve the design and construction of bridge projects: A case study of a long-span steel-box arch bridge project. *International Journal of Advanced Robotic Systems,* 11:125. DOI: 10.5772/58442

Mills, W.H. (2015). *Railway Construction.* An eBook released in December 2015. At https://www.gutenberg.org/files/50696/50696-h50696-h.htm

Wang, G. (2021). *CMGT 3710/3711-Infrastructure Construction Introduction and Highway Materials Course Pack.* East Carolina University, Greenville, NC.

Chapter 15

Innovative Technology Utilization in Infrastructure Construction

15.1 Introduction

Historically, designing a roadway, bridge, or other transportation feature was a distinctly separate operation from the construction of that facility, and maintenance was rarely a consideration in either design or construction. Numerous design revisions due to changed or unanticipated field conditions would be warranted once construction began, causing hours of rework for designers and delays to construction. Often, field construction modifications, such as moving a drainage structure, would create unintended consequences for maintenance crews, making maintaining the roadway difficult. Contractors would often recreate plans in their systems in order to declutter the plan sheets, making it easier for crews to focus on their operations. All of this work and rework led transportation agencies to realize that there had to be a better way. Better communication between designers, contractors, inspectors, and maintenance groups could improve the end product and enhance public perception, but they struggled with how to achieve it.

States began holding planning and scoping meetings at various stages of design, bringing together designers, construction staff, and local maintenance personnel to review the plans and even walk the project limits to allow for greater input and better designs. These meetings were critical in identifying design issues such as bridge clearances, utility line depths, or the need for local construction time restrictions. Construction personnel were able to understand field maintenance staff's needs for ease of access to pipe networks or curb and gutter radius changes to accommodate a local trucking facility (FHWA, 2018).

While all these things were positive improvements, the three groups were still siloed, working with disparate computer systems and paper plans. Designers finished their paper and 2D plans and filed them away. Construction administration staff received printed copies of those 2D plans and made field modifications that, in many cases, never made it back to the designers. Contractors took those paper plans, scanned them into their computer systems, and created their own versions to work from.

At the end of construction, a set of paper redlined "as-built" drawings was delivered to the maintenance staff, and the project was checked off as completed. These hard copy plan sheets were bulky and cumbersome and usually lost over the course of time, leaving the maintenance staff with no good source of information about the roadway they were tasked to maintain.

Today, digital project delivery has undergone significant evolution, and driven by emerging technologies, it has vastly altered the work of infrastructure project delivery.

Cutting-edge technologies like building information modeling (BIM), augmented reality (AR), and virtual reality (VR) offer immersive experiences for design reviews and site visualization.

DOI: 10.1201/9781003197768-15

Unmanned aircraft systems (UAS) facilitate site surveys and data collection, while 3D printing accelerates construction with custom components. The Internet of Things (IoT) enables real-time data collection and analysis. AI and machine learning support predictive analytics, and cloud-based tools enhance collaboration.

These technologies, alongside laser scanning, blockchain, robotics, mobile apps, and 5G connectivity, revolutionize project delivery.

15.2 Building Information Modeling

Building information modeling is a 3D parametric modeling process that provides architects, engineers, and construction professionals with a digital representation of a building's physical and functional characteristics. It involves creating and managing a detailed digital representation of an infrastructure project, including both above-ground and below ground features. For example, it can create a 3D model of a highway bridge showing the underground footings and piles as well as storm drainage, signs, and guardrail features in the area of the bridge. It encompasses not only 3D geometric data but also information about the project's physical and functional characteristics. BIM provides a collaborative platform for stakeholders to work together and share data throughout a project's lifecycle, from design and construction to operation and maintenance. BIM helps improve collaboration, reduce errors, and enhance project visualization (FHWA, 2022).

The Federal Highway Administration recognized the benefits of BIM, saying, "Building Information Modeling, as applied to highway infrastructure (BIM for Infrastructure), is a collaborative work method for structuring, managing, and using data and information about transportation assets throughout their lifecycle." Managing data involves creating data (i.e., supplying data using data models), preserving data (i.e., storing, archiving, securing, and retrieving data), and provisioning, exchanging, or sharing data for use during a variety of business operations. The integration of data sources from multiple business siloes creates a digital twin and increases data accessibility for better decision-making.[1] Life-cycle phases, as shown in Figure 15.1, encompass the entire life of the project from design to construction to operations and maintenance.

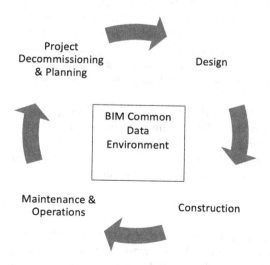

Figure 15.1 BIM Workflow Diagram

15.2.1 3D Modeling and Visualization

BIM starts with the creation of 3D models, which serve as a digital representation of the physical roadway or structure. These models include information about the features' geometry, materials, spatial relationships, and how they interact with its surroundings.

Every asset on the infrastructure project contains a significant amount of data collected at every phase of the project. Some of the data are used for design, construction procurement, fabrication, and installation, while other data are used during operations, maintenance, and asset management. The 3D model provides stakeholders with a visualization of every element of the project, represented in virtual, 3D space. Every element contains its design, procurement, fabrication, installation, and asset management data, all wrapped up in a convenient illustration that gives everyone the ability to better understand the project (FHWA, 2022).

15.2.2 Data Integration and Data Sharing

BIM integrates various data types, including architectural, roadway, structural, and even MEP (mechanical, electrical, and plumbing) data from different design systems to create one common data environment. This collaborative work environment allows all users to access the design, construction, and as built data as needed. This method for structuring, managing, and using data and information about transportation assets throughout their lifecycle is game-changing for transportation professionals. The integration of data sources from multiple business silos creates a digital twin and increases data accessibility for making better decisions. This approach allows different professionals to collaborate effectively in real-time data among project participants, facilitating communication, coordination, and decision-making.

15.2.3 Clash Detection

BIM software can detect clashes or conflicts in the design phase, preventing costly errors and rework during construction. In large highway construction projects, it is not unusual to have many different design firms involved with hundreds of different engineers working on a final set of plans. With all these teams working in a common data environment, BIM allows them to visualize and get a clear overview of the model. This aids in identifying any potential disputes and resolving them early in the design stage, which results in a more accurate and precise project design.

It also improves collaboration amongst multi-disciplinary project teams and models, helping eliminate not only wasted time, but also material waste, installation problems, and element conflicts.

Clash detection aids in BIM, speeding up the construction process by identifying clashes between multi-trade models during the early design stage. For example, a conflict between the phased construction of a bridge bent by the structural design firm and a temporary traffic shift designed by the firm creating the traffic control plan could be caught early in the design process, resulting in lower overall costs, less material waste, and faster project completion.

15.2.4 Life-Cycle Management

BIM extends beyond the construction phase to include information relevant to the operations, maintenance, and renovation of the assets. At the completion of the construction project, during project closeout, the as-built 3D model becomes available and is accessible to the maintenance and operations group charged with keeping the new roadway assets in working order. If the Department of Transportation has a robust asset management system, the data housed in the common data environment can be transferred directly into the asset management system with all its metadata. Data about every element of the bridge, including construction materials and testing data, can be accessible to the bridge maintenance personnel. Pipe networks can be shown in relation to other underground features such as utilities, giving maintenance staff advanced knowledge of conflicts before they start digging.

15.3 Augmented Reality and Virtual Reality

AR and VR technologies are used to create immersive experiences for project stakeholders. They are beneficial for design reviews, virtual walkthroughs, and training purposes. AR and VR can also be used for on-site visualization and as aids for maintenance and operations (Delgado, 2020).

15.3.1 Augmented Reality

AR can offer several benefits to highway construction projects by enhancing the design, planning, and execution phases. It can help construction teams and designers visualize the final project in a real-world context, aiding in making informed decisions about design elements, such as lane placement, signage, and landscape features. It enables stakeholders to see the proposed design superimposed on the actual terrain, which can lead to more efficient and aesthetically pleasing designs, and better public understanding and acceptance of the project.

AR has various applications, from assisting in workplace safety planning to training new workers by familiarizing them with the construction site. It can also be used to monitor the progress of construction by overlaying the planned timeline on the actual site. This allows project managers to track construction milestones and ensure that the project is on schedule (Huang, 2019; 2020).

15.3.2 Virtual Reality

VR technology enables design engineers to depict how facilities like bridges, roadways, sidewalks, and buildings will look after they are completed, and maintenance engineers to examine how the project was built to get a better idea of maintenance needs. 360-degree images are displayed on computer screens, and the engineer wears a headset allowing for the facility to be seen in VR. The engineer can walk or turn and view details of the design up close.

To do this, the engineers match an aerial orthoimage of the construction location with a computer-generated 3D model of the facility. What emerges is a composite view of the proposed highway facility superimposed on top of the photograph of the construction location.

Although the processes used in VR have been available for some time, the technology has become more widespread in recent years thanks to rapid advances and falling costs in computer technology. Today, with high-end graphics workstations and software available at reasonable costs, photo simulations and 3D renderings can be produced quickly and with high accuracy.

VR technology not only helps engineers during their planning process, but it also it serves as an effective way to clearly communicate design implications to non-engineers. Thanks to this technology, highway engineers are now able to take their composite drawings to public meetings and present them to non-engineers with ease (Huang, 2019; 2020).

15.4 Unmanned Aircraft Systems

Unmanned aerial systems (UAS), often simply referred to as drones, are multi-use aircraft systems controlled by a certified operator on the ground and are used for site surveys, inspections, progress monitoring, and data collection. Drones can capture high-resolution images, videos, and data that aid in project planning and management.

The benefits of UAS are wide-ranging and impact nearly all aspects of highway transportation, in some cases replacing human inspections, increasing safety and accuracy, speeding up data collection, and providing access to hard-to-reach locations.

UAS provides high-quality survey and mapping data that can be collected automatically or remotely. Large areas can be mapped relatively quickly in comparison to traditional survey and mapping practices. They can supplement conventional activities, such as bridge safety inspections and routine construction inspections, increasing safety for inspectors and collecting data from otherwise unattainable perspectives. UAS can get closer to bridge bearings than a human inspector, allowing the human inspector to zoom in and look at a portion of the assembly that might have otherwise been missed. Other uses include survey and imagery as part of emergency response events, where traditional surveying and mapping practices may be inadequate or sites impossible to access.

The North Carolina Department of Transportation uses UAS to survey flooded roadways, such as the aftermath of hurricane events. In 2019, in the immediate aftermath of Hurricane Florence, the NCDOT used UAS in an unprecedented manner to monitor and document flooding, road conditions, and traffic impacts. NCDOT earned a regional award in the 2019 America's Transportation Awards competition from the AASHTO, in the "Best Use of Technology and Innovation, Small Project" category.

NCDOT flew more than 200 missions and captured 8,000 pictures and videos of the damage and flooding left behind by Hurricane Florence. This helped state and federal agencies make real-time decisions around aiding emergency response, planning detour routes, assessing future repair needs, expanding disaster declarations, and warning the public of the dangers faced on North Carolina's roads. Lumberton flooded after Hurricane Florence is shown in Figure 15.2.

Construction inspection with UAS allows for a bird's eye view of a project's progress and for the development of digital terrain models (DTMs) that document the construction process and assist in the assessment of earthwork quantity measurement. It can be used for routine inspections, such as flying a programmed path over silt fencing after a rain event to check for sediment buildup, and high-risk inspections, such as crane or falsework construction.

Minnesota DOT has put significant effort into the development of UAS use for bridge inspection and has produced several informative reports that advance the understanding of potentially vital applications and current limitations. New Jersey is currently using UAS to support structural inspections, real-time construction project monitoring, traffic incident management, aerial 3D corridor mapping, emergency response assessments, and traffic congestion assessments. Colorado is using UAS to monitor geohazards in more than 40 mountainous corridors with highly accurate data collection. The accuracy of the data and the lower cost of acquiring it have led to a better understanding and ability to mitigate wide-ranging safety risks.

15.5 3D Printing

While it is not yet a widespread technology, 3D printing, also known as additive manufacturing, is finding innovative applications in highway construction. It has the potential to revolutionize the industry in several ways.

Figure 15.2 I-95 Lumberton Flooded after Hurricane Florence
Source: Photo courtesy of NCDOT

- Rapid Prototyping: 3D printing can be used to create rapid prototypes of highway components and structures. This allows engineers to test and refine designs quickly, reducing the time and cost involved in the design phase.
- Custom Components: 3D printing enables the production of highly customized components for highway construction. For example, it can be used to create unique, complex shapes for barriers, sound walls, or bridge components that are difficult to manufacture using traditional methods.
- Prefabricated Components: Prefabrication is a key benefit of 3D printing. Entire sections of highway infrastructure, such as bridge components or retaining walls, can be 3D printed off-site and then transported to the construction location. This can speed up construction and reduce on-site labor requirements.
- Reinforcement Materials: 3D printing can be used to produce fiber-reinforced concrete, which is stronger and more durable than conventional concrete. This can lead to longer-lasting high-way structures with reduced maintenance needs.
- Lightweight Structures: 3D printing allows for the creation of lightweight structures while maintaining their strength. This is especially useful for designing overpasses, bridges, and other highway components.

- Custom Drainage Systems: Drainage systems, culverts, and stormwater management components can be 3D printed to exact specifications, improving water flow and reducing erosion risks.
- Repair and Maintenance: 3D printing can be used to manufacture replacement parts for highway equipment and vehicles. This can be cost-effective and reduce downtime for maintenance.
- Noise Barriers and Sound Walls: 3D printing can create innovative and customized noise barriers that not only provide noise reduction but also add aesthetic value to highways.
- Barrier Systems: 3D-printed crash barriers and guardrails can be designed to be more efficient at absorbing impact forces and protecting motorists.
- Signage and Wayfinding: 3D printing can be used to produce durable and custom signage and wayfinding markers for highways.

While 3D printing holds significant promise in highway construction, there are also challenges and limitations, such as the need for large-scale printers, material limitations, and cost considerations. Additionally, the technology is still evolving, and regulations and standards need to be adapted to accommodate its use in construction. 3D printing has the potential to improve efficiency, reduce costs, and create innovative, sustainable, and long-lasting highway infrastructure components (Hart, et al., 2022; Gong, et al., 2023).

15.6 Laser Scanning

The term "remote sensing" is used to describe 3D remote data acquisition using technologies such as light detection and ranging (LiDAR, known as terrestrial laser scanning in construction), ground penetrating radar (GPR), road profilers/scanners, and other sensors (i.e., sign retroreflectivity), and the accuracy and precision vary among the different technologies. For example, 3D laser scanners can measure millions of data points per second and generate a very detailed point cloud dataset.

Benefits of this technology began with the time and cost savings in surveying data collection, followed by increased productivity (e.g., less rework) throughout the entire project delivery process. There has been improved quality with the increased level of detail, accuracy, and scalability in these technologies. For example, when high-definition surveys were provided to contractors during the pre-bid stages, the increased accuracy and detail reduced uncertainty and allowed the contractors to submit more competitive bids.

Another benefit of using remote sensing has been improved safety because the non-contact technologies minimized or eliminated the time field crews were exposed to traffic and other dangerous conditions. Also, non-contact technologies minimized or eliminated impacts on environmentally sensitive areas. Finally, the use of remote sensing technology during surveying phases has provided a building block for information modeling in design and as-built construction documentation.

15.7 Artificial Intelligence

Artificial intelligence (AI) is being increasingly integrated into highway construction to improve efficiency, safety, and decision-making processes. AI algorithms can analyze traffic data to optimize road design and lane configurations, thereby improving traffic flow and reducing congestion (Baduge, et al, 2022).

It can help assess and minimize the environmental impact of construction by optimizing designs to reduce carbon emissions and pollution. In the conceptual phases of a project, AI algorithms can process geospatial data to identify suitable construction sites, analyze terrain, and optimize the placement of highway infrastructure.

During construction, AI-based computer vision systems can inspect materials, structures, and infrastructure components for defects and adherence to design specifications. They can help optimize the procurement, storage, and use of construction materials, reducing waste and costs. They can analyze test reports, looking for trends that can lead to shortened asset life in the future.

Contractors have found ways to use AI to predict equipment maintenance needs, reduce downtime, and extend the life of construction machinery, monitor construction sites for safety compliance, and detect potential safety hazards, allowing for quick corrective actions.

AI can analyze historical project data, helping contractors create more accurate cost estimates and bid packages, reducing the risk of budget overruns. Stakeholders can make informed decisions by processing large datasets and identifying trends and patterns in project data quickly, producing a better overall project outcome.

Chatbots and virtual assistants can engage with the public to answer questions, provide project updates, and address concerns about construction projects.

15.8 Robotics and Automation

In the realm of highway construction, robotics and automation have ushered in a new era of innovation, efficiency, and safety. The application of cutting-edge technology has brought about significant advancements in the way roads are planned, built, and maintained. From precision-guided robotic pavers ensuring perfectly even road surfaces to autonomous surveying and drones mapping vast areas in record time, the integration of robotics and automation is revolutionizing the construction industry. Advances like intelligent traffic management systems, self-driving construction vehicles, and automated material handling are contributing to safer, faster, and more cost-effective highway projects. Some examples of these remarkable advancements are:

15.8.1 Robotic Pavers

Automation and robotics have generated much interest in the transportation construction community, and combined with high accident rates in work zones, much research has been conducted attempting to develop autonomous robot pavers. To date, these paver prototypes are still being developed and are not used in mainstream construction. Results are promising that these paver robots can improve productivity, safety, efficiency, and smoothness (Bryson, et al., 2005).

15.8.2 Automated Surveying

Automated surveying technology has emerged as a game-changer in the field of land surveying and mapping. Integrating cutting-edge instruments, satellite-based positioning systems, and advanced software has transformed the way construction project data is collected and analyzed. Automated surveying technology offers increased accuracy, efficiency, and safety, making it invaluable.

Automated surveying heavily relies on global navigation satellite systems (GNSS), such as global positioning system (GPS) and global navigation satellite system (GLONASS). These satellite networks provide precise positioning and navigation data, enabling surveyors to determine accurate coordinates for various points on the Earth's surface. GNSS technology has significantly improved surveying efficiency, allowing surveyors to work faster and with increased precision, with fewer personnel and, in many cases, away from traffic.

Robotic total stations are advanced surveying instruments with robotic technology and motorized telescopes. They can automatically track and measure target points, reducing the need for surveyors to operate the instrument manually. Robotic total stations enhance accuracy and productivity, especially in construction and land development projects.

Automated surveying technology has revolutionized highway construction projects, allowing for precise project layout and ensuring that structures such as drainage systems and bridges are built in the correct locations and with the proper orientation. This technology also aids in excavation and earthwork volume calculations, eliminating the need for laborious cross-section calculations, thereby streamlining the construction process and minimizing errors. Figure 15.3 shows a Chironix surveying robot.

15.8.3 Automated Grading Equipment

The global navigation satellite system (GNSS) works by using satellites to send signals to GPS/GLONASS receivers on Earth. GNSS technology in the construction industry is used to organize sites, keep track of equipment, survey, grade, excavate, drill, pile drive, and compact. Applications of GNSS in the construction industry can help contractors more efficiently and accurately complete projects, especially those that are larger-scale and traditionally require larger crews.

Figure 15.3 A Chironix Surveying Robot

Source: Courtesy of Chironix

Putting these GPS/GLONASS receivers on grading equipment like bulldozers and graders achieves precise grading and earthmoving, eliminating the need for repetitive surveying and grade checking, and reducing the need for rework when over grading is done.

15.8.4 Autonomous Vehicles and Robots

Researchers and leading contractors are increasingly interested in autonomous vehicles and robots because of their unprecedented agility and advanced autonomy. Unmanned ground vehicles (UGV) and quadruped robots have exceptional spatial mapping and obstacle avoidance capabilities with their integrated payloads, including laser scanners, 360-degree cameras, depth cameras, and infrared cameras. In addition, various environmental sensors allow autonomous vehicles and robots to measure data in potentially hazardous construction sites, such as temperature, particulate levels, and gas leaks. Figure 15.4 shows a Boston Dynamics Spot with various payloads.

15.9 Mobile Apps and Field Data Collection

Highway construction projects are complex undertakings with hundreds of activities and pay items. They require precise planning, efficient execution, and thorough data management. In recent years, mobile applications and automated field data collection tools have become invaluable assets by streamlining data collection, improving communication among project stakeholders, and enhancing overall project efficiency.

These tools improve accuracy and reduce human error in data collection and reporting by collecting the location and quantity of assets at the source, leading to more accurate project information. Because they streamline data collection and eliminate manual paperwork, they save time by reducing administrative overhead. This efficiency saves time and reduces project costs.

Figure 15.4 Boston Dynamics Spot with Various Payloads
Source: Courtesy of Boston Dynamics

They provide enhanced communication by facilitating real-time collaboration among project stakeholders, including contractors, engineers, and project managers, sharing pictures and conditions, leading to better decision-making. Automated data collection tools allow for remote monitoring of construction sites, providing real-time insights into project progress and challenges.

Data collected in the field during construction becomes available to future maintenance personnel responsible for maintaining the assets. This provides valuable insights for understanding the placement, positioning, and specifications of the assets, enabling them to make informed decisions without delay. Some of these tools include:

15.9.1 Asset Management

These tools include various types of data collectors loaded with construction plans and project specifications. These are used to mark the location of features like pipes, guardrails, and sign structures, including any field changes made during construction. Construction personnel develop an inventory of construction assets, including materials, equipment, and labor, helping to track and optimize resource allocation as well as creating a permanent record for future use.

15.9.2 Quality Control and Inspection

Automated field data collection tools assist in quality control and inspection procedures by using preloaded forms to enable real-time data capture and reporting for compliance, quality assurance, and accurate payment.

15.9.3 Project Scheduling and Management

Mobile project management apps enable construction professionals to manage projects more efficiently. They provide access to project plans, schedules, and task lists, ensuring real-time updates and seamless communication among team members. They allow project managers to track progress, assign tasks, and make schedule adjustments on the go, improving project coordination and minimizing delays.

15.10 Innovative Materials Utilization – A Case Study of Harkers Island Bridge Replacement Project in North Carolina

Project Overview: Balfour Beatty was selected to replace two 50-year-old bridges on behalf of the North Carolina Department of Transportation (NCDOT) to increase emergency access and evacuation capacity, and reduce congestion and delays for marine vessels and vehicular traffic.

Construction included the development of a new, single, 3,200-foot-long fixed-span bridge to replace the existing Earl C. Davis Memorial Bridge (Bridge No. 73) and Bridge No. 96 that connected the town of Straits to Harkers Island. The new bridge was constructed over the straits to the east of the existing bridges and provided a direct path for drivers traveling to and from the mainland. With a 45-foot navigational clearance, the new bridge allows boats to pass unimpeded and safely allows motorists to travel without delays from bridge openings.

Bridge No. 73 will be removed, and Carteret County will take ownership of Bridge No. 96. This bridge will be repurposed as a pedestrian bridge to provide access to the center island and a renovated Straits Fishing Pier. Figures 15.5 and 15.6 show the Hankers Bridge under construction and the Birdseye view of the bridge replacement project.

Figure 15.5 Hankers Bridge Under Construction

Source: Photo courtesy of Balfour Beatty US Civils and NCDOT

Figure 15.6 Birdseye View of the Bridge Replacement Project

Source: Photo courtesy of Balfour Beatty US Civils and NCDOT

Material Innovation: Balfour Beatty leveraged the latest technologies, including non-corroding carbon fiber reinforced polymer (CFRP) strands and glass fiber-reinforced polymer (GFRP) bars, a proven advancement in transportation technology that performs comparably to steel in the finished product and is lighter and more durable than traditional materials. The new Harkers Island Bridge is expected to withstand the elements better in the coastal environment, resulting in less maintenance and a longer lifespan for the new bridge. Figures 15.7 and 15.8 show the installation of glass fiber-reinforced polymer bars and non-corroding carbon fiber-reinforced polymer strands used in the bridge replacement project.

Design and Construction: The bridge was designed with the new CFRP and GFRP materials, and strict requirements prohibited the use of ferrous materials in its construction. This required the contractor to innovate certain building processes, which presented challenges to traditional construction methods.

Figure 15.7 Installing Glass Fiber-Reinforced Polymer (GFRP) Bars

Source: Photo courtesy of Balfour Beatty US Civils and NCDOT

Figure 15.8 Non-Corroding, Carbon Fiber-Reinforced Polymer (CFRP) Strands Are Used

Source: Photo courtesy of Balfour Beatty US Civils and NCDOT

Environmental Considerations: A vital aspect of the project is its environmental stewardship. The CFRP and GFRP components will eliminate the risk of corrosion, ultimately increasing the bridge's lifespan. Additionally, the new bridge was constructed while an in-water work moratorium was observed from April 1 through September 30 each year to protect fish spawning.

Community Impact: This bridge is a transportation link and a community landmark. It replaces a 50-year-old steel swing span bridge that required frequent maintenance.

Economic Benefits: The new design with non-corrosive materials will prolong the bridge's lifespan and reduce NCDOT's future maintenance costs.

Project Timeline: The bridge was opened almost a year ahead of its contracted timeline. Following the opening of the new bridge, the existing bridge will be removed in its entirety.

15.11 Questions

1. What is building information modeling (BIM) and how does it aid infrastructure projects?
2. Describe how augmented reality (AR) can benefit highway construction projects.
3. Explain the role of virtual reality (VR) in infrastructure development and its impact on public engagement.
4. How do drones, or unmanned aircraft systems (UAS), contribute to the construction and maintenance of infrastructure?
5. Discuss the advantages of 3D printing in highway construction and the types of components it can produce.
6. What is laser scanning, and how does it enhance the accuracy and efficiency of construction projects?
7. How is artificial intelligence (AI) integrated into highway construction to improve project outcomes?
8. Detail the use of robotics and automation in modern infrastructure construction and provide examples of their applications.
9. What are the benefits of using mobile apps and field data collection tools in highway construction projects?
10. Explain the significance of data integration and sharing in the BIM process.
11. How does clash detection in BIM contribute to cost savings and project efficiency?
12. Describe how life-cycle management is implemented in infrastructure projects using BIM.
13. Discuss the benefits and limitations of 3D printing technology in construction infrastructure.
14. How does the use of AI in infrastructure construction enhance safety and decision-making processes?
15. In what ways are automated surveying technologies changing the landscape of land surveying in construction?
16. What role do augmented and virtual reality technologies play in construction training and operational phases?
17. How does the integration of mobile apps improve communication and project management in construction?
18. Describe a specific example of a successful application of UAS in disaster response and its impact on infrastructure assessment.
19. What are the potential economic benefits of integrating innovative materials like CFRP and GFRP in bridge construction?
20. How has the digital transformation in infrastructure construction affected traditional methods and project delivery timelines?

Note

1 FHWA (2023). Building Information Modeling (BIM) for Infrastructure Overview

References

Baduge, S.K., Thilakarathna, S., Perera, J.S., Arashpour,M., Sharafi, P., Teodosio, B., Shringi, A., & Mendis, P. (2022). *Artificial intelligence and smart vision for building and construction 4.0: Machine and deep learning methods and applications*. https://research.monash.edu/en/publications/artificial-intelligence-and-smart-vision-for-building-and-constru. Monash University, Melbourne, Australia.

Bryson,L. S., Maynard, C., Castro-Lacouture, D., & Williams, R. W. (2005). *Fully autonomous robot for paving operations.* DOI:10.1061/40754(183)37.

Delgado, O. & Demian, B. (2020). *A research agenda for augmented and virtual reality in architecture, engineering, and construction.* DOI:10.1016/j.aei.2020.101122.

FHWA (2018). Automation in highway construction part 1: Implementation challenges at state transportation departments and success stories (FHWA-HRT-16-030, October 2018).https://www.fhwa.dot.gov/publications/research/infrastructure/pavements/16030/16030.pdf. Federal Highway Administration (FHWA), Washington, DC.

FHWA (2022). Building Information Modeling (BIM) for infrastructure. https://www.youtube.com/watch?v=bEI-TCPmmT0. Federal Highway Administration (FHWA), Washington, DC.

FHWA (2022). Design visualization. https://highways.dot.gov/federal-lands/design-visualization. Federal Highway Administration (FHWA), Washington, DC.

Hart, J.A., Chen, W., Quinlan, H., Quinlan, H., & Gerasimidis, S. (2022). *Feasibility of 3D Printing Applications for Highway Infrastructure Construction and Maintenance.* No. 22-029, May 2022. Massachusetts Department of Transportation, Report No. 22-029. Massachusetts Department of Transportation, Office of Transportation Planning, Ten Park Plaza, Suite 4150, Boston, MA 02116.

Gong, F., Cheng, X., Wang, Q., Chen, Y., You, Z., & Liu, Y. (2023). A review on the application of 3D printing technology in pavement maintenance. *Sustainability*, 15(7), 6237. At https://doi.org/10.3390/su15076237

Huang, Y. (2019). A Comparative Review of Construction VR Applications: Functionality,Simulation, and Collaboration. *International Journal of Emerging Engineering Research and Technology*, 7(2), 24-37.

Huang, Y. (2020). Evaluating Mixed Reality Technology for Architectural Design and Construction Layout. *Journal of Civil Engineering and Construction Technology*, 11(1), 1- 12.

Kim, J. H. & An, D. (2018). Development of a mobile application for real-time geospatial data collection on construction sites. *International Journal of Advanced Smart Convergence*, 7(1), 20–26.

Liang, W. & Kamat, M. (2021). Human–robot collaboration in construction: Classification and research trends publication. *Journal of Construction Engineering and Management*, 147(10). https://doi.org/10.1061/(ASCE)CO.1943–7862.0002154

Mohamed, T. (2022). *Exploring the use of mobile technologies for highway construction inspection.* Construction Research Conference, Arlington, Virginia, March 9–12, 2022.

Moselhi, O., Shalaby, A., & Waly, A. (2018). Development of a quality control and inspection application for construction projects. *Journal of Computing in Civil Engineering*, 32(4), 04018027.

Ngwenyama, O., Guergachi, A., & Gutiérrez, J. (2019). Design and evaluation of a mobile application for field construction site safety audits. *Automation in Construction*, 105, 102857.

Ogunrinde, N.A. (2020). *Application of emerging technologies for highway construction quality management: A review.* Construction Research Conference, Tempe, Arizona, March 8–10, 2020.

Petrie, G., Baltsavias, E., Jutzi, B., & Puissant, A. (2012). The development of photogrammetric technology and its current state. *ISPRS Journal of Photogrammetry and Remote Sensing*, 67, 1–19.

Rizos, C. & Lachapelle, G. (2016). GNSS in the challenging environment of civil engineering. *Journal of Geodesy*, 90(2), 101–105.

Shafiq, M., Hafeez, A., Ahmad, R., & Khan, S.A. (2019). Critical success factors for project scheduling and control through mobile-based project management (MBPM) applications in the construction industry. *Arabian Journal for Science and Engineering*, 44(1), 557–576.

Su, Y. H. & Jung, J. (2019). A review of the applications of unmanned aerial vehicles in land surveying and mapping. *Journal of Photogrammetry and Remote Sensing*, 151, 20–28.

Teizer, J., Lee, J. K., & Venugopal, M. (2017). Survey and recommendations for mobile IT-based systems for construction safety and health. *Automation in Construction*, 81, 254–264.

Teo, T.A. (2018). An overview of the use of LiDAR in surveying and mapping. *The Photogrammetric Record*, 33(163), 148–161.

Wang, G., Li, J., Saberian, M., Rahat, M., Massarra, C., Buckhalter, C., Farrington, J., Collins, T., & Johnson, J. (2022). Use of COVID-19 single-use face masks to improve the rutting resistance of asphalt pavement. *Journal of Science of the Total Environment*, 826(2022), 154118. At https://doi.org/10.1016/j.scitotenv.2022.154118

Appendix

Lab Practices for Basic Highway Materials

This appendix provides the lab practice procedures for basic highway materials. ASTM or AASHTO standard test methods are referred to.

A.1 California Bearing Ratio Test (CBR)

ASTM D1883 – Test Method for California Bearing Ratio (CBR) of Laboratory-Compacted Standard Soils

Purpose: To measure the strength and swelling potential of soil.

Theory: The California Bearing Ratio test is one of the most commonly used methods to evaluate the strength of subgrade soil for pavement thickness design. A soil is compacted in a mold with the standard compaction effort at its optimum water content (which is at about 100% of its maximum density, as determined by the standard compaction test). This test simulates the perspective of the actual condition at the surface of the subgrade. A surcharge is placed on the surface to represent the mass of pavement materials above the subgrade. The sample is soaked to simulate its weakest condition in the field. Expansion of the sample is measured during soaking to check for potential swelling. After soaking, the strength is measured by recording the force required to shove a penetration piston into the soil.

Apparatus: Compression machine with penetration piston that is 49.5 mm (1.95 in) in diameter, with an area of 19.35 cm² (3.0 in²); mold that is 152.4 mm (6.0 in) in diameter and 177.8 mm (7.0 in) high, with collar and base; spacer that is 61.4 mm (2.416 in) high to fit mold; standard compaction hammer; surcharge masses, each weighing 2.27 kg (5 lb); swell-measuring apparatus.

Sample: Approximately 5,000 g of aggregate including 2,500±10 g of 19 mm to 12.5 mm (¾–½ in) size and 2,500 ± 10 g of 12.5–9.5 mm (½ in–⅜ in) size.

A.1.1 Procedure

1. Place the spacer in the mold, with the base plate and collar attached.
2. Fill the mold with three layers of soil at its optimum moisture content. Compact each layer with 56 blows of the standard compaction hammer. Remove the collar and level the surface with a straightedge.
3. Remove the perforated base plate, invert the mold, and replace it on the base plate.
4. Remove the spacer, insert the expansion-measuring apparatus, place the surcharge rings on the surface, and take an initial swell-gauge reading.

Note: A minimum of 4.54 kg (10 lb) of surcharge is used, which represents about 125 mm (5 in) of pavement structure.

5. Soak the sample for four days, with water available at the top and bottom.
6. Take a final swell-gauge reading. Remove the sample and mold from the water. Allow the sample to drain for 15 minutes.
7. Place it in the compression machine and seat the penetration piston on the surface with a 44.5 N (10 lb) load.
8. Apply a load on the piston, and then record the total load when the piston has penetrated the soil to a depth of 2.54 mm (0.1 in).

A.1.2 Results

1. Swell-gauge readings (A): Initial _____ Final _____
2. Load required to cause 2.54 mm (0.1 in) penetration: _____ (L)

A.1.3 Calculation

1. Calculate percentage swell (%)

   $$S = A / B \times 100\% \qquad\qquad\qquad \text{Eq. A1.1}$$

   where
   S = percentage swell (%)
   A = amount of swell or height of sample
   B = height of mold (116.4 mm or 4.584 in)

2. Calculate CBR value

   $$CBR = L/S \times 100\% \qquad\qquad\qquad \text{Eq. A1.2}$$

   where
   L = actual load required to cause 2.54 mm (0.1 in) penetration
   S = standard load at 2.54 mm (0.10 in) penetration, 13.3 kN or 3,000 lb

A.2 Los Angeles (LA) Abrasion Test (Coarse Aggregate)

ASTM C131 (AASHTO T96-02) Standard Test Method for Resistance to Degradation of Small-Size Coarse Aggregate by Abrasion and Impact in the Los Angeles Machine

The full references are ASTM C535-09 Standard Test Method for Resistance to Degradation of Large-Size Coarse Aggregate by Abrasion and Impact in the Los Angeles Machine.

Purpose: To determine the hardness of aggregates.

Theory: To measure the hardness of aggregates, a sample is placed in a drum with steel balls. The drum is rotated, and the balls grind down the aggregate particles. Soft aggregates are quickly ground to dust, while hard aggregates lose little mass.

Apparatus: Los Angeles abrasion machine; Sieve; Balance (accurate to 0.01 g).

Sample: Approximately 5,000 g of aggregate including 2,500 ± 10 g of 19–12.5 mm (¾–in ½ in) size and 2,500 ± 10 g of 12.5–9.5 mm (½–⅜ in) size.

A.2.1 Procedure

1. Wash, dry, and obtain mass of the sample.
2. Place it in the abrasion machine.
3. Add 11 standard steel balls.
4. Rotate the drum for 500 revolutions at 30–33 rpm.
5. Remove the sample. Sieve it on a 1.70 mm sieve. Wash the sample retained. Obtain the mass

A.2.2 Results

I.   Mass of original sample _____ g (A)
II. Mass of final sample _____ g (B)
III. Loss _____ g (A – B)

A.2.3 Calculation

% Loss = (A – B) /A × 100 _____ %   Eq. A1.3

where

A = mass of original sample, g
B = mass of final sample, g

The LA abrasion test is a common test method used to indicate aggregate toughness and abrasion characteristics. Aggregate abrasion characteristics are important because the constituent aggregate in HMA must resist crushing, degradation, and disintegration to produce a high-quality HMA.

The standard LA abrasion test subjects a coarse aggregate sample (retained on the No. 12 (1.70 mm) sieve) to abrasion, impact, and grinding in a rotating steel drum containing a specified number of steel spheres. After being subjected to the rotating drum, the weight of aggregate that is retained on a No. 12 (1.70 mm) sieve is subtracted from the original weight to obtain a percentage of the total aggregate weight that has broken down and passed through the No. 12 (1.70 mm) sieve. Therefore, an LA abrasion loss value of 40 indicates that 40% of the original sample passed through the No. 12 (1.70 mm) sieve.

Aggregates undergo substantial wear and tear throughout their life. They should be hard enough to resist crushing, degradation, and disintegration from any associated activities, including manufacturing, stockpiling, production, placing, and compaction. Furthermore, they must be able to adequately transmit loads from the pavement surface to the underlying layers and, eventually, the subgrade. These properties are especially critical for open or gap-graded HMA, which do not benefit from the cushioning effect of fine aggregate and where coarse particles are subjected to high contact stresses. Aggregates not adequately resistant to abrasion and polishing may cause premature structural failure and/or a loss of skid resistance. Furthermore, poor abrasion resistance can produce excessive dust during HMA production, resulting in possible environmental and mixture control problems. Because of the preceding issues, a test to predict aggregate toughness and abrasion resistance is valuable. The LA abrasion test is the predominant test in the United States; 47 states use it.

The LA abrasion test measures the degradation of a coarse aggregate sample placed in a rotating drum with steel spheres. As the drum rotates, the aggregate degrades by abrasion and impact with other aggregate particles and the steel spheres (called the "charge"). Once the test is complete, the calculated mass of aggregate that has broken apart to smaller sizes is expressed as a percentage of

Figure A.1 LA Abrasion Equipment

the total mass of aggregate. Therefore, lower LA abrasion loss values indicate aggregates that is tougher and more resistant to abrasion.

A.3 Micro-Deval Test

ASTM D6928 – Standard Test Method for Resistance of Coarse Aggregate to Degradation by Abrasion in the Micro-Deval Apparatus

The Micro-Deval test has gained acceptance and popularity in recent years as an economical and accurate procedure for aggregate abrasion testing. The basis for this method was developed in France during the 1960s and provides a measure of toughness, abrasion resistance, and durability of mineral aggregates as they are ground with steel balls in the presence of water. Materials yielding a low loss in the test are unlikely to show significant degradation during handling, mixing, or placing, and will allow better long-term performance of pavements. There are separate

Figure A.2 Micro-Deval Test Equipment

test methods for fine or coarse aggregate materials. Abrasion loss is determined by measuring the amount of degraded material passing a 1.18 mm (No. 16) or 75 μm (No. 200) sieve after the test.

A.3.1 How Does the Micro-Deval Test Work?

A sample is prepared by separating it into individual size fractions of the required masses. Typical prepared sample sizes are 500 g for fine aggregate and 1,500 g for coarse aggregate.

1. The sample is immersed in tap water for a minimum of 1 hour in the Micro-Deval jar or other suitable container.
2. An abrasive charge of magnetic stainless steel balls is added to the prepared test sample with the water. The operator then secures the cover and places the Micro-Deval jar on the machine.

Figure A1.3 Micro-Deval Test Equipment

3. The Micro-Deval machine is set to rotate the jars at 100 rpm for a specified length of time or a specified total number of revolutions.
4. At completion, the sample is carefully washed over a specified sieve, and the percentage loss is determined by comparing the oven-dried mass of the retained sample to the original total sample weight. The 75 μm (No. 200) sieve is used for fine aggregates, and a 1.18 mm (No. 16) sieve is used for coarse fractions.
5. Micro-Deval testing requires a supply of calibration aggregate, which may be developed from a local source.

A.3.2 What Are the Benefits?

Both the LA abrasion test and the Micro-Deval test offer unique benefits. The Micro-Deval test has a smaller equipment size, lower sample quantities, and a simple procedure. Operators can also run two samples through a Micro-Deval apparatus simultaneously for increased efficiency and cost-effectiveness.

A.4 Field Density Test Using a Density Gauge

ASTM D6938 – Standard Test Methods for In-Place Density and Water Content of Soil and Soil-Aggregate by Nuclear Methods (Shallow Depth)

AASHTO T 3101 – In-Place Density and Moisture Content of Soil and Soil-Aggregate by Nuclear Methods (Shallow Depth)

Purpose: To determine the field density of soil, granular base, and asphalt concrete.

Theory: Nuclear moisture/density gauges are testing devices that use low-level radiation to measure the wet density, dry density, and moisture content of soil, granular construction materials, and asphalt concrete (pavement). When the appropriate safety practices are followed, the nuclear gauges pose no danger of radiation exposure to the operator.

Density testing on HMA is performed using a method referred to in the standards as backscatter. Backscatter means that the gamma detector picks up radiation emitted from the source at the tip of the rod when the source is flush with the pavement's surface. When this testing method is used, the particles emitted from the radioactive source are scattered into the material under test. In the Backscatter Method for density testing, the nuclear gauge must be seated in contact with the surface of the material being tested. No air gaps may be under the gauge caused by surface debris or roughness. The long dimension of the indicator must be parallel to the compaction equipment's travel direction. When the source rod handle is depressed, the radiation source is lowered to just above the material's surface, and gamma photons are emitted.

The gamma photons that measure density penetrate far into the material, but over 70% of the photons are scattered back to the detector tubes in the first two inches of material being tested. Ninety-five percent of the photons are scattered back to the detector tubes from the top three inches. Not many gamma photons are left to be scattered below the third inch. Therefore, virtually no density information is obtained below three inches with the Backscatter Method. In the Backscatter Method, in-place wet density is obtained by testing the material's surface. This method is usually

Figure A.4 Nuclear Density Gauge

used when determining the density of granular materials. Generally, the Backscatter Method is not used on soils except when the soil is very loose and granular. Whenever backscatter is used, the gauge's bottom surface must be clean. The backscatter test is required to be a 4-minute reading.

Apparatus: Troxler Nuclear Gauge.

Test Site Selection: Test sites are required to represent the area being tested. Proper seating of the gauge, without air gaps, is necessary to ensure reliable readings. Most materials are compacted with vibratory, pneumatic, or steel wheel rollers, leaving the surface smooth enough to test without special preparation. However, the material must be tested as soon after compaction as possible to avoid any unnecessary surface drying and shrinkage. If the surface has already dried, gently remove some of the dry surface material with a stiff brush until signs of moisture are visible. Extreme care is taken when scraping or brooming granular materials, as these materials may tend to loosen up when disturbed, and a reliable test is challenging to obtain.

A.4.1 Procedure

The procedure using the backscatter method is as follows:

Refer to the information issued with each gauge for supplemental information on conducting backscatter density testing.

The more tests taken, the more accurate the results. If several tests are conducted on a lot of material and the test results are averaged for final acceptance, more representative results are obtained.

Follow the procedures in AASHTO T 310.

A.4.2 Results

Given directly by the gauge.

A.5 Measuring Flat and Elongated Particles

ASTM D4791 – Standard Test Method for Flat Particles, Elongated Particles, or Flat and Elongated Particles in Coarse Aggregate

A.5.1 Purpose

For Superpave HMA concrete design, aggregates should meet four (4) requirements: i.e., (i) coarse aggregate angularity; (ii) fine aggregate angularity; (iii) flat and elongated particles; and (iv) clay content in fine aggregate.

This test measures the flat and elongated particles.

A.5.2 Apparatus

The Proportional Caliper Device determines (i) flat particles; (ii) elongated particles; or (iii) flat and elongated particles in coarse aggregates. Specific procedures for conducting this test are described in ASTM D 4791 – Standard Test Method for Flat Particles, Elongated Particles, or Flat and Elongated Particles in Coarse Aggregate.

The Proportional Caliper Device is constructed from steel for durability and strength and is plated for corrosion resistance. The 6 in × 16 in base plate has rubber feet attached for added stability and use in tabletop testing. Four positions control the various ratios. The desired ratio is obtained by moving the wing nut to the corresponding number stamped on the pivot block, i.e., 2 = 1:2, 3 = 1:3, 4 = 1:4, and 5 = 1:5.

A.5.3 Procedure

Aggregate particles are defined as follows:

1. Flat or elongated particles of aggregate – particles having a ratio of width to thickness or length to width greater than a specified value
2. Flat and elongated particles of aggregate – particles having a ratio of length to thickness greater than a specified value (for Superpave HMA, the required ratio is 1:5) – this is a Superpave requirement

Individual particles of aggregate of specific sieve sizes are measured to determine the ratios of width to thickness, length to width, or length to thickness.

In the Flat Particle test, the particle is classified as flat if its thickness can be placed in the smaller opening without adjusting the caliper.

In the Elongated Particle test, the particle is elongated if the width can be placed within the smaller opening without readjusting the caliper. Using the caliper positioned at the proper ratio, set the larger opening equal to the particle length.

If the particle does not meet either the classification of flat or elongated, it is classified as neither flat nor elongated. After particles are classified, determine the proportion of the sample in each group by either counting the number of particles or by mass, as required.

In the Flat and Elongated Particle test (Superpave mix design requirement), the collected particles are grouped as either (i) flat and elongated or (ii) not flat and elongated. The Proportional Caliper, set at the desired ratio (1:5, required by Superpave mix design procedures), is used with the more extensive opening set equal to the particle length. The particle is flat and elongated if the thickness can be placed in the smaller opening; otherwise, it is classified as not flat and elongated. After particles are classified into groups, determine the proportion of the sample in each group by either count of the number of particles or by mass.

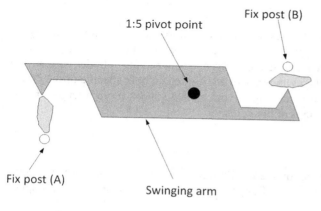

Figure A.5 Measuring Flat and Elongated Particles (A)

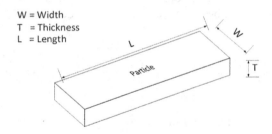

W = Width
T = Thickness
L = Length

Flat : W/T > 3
Elongated: L/W > 3
Flat and Elongated: meets both criteria

Figure A.6 Measuring Flat and Elongated Particles (B)

The aggregate particle is first placed with its largest dimension between the swinging arm and the fixed post at position A. The swinging arm then remains stationary while the aggregate is placed between the swinging arm and the fixed post at position B. If the aggregate fits in (or can pass through) this gap, then it is counted as a flat or elongated particle.

A.6 Fine Aggregate Angularity by Measuring the Loose Uncompacted Void Content of a Sample of Fine Aggregate

ASTM C1252-06 – Standard Test Methods for Uncompacted Void Content of Fine Aggregate (as Influenced by Particle Shape, Surface Texture, and Grading)

A.6.1 Purpose

Determine the fine aggregate angularity by measuring the loose, uncompacted void content of a sample of fine aggregate, which indicates the aggregate's angularity.

A.6.2 Theory

A small, fine, washed, and dried aggregate is poured into a small calibrated cylinder through a standard funnel. By measuring the mass of fine aggregate (W) in the filled cylinder of known volume (V), the void content can be calculated as the difference between the cylinder volume and the fine aggregate volume collected in the cylinder. The fine aggregate bulk specific gravity (G_{sb}) is used to compute the fine aggregate volume. Higher void contents mean more fractured faces. Superpave requires greater than 40% or 45% of the voids.

A.6.3 Apparatus

Cylindrical Measure: A right cylinder of approximately 100-mL capacity having an inside diameter of approximately 39 mm and an inside height of approximately 86 mm made of drawn copper water tube.

Table A1.1 Superpave Aggregate Consensus Property Requirements

Superpave Aggregate Consensus Property Requirements

Design ESALs[1] (million)	Coarse Aggregate Angularity (Percent), minimum		Uncompacted Void Content of Fine Aggregate (Percent), minimum		Sand Equivalent (Percent), minimum	Flat and Elongated[3] (Percent), maximum
	≤ 100 mm	> 100 mm	≤ 100 mm	> 100 mm		
< 0.3	55/-	-/-	-	-	40	-
0.3 to < 3	75/-	50/-	40	40	40	10
3 to < 10	85/80[2]	60/-	45	40	45	10
10 to < 30	95/90	80/75	45	40	45	10
≥ 30	100/100	100/100	45	45	50	10

1. Design ESALs are the anticipated project traffic level expected on the design lane over a 20 year period. Regardless of the actual design life of the roadway determine the design ESALs for 20 years and choose the appropriate N_{design} level.

2. 85/80 denotes that 85% of the coarse aggregate has one fractured face and 80% has two or more fractured faces.

3. Criterion based upon a 5:1 maximum-to-minimum ratio.

(If less than 25% of a layer is within 100 mm of the surface, the layer may be considered to be below 100 mm for mixture design purposes.)

Funnel: The lateral surface of the right frustum of a cone sloped 4° from the horizontal with an opening of 0.6 mm in diameter. The funnel section shall be a piece of metal, smooth on the inside and at least 38 mm high. It shall have a volume of at least 200 mL or shall be provided with a supplemental glass or metal container to provide the required volume.

Funnel Stand: A three- or four-legged support capable of holding the funnel firmly in position. The funnel opening shall be 2 mm above the top of the cylinder.

Glass Plate: A square glass plate approximately 60 by 60 mm with a minimum 4-mm thickness is used to calibrate the cylindrical measure.

Pan: A metal or plastic pan of sufficient size to contain the funnel stand and to prevent the loss of material. The purpose of the pan is to catch and retain fine aggregate particles that overflow the measure during filling and striking off.

Metal Spatula (with a Blade and Straight Edges): The end shall be cut at a right angle to the edges. The straight edge of the spatula blade is used to strike off the fine aggregate.

Scale or Balance: This is capable of weighing the cylindrical measure and its contents.

A.6.4 Procedure

The sample is to flow through a funnel from a fixed height into the measure. The fine aggregate is struck off, and its mass is determined by weighing. Uncompacted void content is calculated as the difference between the volume of the cylindrical measure and the absolute volume of the fine aggregate collected in the measure. Uncompacted void content is calculated using the fine aggregate's dry relative density (specific gravity). Two runs are made on each sample, and the results are averaged.

A.7 Superpave Mix Design – Lab procedures

ASTM D6925 – Standard Test Method for Preparation and Determination of the Relative Density of Asphalt Mix Specimens by Means of the Superpave Gyratory Compactor

Testing and design procedures of Superpave closely simulate the actual loading and climatic conditions that pavements endure.

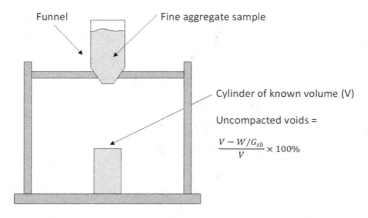

Funnel

Fine aggregate sample

Cylinder of known volume (V)

Uncompacted voids =

$$\frac{V - W/G_{sb}}{V} \times 100\%$$

Figure A.7 Fine Aggregate Angularity Testing

Figure A.8 Fine Aggregate Angularity Apparatus

The mix design procedures are as follows:

The Superpave mix design method is a performance-based method, in which the testing and evaluation procedures simulate actual field conditions. Three levels of mix design are outlined. Level 1 is suggested for low-volume roads, up to 1 million total ESAL; Level 2 is recommended for medium-volume traffic, ESAL from 1 to 10 million; and Level 3 for higher volumes and critical situations, such as exposure to extreme temperatures or where the risk of pavement failure must be reduced to a minimum. Acceptance for Level 1 standards is the first requirement for all levels. Levels 2 and 3 use samples developed in the Level 1 testing. In Level 1, aggregates and binders are tested and checked for acceptance. The trial mixes produced and tested in Level 1 are compacted in the Superpave gyratory compactor at a gyration angle of 1.25°. The vertical pressure for compaction is 600 kPa (87 psi). The number of compaction blows and the temperature of the material during compaction vary according to the design requirements and will be illustrated in the description of the following procedure.

The following steps in the Level 1 mix design procedure are adapted from the Asphalt Institute's Level I Mix Design Manual – SP-2.

1. Determine the design traffic load (total ESALs), average design high air temperature, and the maximum size of the aggregate to be used.
2. Select the asphalt binder grade.
3. Select the aggregate blend. Aggregates must meet the gradation and quality requirements. Three blends of aggregates are selected.

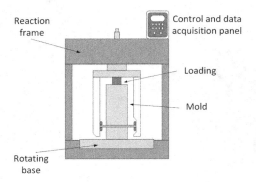

Figure A.9 Superpave Gyratory Compactor (SGC)

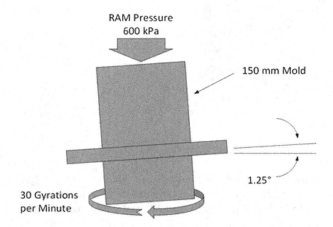

Figure A.10 Illustration of setting and work conditiosn of Superpave Gyratory Compactor

4. Select the trial asphalt binder content. The asphalt content includes both the absorbed and the effective asphalt. These quantities can be calculated using relationships involving the bulk and apparent relative densities of the aggregate, the relative density of the asphalt, the approximate proportions of these materials in the mix, and the size of the aggregate, as outlined in the Asphalt Institute Manual. However, for normal aggregates, the main factors are the aggregate size and its asphalt absorption. The asphalt absorption can be assumed to be 60–80% of the water absorption, which is usually known.

5. Determine the mixing and compacting temperatures from temperature–viscosity charts or, for modified asphalt binders, from the manufacturer''s recommendations. Mixing viscosities are between 0.15 and 0.19 Pa-s, and compacting viscosities are between 0.25 and 0.31 Pa-s.

6. Prepare aggregate samples. Two briquettes, each requiring 4700 g, are recommended for each blend. Another sample of 1, 1.5, 2, or 2.5 kg is needed to determine the maximum density of each mix for 9.5, 12.5, 19, and 25 mm aggregates, respectively. About 3700 g is needed for the moisture-sensitivity test.

7. Determine the compactive effort to be used. The briquettes must be compacted with N_{max} (maximum) number of gyrations in the compactor. The N_{des} (design) data is used for the calculation of volumetric properties. The N_{ini} (initial) and N_{max} values measure the compactibility of the mix.

8. Heat aggregates and asphalt binder to the mixing temperature and mix.

9. Place mixed samples in an oven at 135°C for short-term aging. Remove after four hours.

10. Bring the samples to the compacting temperature and compact with N_{max} gyrations. The height of the briquette is recorded for each gyration.

11. Calculate the density of each briquette. Using this density and the height of the briquette at N_{max} gyrations, calculate the densities at N_{des} and N_{ini}.

12. Determine the maximum bulk density of the mix using the sample of loose material, and determine corrected densities at N_{des} and N_{ini}.

13. Calculate air voids content, VMA, and VFA at the N_{des} density.

14. Estimate asphalt content at 4% air voids by subtracting $0.4 \times (4 - AV\%)$ from percent asphalt as mixed.

15. Correct the values for VMA and VFA. The corrected value of VMA is the initial value $+ 0.1$ $(4 - AV\%)$ for air voids content under 4.0% and $+ 0.2 (4 - AV\%)$ for other values. VFA values can be found from $(VMA - 4.0)/VMA$.

16. Check corrected VMA and VFA results for each blend for compliance with the specifications.

17. The density at N_{ini} must be less than 89% of the maximum density to ensure that there is a strong structure to resist rutting. The density at N_{max} must be less than 98% of the maximum density to ensure that a minimum of 2% air voids remain in the pavement in its densest condition.

18. The sample for moisture sensitivity is tested. The tensile strength of a saturated sample must be a minimum of 80% of the tensile strength of the control sample.

19. A value for the dust content is also calculated. This is the percentage of the aggregate passing 0.075 mm divided by the effective asphalt content as a percentage of the total mix. The dust proportion must be between 0.6 and 1.2.

20. If none of the aggregate blends tested meet the specifications adequately, select other blends or aggregates for further trials.

21. When an acceptable blend has been determined, new samples are made at four asphalt contents: the estimated content for 4% air voids, and contents that are 0.5% less, and 0.5% and 1.0% greater, than the estimated.

22. Final air voids, VMA, and VFA results are plotted, and the asphalt content at 4% air voids, if acceptable for other properties, is used for production or Level 2 and 3 tests.

A.8 Bulk and Maximum Specific Gravity of HMA

ASTM D2041 – Standard Test Method for Theoretical Maximum Specific Gravity and Density of Bituminous Paving Mixtures

AASHTO T 166 – Bulk Specific Gravity of Compacted Asphalt Mixtures Using Saturated Surface Dry Specimens

A.8.1 Introduction

This procedure is used to determine the maximum specific gravity (G_{mm}) of uncompacted asphalt paving mixtures as well as the bulk specific gravity (G_{mb}) of compacted asphalt mixtures.

A.8.2 Equipment and Materials

- Vacuum assembly
- Scale
- Thermometer
- Water bath
- Suspension apparatus
- Loose asphalt mixture
- Compacted asphalt specimen

A.8.3 Procedure

A.8.3.1 Rice Test – Maximum Specific Gravity (G_{mm})

1. Retrieve the loose asphalt mixture from the low-temperature oven and break up the mix, taking care not to fracture the mineral particles, so that the particles of the fine aggregate portion are not larger than ¼ inch. (1,500–1,700 g depending on max aggregate size).
2. Place the sample in the vacuum bowl (without the lid) and weigh it. Record the weight of the sample (subtracting the bowl's dry weight).
3. Add 77°F water until the sample is completely covered.
4. Remove entrapped air by subjecting the contents of the container to a partial vacuum of 27.5 ± 2.5 mm Hg, absolute pressure for 15 ± 2 minutes.
5. Gradually release the vacuum pressure using the bleeder valve.
6. Suspend the bowl and contents (without the lid) in 77°F for 10 ± 1 minutes. Record the weight of the suspended sample (subtracting bowl wet weight).
7. Calculate the maximum specific gravity (G_{mm}) of the test sample to three significant digits.

$$G_{mm} = \frac{A}{A-C}$$

Eq. A1.4

where
   A = weight of oven dry sample in air, g
   C = weight of the sample in water after vacuum, g

8. Record the calculations.

A.8.4 Bulk Specific Gravity (G_{mb})

1. Remove the compacted specimen from the low-temperature oven and weigh it. Record the weight.
2. Suspend the specimen in water at 77°F for 4 ± 1 minutes and record the immersed weight.
3. Remove the specimen from the water, surface dry by blotting with a damp cloth towel as quickly as possible, and determine the surface-dry weight.

Note: Completely immerse the entire towel in water and wring it out. Damp is considered to be when no water can be wrung from the towel.

4. Calculate the specimen bulk gravity (G_{mb}) as follows:

$$G_{mb} = \frac{A}{B-C}$$

Eq. A1.5

where

A = weight of the dry specimen in air, g
B = weight of the saturated surface-dry specimen, g
C = weight of the specimen in water, g

5. Record the calculations.

A.8.5 Air Voids Calculation

$$\%Voids = \frac{G_{mm} - G_{mb}}{G_{mm}} \times 100$$

Eq. A1.6

A.9 Aggregate Soundness Test

ASTM C88 – Standard Test Method for Soundness of Aggregates by Use of Sodium Sulfate or Magnesium Sulfate

Purpose: To measure the resistance of aggregate's soundness.

Theory: In the soundness test, aggregates are soaked in a solution of $MgSO_4$ or $NaSO_4$ (magnesium or sodium sulfate). The salt solution soaks into the pores of the aggregate. The sample is removed from the solution, drained, and then dried. During drying, crystals form in the pores, just as ice crystals form in aggregates exposed to weathering. This soaking and drying operation is carried on for a number of cycles. At the end of the test, the amount of material that has broken down is found, and the percentage loss is calculated. (The reliability of this test to simulate freezing and thawing damage is questionable.)

Apparatus: Saturated solution of $MgSO_4$; containers for soaking samples; sieves; balance (accurate to 0.01 g).

Sample: Approximately 5,000 g of aggregate including 2,500 ± 10 g of 19–12.5 mm (¾–½ in) size and 2,500 ± 10 g of 12.5 mm to 9.5 mm (½–⅜ in) size.

A.9.1 Procedure

1. Wash, dry, and obtain the mass of the test sample; approximately 1000 g if the size range is 19–9.5 mm (¾–⅜ in).
2. Place in solution for 16–18 hours.
3. Remove, drain, and place in the oven for about six hours.
4. Remove when dry; cool.
5. Repeat steps 2, 3, and 4 for five cycles.
6. Wash the sample thoroughly; dry.
7. Sieve the sample over an 8 mm (⁵/₁₆ in) sieve and measure the mass retained.

A.9.2 Results

A. Mass of original sample _____ g (A)
B. Mass of final sample _____ g (B)
C. Loss _____ g (A − B)

A.9.3 Calculation

$$\% \text{ Loss} = (A - B)/A \times 100 \underline{\hspace{3cm}} \%$$ Eq. A1.7

where
A = mass of original sample
B = mass of final sample

A.10 Sand Equivalent Value of Fine Aggregate

ASTM D2419 – Standard Test Methods for Sand Equivalent Value of Soils and Fine Aggregate
    Purpose: To qualify aggregates for applications where sand is desirable, but fines and dust are not. A higher sand equivalent value indicates less clay-like material in a sample.
    Theory: During the test, material from the test specimen that can pass through a No. 4 sieve is mixed with calcium chloride, formaldehyde, and glycerin solutions in a cylinder. The content is then left for sedimentation. After about 20 minutes, the level of clay suspension and the sand level are read on the cylinder scale.

A.10.1 Sample Preparation

Thoroughly mix the sample and reduce it using the applicable procedures in practice, obtaining at least 1,500 g of material passing the 4.75 mm (No. 4) sieve. Break down any lumps of material in the coarse fraction to pass the 4.75 mm (No. 4) sieve. Use a mortar and rubber-covered pestle or any other means that will not cause appreciable degradation of the aggregate.

- If necessary, dampen the material to avoid segregation or loss of fines during the splitting or quartering operations. Use care in adding moisture to the sample to retain a free-flowing condition of the material.
- Using the measuring tin, dip four of these measures from the sample. Each time a measure full of the material is dipped from the sample, tap the bottom edge on a work table or other hard surface at least four times and jog it slightly to produce a consolidated material that is level-full or rounded somewhat above the brim.
- Determine and record the amount of material contained in these four measures, either by mass or by volume, in a dry plastic cylinder.
- Return this material to the sample, proceed to split or quarter the sample, and make the necessary adjustments to obtain the predetermined mass or volume. When this mass or volume is received, two successive split or quartering operations without adjustment should provide the proper amount of material to fill the measure and, therefore, offer one test specimen.
- Dry the test specimen to a constant mass at 230 ± 9°F (110 ± 5°C) and cool to room temperature before testing.

A.10.2 Apparatus

Fit the siphon assembly to a 1.0 gal (3.8 L) working calcium chloride solution bottle. Place the bottle on a shelf 36 ± 2 in (90 ± 5 cm) above the working surface. Start the siphon by blowing into the solution bottle's top through a short piece of tubing piece while the pinch clamp is open.

A.10.3 Procedure

- Siphon 4 ± 0.1 in (102 ± 0.3 mm) of working calcium chloride solution into the plastic cylinder.
- Pour one of the test specimens into the plastic cylinder using the funnel to avoid spillage.
- Tap the bottom of the cylinder sharply on the heel of the hand several times to release air bubbles and promote thorough wetting of the specimen.
- Allow the wetted specimen and cylinder to stand undisturbed for 10 ± 1 min.
- At the end of the 10-minute soaking period, stopper the cylinder and loosen the material from the bottom by partially inverting the cylinder and shaking it simultaneously.
- Allow the cylinder and contents to stand undisturbed for 20 min. Start the timing immediately after withdrawing the irrigator tube.
- After the suspension reading has been taken, place the weighted foot assembly over the cylinder and gently lower it until it rests on the sand. Do not allow the indicator to hit the mouth of the cylinder as the assembly is being lowered.
- As the weighted foot comes to rest on the sand, tip the assembly toward the graduations on the cylinder until the indicator touches the inside of the cylinder. Subtract 10 in (25.4 cm) from the level indicated by the extreme top edge of the indicator and record this value as the "sand reading."

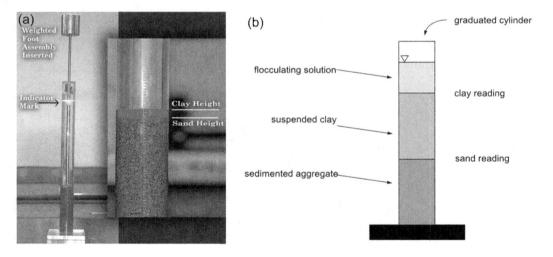

Figure A.11 Sand Equivalent Test (left), Clay Reading, and Sand Reading (right)

A.10.4 Calculation

Calculate the sand equivalent to the nearest 0.1 % as follows:

$$SE = S/C \times 100\%$$ Eq. A1.8

where
SE = sand equivalent value (%)
S = sand reading
C = clay reading

Index